Molecular Targets in Cancer Therapy

Anand Shankar Ramteke
Suresh Kumar Ramakrishnan
Rajesh Agarwal
Editors

Molecular Targets in Cancer Therapy

Understanding Cellular Pathways for Therapeutic Interventions

Springer

Editors
Anand Shankar Ramteke
Department of Molecular Biology and Biotechnology
Tezpur University
Tezpur, Assam, India

Suresh Kumar Ramakrishnan
Molecular Biology Division
ICMR-National Institute of Cancer Prevention and Research
Noida, Uttar Pradesh, India

Rajesh Agarwal
Department of Pharmaceutical Sciences
University of Colorado Cancer Center
Colorado, CO, USA

ISBN 978-981-96-7335-3 ISBN 978-981-96-7336-0 (eBook)
https://doi.org/10.1007/978-981-96-7336-0

This Springer imprint is published by the registered company Springer Nature Singapore Pte Ltd.
The registered company address is: 152 Beach Road, #21-01/04 Gateway East, Singapore 189721, Singapore

Dedicated to
All our teachers who enlightened our path
through their knowledge and wisdom

Foreword

Recent advances in the technology, particularly the "Multi-omics" and high-throughput screens, have generated a vast amount of knowledge in cancer biology that led to generation of new tools, novel approaches in treatment modalities for the age-old unresolved problem of cancer disease. The knowledge we acquired in the recent past have charted the novel potential targets and their role in carcinogenesis process. Knowing those new developments in the field of cancer biology is important for the present and many more coming generations.

This book *Molecular Targets in Cancer Therapy* introduces the various cellular physiological processes and targets that have been successfully implemented for cancer therapies. The readers will find here the latest information on molecular pathways and targets involved in cancers related to tyrosine kinase receptors, cell death regulators, epigenetic modifications, angiogenesis, epithelial-mesenchymal transition, inflammation, metabolic reprograming, genomic instability, and multi-drug resistance that can be used as targets for cancer therapy. These topics include basic and up-to-date information on the various pathways, which would be beneficial not only for the experts in the field but also for early learners to understand the subject.

I congratulate the editors, Drs. Anand Shankar Ramteke, Suresh Kumar Ramakrishnan, and Rajesh Agarwal, along with the contributors of the chapters, for bringing out this important book that would bridge the knowledge gap and also advance our understanding of molecular targets in cancer research and therapy.

Raosaheb K. Kale
Former Professor and Dean
School of Life Sciences
Jawaharlal Nehru University
New Delhi, India

Former Vice Chancellor
Central University of Gujarat
Gandhinagar, Gujarat, India

Preface

The global cancer statistics for the year 2022 based on estimates from the International Agency for Research on Cancer (IARC) showed 20 million new cancer cases and 9.7 million deaths from the disease. The demographic prediction suggests that new cancer cases will increase to 35 million, i.e., 77% increase by 2050, indicating the need for global efforts for targeted strategies and programs to control cancer. There were limited treatment options available for patients for many decades, but the recent advancements in the omics platforms and high-throughput screens led to the generation of new tools and methodologies for the diagnosis and therapy of cancer. The application of multi-omics and CRISPR-Cas9 screen technology have added newer hallmarks in cancer which have impacted the development of modern treatment modalities, making possible the addition of novel potential targets for cancer therapy. This book is a compilation of up-to-date information on the various cellular pathways and major targets that could be successfully exploited in cancer therapies. There are 12 chapters, each focusing on the different mechanisms and pathways that cancer cells exploit for their growth and survival, and these pathways are targeted for therapy in clinics as well as research laboratories for developing better treatment regimens. The major pathways that are covered include tyrosine kinase receptors, cell death mechanisms, telomerase activity, epigenetic regulations, inflammatory response, epithelial-mesenchymal changes, angiogenic pathways, metabolism, genomic instability, and multi-drug resistance. Each chapter elaborates on the above topics providing basic and in-depth information on the roles of various pathways on the progression and development of cancer and exploring molecules that are being targeted for present therapy or that could be a suitable potential target in the future. The readers will find this book very useful for understanding the concepts of the cellular pathways that are involved in the carcinogenesis process, how these pathways can used as targets in cancer therapy and the strategies to overcome challenges. The latest information on cancer targets that can be used for developing personalized therapies have also been discussed.

This book is an outcome of the sincere efforts of all contributors; we would like to express our gratitude to all those who have contributed to the completion of this important book which would advance our understanding of molecular targets and how they could be explored for developing better therapeutic regimens for the management of cancer.

Tezpur, Assam, India — Anand Shankar Ramteke
Noida, Uttar Pradesh, India — Suresh Kumar Ramakrishnan
Colorado, CO, USA — Rajesh Agarwal

Acknowledgments

We would like to express our profound gratitude to everyone who supported us throughout the journey of publishing this book.

First and foremost, we extend our sincere gratitude and thanks to Professor Shambhu Nath Singh, Vice Chancellor, Tezpur University, Tezpur, Assam, India, for his immense encouragement and support for bringing up this book on "Molecular Targets in Cancer Therapy."

We are grateful to Dr. Shalini Singh, Director, ICMR-National Institute of Cancer Prevention and Research, Noida, Uttar Pradesh, India, for her support and keen interest throughout in publishing this book.

The editors extend their deepest gratitude and thanks to all the contributors for sharing their expertise in the field of cancer biology and relentless efforts, which got transformed into this book. Special thanks to all our peers for their invaluable feedback and constructive critiques during the review process, which has enhanced the quality of the manuscripts.

The editors, Drs. Anand Shankar Ramteke and Suresh Kumar Ramakrishnan, take this opportunity to express long-lasting gratitude to their teachers, Prof (Dr.) Raosaheb K. Kale, Prof (Dr.) Pramod K. Yadava, and late Prof (Dr.) A. Ramesha Rao, School of Life Sciences, Jawaharlal Nehru University, New Delhi, for instilling their knowledge, thoughts, and wisdom through their teaching and discussions.

We also gratefully acknowledge the contribution of the team at Springer Nature, for their insights and feedback, which has greatly enhanced the presentation of the work. The contribution of all the professional staff at Springer Nature is also acknowledged for transforming this book, ready for readers.

Thank you to everyone who remained with us throughout this journey for your unwavering support and kindness in making this book a reality.

Anand Shankar Ramteke
Suresh Kumar Ramakrishnan
Rajesh Agarwal

Contents

Editors and Contributors

About the Editors

Anand Shankar Ramteke did his doctorate in Life Sciences at Jawaharlal Nehru University, New Delhi, India, with a specialization in the area of radiation and cancer biology. He has been working in the area of exosome biology, tumor microenvironment, and drug resistance in head and neck, breast, and prostate cancers. His research interest also includes developing a nano-bio system for targeted drug delivery for cancer therapy. Presently, he is working as a Professor at Tezpur University, Tezpur, Assam, India, and has published high-impact papers in the field. He was also a visiting Professor/Faculty at the University of Oklahoma Health Sciences, Oklahoma, USA, and the University of Colorado, Denver, USA.

Suresh Kumar Ramakrishnan did his doctorate in Life Sciences at Jawaharlal Nehru University, New Delhi, India. He worked in the field of non-canonical function of telomerase in cancer cells and unraveled telomerase RNA function in controlling cellular physiology. His interest also lies in epigenetic modifications, drug resistance, and chemoprevention in cancers. He was a visiting scientist at The Wistar Institute, Philadelphia, USA, and worked in the field of non-coding RNAs in cancer. Presently, he is working as a Scientist F at ICMR-National Institute of Cancer Prevention and Research, Noida, Uttar Pradesh, India.

Rajesh Agarwal graduated from Lucknow University, Lucknow, India, and worked at several positions in India and Case Western Reserve University (CWRU), Cleveland, USA. Presently he is working as a Professor in the Department of Pharmaceutical Sciences, School of Pharmacy, University of Colorado, Anschutz Medical Campus (UC-AMC), Aurora, CO, USA. He is a pioneer in cancer pharmacology and chemoprevention and has published more than 400 research papers in high-impact journals. He is a fellow of the American Association for the Advancement of Science, and a recipient of several national and international awards.

Contributors

Sourin Adhikary Food Toxicology Laboratory, FEST Division, CSIR-Indian Institute of Toxicology Research, Lucknow, Uttar Pradesh, India

Academy of Scientific and Innovative Research (AcSIR), Ghaziabad, Uttar Pradesh, India

Farzaneh Amirmahani Department of Oncology Science, University of Oklahoma Health Sciences Center, Oklahoma City, OK, USA

Saria Anjum Food Toxicology Laboratory, FEST Division, CSIR-Indian Institute of Toxicology Research, Lucknow, Uttar Pradesh, India

Academy of Scientific and Innovative Research (AcSIR), Ghaziabad, Uttar Pradesh, India

Kausar M. Ansari Food Toxicology Laboratory, FEST Division, CSIR-Indian Institute of Toxicology Research, Lucknow, Uttar Pradesh, India

Academy of Scientific and Innovative Research (AcSIR), Ghaziabad, Uttar Pradesh, India

R. S. Arya Laboratory of Immunopharmacology and Experimental Therapeutics, Division of Cancer Research, Regional Cancer Centre (Research Centre, University of Kerala), Thiruvananthapuram, Kerala, India

I. S. Aswathy Laboratory of Immunopharmacology and Experimental Therapeutics, Division of Cancer Research, Regional Cancer Centre (Research Centre, University of Kerala), Thiruvananthapuram, Kerala, India

Noble T. Baby Laboratory of Immunopharmacology and Experimental Therapeutics, Division of Cancer Research, Regional Cancer Centre (Research Centre, University of Kerala), Thiruvananthapuram, Kerala, India

M. Caroline Diana Sherly Laboratory of Immunopharmacology and Experimental Therapeutics, Division of Cancer Research, Regional Cancer Centre (Research Centre, University of Kerala), Thiruvananthapuram, Kerala, India

Santanu Chatterjee SRM Institute of Science and Technology, Kattankulathur Campus, Chengalpattu, Tamil Nadu, India

Ujjayita Chowdhury Stem cell and Tissue Homeostasis Lab, Cancer Research Institute, ACTREC-Tata Memorial Centre, Navi Mumbai, India

Homi Bhabha National Institute, Mumbai, India

Gagan Deep Department of Internal Medicine-Gerontology and Geriatric Medicine, Wake Forest University School of Medicine, Winston-Salem, NC, USA

Department of Cancer Biology, Wake Forest University School of Medicine, Winston-Salem, NC, USA

Sanchayeta Dey Department of Molecular Biology and Biotechnology, Tezpur University, Tezpur, Sonitpur, Assam, India

Reetika Ghosh Department of Molecular Biology and Biotechnology, Tezpur University, Tezpur, Sonitpur, Assam, India

Chandrashekaran Guruvayoorappan Laboratory of Immunopharmacology and Experimental Therapeutics, Division of Cancer Research, Regional Cancer Centre (Research Centre, University of Kerala), Thiruvananthapuram, Kerala, India

Abhirami Harjith Laboratory of Immunopharmacology and Experimental Therapeutics, Division of Cancer Research, Regional Cancer Centre (Research Centre, University of Kerala), Thiruvananthapuram, Kerala, India

Anowar Hussain Department of Biophysics, Pub Kamrup College, Baihata Chariali, Kamrup, Assam, India

K. S. Janasevakan Laboratory of Immunopharmacology and Experimental Therapeutics, Division of Cancer Research, Regional Cancer Centre (Research Centre, University of Kerala), Thiruvananthapuram, Kerala, India

Deepak Kalyanasundaram School of Life Sciences, Bharathidasan University, Palkalaiperur Campus, Tiruchirappalli, Tamil Nadu, India

Rohan J. Khadilkar Stem cell and Tissue Homeostasis Lab, Cancer Research Institute, ACTREC-Tata Memorial Centre, Navi Mumbai, India

Homi Bhabha National Institute, Mumbai, India

Praveen K. Kujur Cancer Biology Laboratory, School of Life Sciences, Jawaharlal Nehru University, New Delhi, India

Niti Kumari Biomedical Sciences Department, Creighton University School of Medicine, Omaha, NE, USA

Rahul Kumar Dr B. R. A. Institute Rotary Cancer Hospital, All India Institute of Medical Sciences, New Delhi, India

Saurav Kumar Department of Oncology Science, University of Oklahoma Health Sciences Center, Oklahoma City, OK, USA

Ajaikumar B. Kunnumakkara Cancer Biology Laboratory, Department of Biosciences and Bioengineering, Indian Institute of Technology Guwahati (IITG), Guwahati, Assam, India

Mantu Lal Radiation and Cancer Therapeutics Lab, School of Life Sciences, Jawaharlal Nehru University, New Delhi, India

Mukesh Kumar Manickasamy Cancer Biology Laboratory, Department of Biosciences and Bioengineering, Indian Institute of Technology Guwahati (IITG), Guwahati, Assam, India

Amy S. Mathew Laboratory of Immunopharmacology and Experimental Therapeutics, Division of Cancer Research, Regional Cancer Centre (Research Centre, University of Kerala), Thiruvananthapuram, Kerala, India

Sree Deepthi Muthukrishnan Department of Oncology Science, University of Oklahoma Health Sciences Center, Oklahoma City, OK, USA

Saurabh Pal Food Toxicology Laboratory, FEST Division, CSIR-Indian Institute of Toxicology Research, Lucknow, Uttar Pradesh, India

Soumitro Pal Division of Nephrology, Boston Children's Hospital, Boston, MA, USA

Harvard Medical School, Boston, MA, USA

Ram Raj Prasad Cardiovascular Pulmonary Research Laboratories, Anschutz Medical Campus, University of Colorado, Aurora, Denver, CO, USA

Seydur Rahman Department of Zoology, Faculty of Science, Assam Down Town University, Guwahati, Kamrup (M), Assam, India

Suresh Kumar Ramakrishnan ICMR-National Institute of Cancer Prevention and Research, Noida, Uttar Pradesh, India

Anand Shankar Ramteke Department of Molecular Biology and Biotechnology, Tezpur University, Tezpur, Sonitpur, Assam, India

Akash Sabarwal Division of Nephrology, Boston Children's Hospital, Boston, MA, USA

Harvard Medical School, Boston, MA, USA

Anjana Sajeev Cancer Biology Laboratory, Department of Biosciences and Bioengineering, Indian Institute of Technology Guwahati (IITG), Guwahati, Assam, India

Rana P. Singh Cancer Biology Laboratory, School of Life Sciences, Jawaharlal Nehru University, New Delhi, India

Department of Pharmaceutical Sciences, Skaggs School of Pharmacy and Pharmaceutical Sciences, University of Colorado, Anschutz Medical Campus, Aurora, CO, USA

Gautam Buddha University, Greater Noida, Uttar Pradesh, India

Aleena Sinha Cancer Biology Laboratory, School of Life Sciences, Jawaharlal Nehru University, New Delhi, India

Pranay Tanwar Dr B. R. A. Institute Rotary Cancer Hospital, All India Institute of Medical Sciences, New Delhi, India

Ravi Thakur Department of Oncology Science, University of Oklahoma Health Sciences Center, Oklahoma City, OK, USA

Department of Life Sciences, Christ University, Bengaluru, India

Ashu Bhan Tiku Radiation and Cancer Therapeutics Lab, School of Life Sciences, Jawaharlal Nehru University, New Delhi, India

Akhil Varshney Center for Advanced Vision Science, Department of Ophthalmology, University of Virginia, Charlottesville, VA, USA

Sabu Vidya Laboratory of Immunopharmacology and Experimental Therapeutics, Division of Cancer Research, Regional Cancer Centre (Research Centre, University of Kerala), Thiruvananthapuram, Kerala, India

Ravichandran Vishwa Cancer Biology Laboratory, Department of Biosciences and Bioengineering, Indian Institute of Technology Guwahati (IITG), Guwahati, Assam, India

Pramod K. Yadava School of Life Sciences, Jawaharlal Nehru University, New Delhi, India

Monika Yadav Laboratory of Nanotechnology and Chemical Biology, Regional Centre for Biotechnology, Faridabad, Haryana, India

Structural Features and Signaling Mechanisms of Receptor Tyrosine Kinases in Cancer

Rahul Kumar, Monika Yadav, Pranay Tanwar, Soumitro Pal, and Akash Sabarwal

Abbreviations

AKT	Protein Kinase B
ALK	Anaplastic Lymphoma Kinase
BCR–ABL	Breakpoint Cluster Region-Abelson Murine Leukemia Viral Oncogene Homolog
CI	Confidence Interval
CNS	Central Nervous System
CSF-1	Colony Stimulating Factor 1
ECD	Extracellular Domain
EGFR	Epidermal Growth Factor Receptor
ERK	Extracellular Signal-Regulated Kinase
FDA	Food and Drug Administration
FGFR	Fibroblast Growth Factor Receptor
FLT-3	FMS-like Tyrosine Kinase 3
Fn3	Fibronectin type 3
GBM	Glioblastoma Multiforme
GIST	Gastrointestinal Stromal Tumor
Her2	Human Epidermal Growth Factor Receptor 2

R. Kumar · P. Tanwar
Dr B. R. A. Institute Rotary Cancer Hospital, All India Institute of Medical Sciences, New Delhi, India

M. Yadav
Laboratory of Nanotechnology and Chemical Biology, Regional Centre for Biotechnology, Faridabad, Haryana, India

S. Pal · A. Sabarwal (✉)
Division of Nephrology, Boston Children's Hospital, Boston, MA, USA

Harvard Medical School, Boston, MA, USA
e-mail: akash.sabarwal@childrens.harvard.edu

A. S. Ramteke et al. (eds.), *Molecular Targets in Cancer Therapy*,
https://doi.org/10.1007/978-981-96-7336-0_1

HGFR	Hepatocyte Growth Factor Receptor
Ig	Immunoglobulin-like
Ins	Insulin
IR	Insulin Receptor
IRR	Insulin Receptor-Related Receptor
JAK	Janus Kinase
JAK-STAT	Janus Kinase-Signal Transducer and Activator of Transcription
JMD	Juxta Membrane Domain
JNJ	Johnson & Johnson
KIT	KIT Proto-Oncogene Receptor Tyrosine Kinase
MAPK	Mitogen-Activated Protein Kinase
MEK	Mitogen-Activated Protein Kinase Kinase
MET	Mesenchymal-Epithelial Transition
miRNAs	MicroRNAs
mOS	Median Overall Survival
mPFS	Median Progression-Free Survival
mTOR	Mammalian Target of Rapamycin
NGF	Nerve Growth Factor
NMPA	National Medical Products Administration
NSCLC	Non-Small Cell Lung Cancer
ORR	Objective Response Rate
PD-1	Programmed Cell Death Protein 1
PDGFR	Platelet-Derived Growth Factor Receptor
PDGFRA	Platelet-Derived Growth Factor Receptor Alpha
PDGFR-β	Platelet-Derived Growth Factor Receptor Beta
PI3K	Phosphoinositide 3-Kinase
PLCγ	Phospholipase C Gamma
PSC	Pulmonary Sarcomatoid Carcinoma
PTB	Phosphotyrosine-Binding
PTEN	Phosphatase and Tensin Homolog
RAF	Rapidly Accelerated Fibrosarcoma
RET	Rearranged during Transfection
ROS1	c-ros Oncogene 1
RTK	Receptor Tyrosine Kinase
SCLC	Small Cell Lung Cancer
SH2	Src Homology 2
sssABL-BCR	Abelson-Breakpoint Cluster Region
STAT	Signal Transducer and Activator of Transcription
TKD	Tyrosine Kinase Domain
TKIs	Tyrosine Kinase Inhibitors
TMD	Transmembrane Domain
TRAEs	Treatment-Related Adverse Events
VEGFR	Vascular Endothelial Growth Factor Receptor

1 Introduction

Over the past few decades, cell signaling has evolved to illustrate the physiology and pathology of the cell. It is a complex web of molecular interaction that enables cells to coordinate their function after perceiving external stimuli. The information is transmitted using various signaling molecules (first messenger) binding with the receptors (primary effectors) to initiate intracellular signaling cascades (secondary effectors) and the effector molecules that execute specific responses [1, 2]. The cell-surface receptor family member receptor tyrosine kinase (RTK) is one such pathway implicated in the cell's growth, proliferation, differentiation, and survival [3, 4]. The discovery of the RTK was started in the mid-nineteenth century and can be attributed to the work of several research groups. The pioneer discoveries of nerve growth factor (NGF) and epidermal growth factor (EGF) in the 1950s and 1960s have established several paradigms associated with growth factor-mediated signaling. These growth factors are found to bind with the cell surface receptor [5–7]. Subsequently, a large number of growth factors, such as FGF and PDGF were identified and studied for their role in several normal cellular processes [8–11]. However, it's their role in various pathologies, such as cancer, that has brought RTKs into sharp focus. In 1984, Ullrich et al. did a molecular characterization study and were able to shed light on the primary structure of the EGF receptor (EGFR) [12]. Further investigations revealed that RTKs constitute a varied group of cell surface receptors distinguished by their extracellular ligand-binding domain, a singular transmembrane helix, and an intracellular tyrosine kinase domain.

The structural elements of RTKs encompass various domains such as extracellular ligand-binding domain, transmembrane helix, juxta membrane domain (JMD), tyrosine kinase domain (TKD), C-terminal tail, and dimerization interface [13, 14]. These components collectively enable RTKs to transmit extracellular signals across the cell membrane, followed by activation of intracellular signaling pathways and regulate several fundamental cellular processes. It was found that structural changes of RTKs were observed upon ligand binding, and a series of changes, including auto-phosphorylation and dimerization, were undergone to activate downstream signaling cascades. In addition to being transmembrane cell surface receptors, the RTKs are enzymes with kinase activity. Numerous studies have shown that mutations in RTK domains are linked to various diseases, most notably cancer [15]. This underscores the significance of RTKs as potential therapeutic targets, a role that is further explored in this chapter. The present chapter provides a comprehensive understanding of the structured framework of RTKs and their biology, ranging from classification, activation, and regulation mechanisms to pathological implications in cancer. It also covers monoclonal antibodies and small molecule inhibitors that are available for use as targeted therapeutics against RTKs.

2 Classification of Receptor Tyrosine Kinases

The classification of RTKs was mainly based on the structure and sequence of extracellular domain (ECD). Each RTKs subfamily has its own set of distinct protein motifs, including immunoglobulin-like (Ig) domain, Cys-rich domain (CR domain), Leu domain (L domain), Fibronectin type 3 (Fn3)-like domain, or α and β chains which determines its ligand specificity [16]. The ECD's structural characteristics help us understand the diversity of RTKs and their ligands. The human genome has 90 different kinase genes encodes 58 types of receptors, which are distributed among 20 subfamilies [17]. Each subfamily has its own specialized function, but they are similar in structure with distinct domains and share the conserved architecture of the kinase domain. Of these 20 subfamilies, some families (Fig. 1) such as epidermal growth factor receptor (EGFR), platelet-derived growth factor receptor (PDGFR), vascular endothelial growth factor receptor (VEGFR), fibroblast growth factor receptor (FGFR), and hepatocyte growth factor receptor (HGFR) are found to be significantly associated with the development and progression of cancer [18–20] (Table 1). The (EGF/ErbB) receptor family is often found to have high expression levels in tumors arising from epithelial cells, including head and neck squamous cell

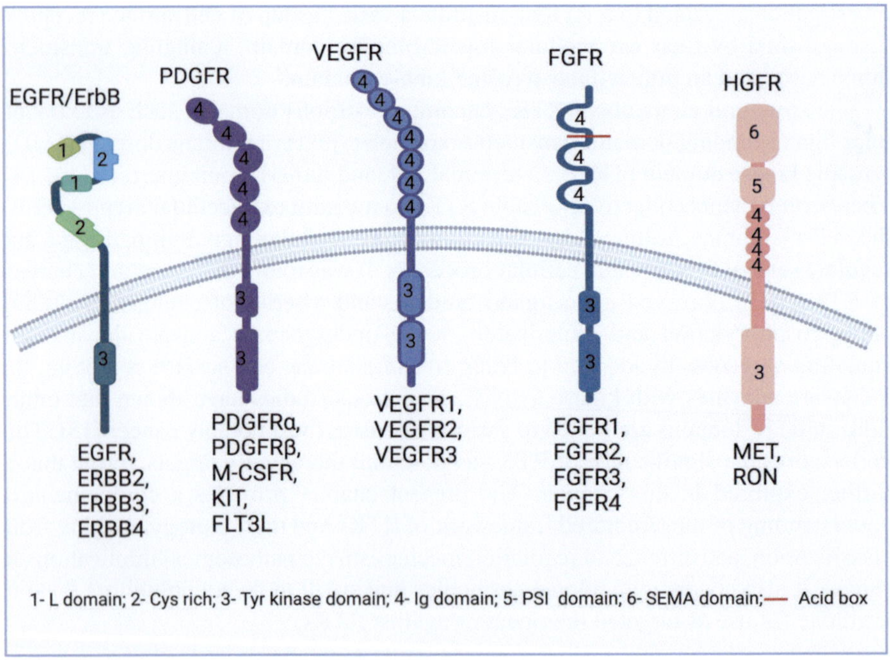

Fig. 1 The basic structure of various receptor tyrosine kinases (RTKs). Epidermal growth factor receptor (EGFR), Platelet-derived growth factor receptor (PDGFR), vascular endothelial growth factor receptor (VEGFR), fibroblast growth factor receptor (FGFR), and hepatocyte growth factor receptor (HGFR)

Table 1 Major receptor tyrosine kinase (RTK) subfamily involved in cancer

Subfamily	Members	Ligand	Structural domain of the ECD
EGFR/ErbB	ERBB1 (EGFR), ERBB2 (HER2), ERBB3 (HER3), ERBB4 (HER4)	EGF, TGFα, EREG, BTC, EPGN, AREG, HB-EGF, NRG1, NRG2, NRG3, NRG4	2 Cysteine rich domains, 2 Leucine domains
PDGFR	PDGFRα, PDGFRβ, CSF1R, KIT, FLT3L	PDGF-A, PDGF-B, PDGF-C, PDGF-D, SCF, FLT3L	5 Ig-like domains
VEGFR	VEGFR1, VEGFR2, VEGFR3	PIGF, VEGFB, VEGFC, VEGFD, VEGFE	7 Ig-like domains
FGFR	FGFR1, FGFR2, FGFR3, FGFR4	FGF1–22	3 Ig-like domains, 1 acidic box
HGFR	MET, RON	HGF, MSP	3 Ig-like domains, 1 PSI domain, 1 SEMA domain

carcinoma, colon and rectum, kidney, bladder, liver, pancreas, cervix, non-small cell lung cancer (NSCLC), prostate, glioblastoma, ovary, breast, and esophagogastric. The overexpression of these receptors is associated with increased cell proliferation, survival, and migration, which are key factors in cancer development and progression [21, 22]. The PDGFR family members are essential in various cellular processes, including blood vessel formation [23]. Many malignant tumors, including gliomas, neuroendocrine tumors, NSCLC, breast cancer, gastrointestinal stromal tumor (GIST), hepatocellular carcinoma, pancreatic cancer, renal cell carcinoma, and ovarian carcinoma, exhibit an increase in PDGF and its receptor expression. This increase has been associated with therapeutic resistance, while its inhibition can significantly reduce cancer cell proliferation and progression [24, 25]. Under normal physiological conditions, VEGFR members promote the growth of hematopoietic cells, lymphatic vessels, and endothelial cells and regulate metabolic homeostasis [26]. Lymphangiogenesis or angiogenesis may be regarded as an essential precursor for the progression of a tumor and its metastatic dissemination. The high expression level of VEGF was observed in many metastatic cancer cases and is correlated with highly unfavorable or decreased survival [27, 28]. The FGFR family comprises four members and actively participates in cell proliferation, differentiation, and survival. This biological function can be achieved through the four major downstream intracellular signaling pathways, such as RAS-RAF-MEK-MAPK, PLCγ, JAK-STAT, and PI3K-AKT-mTOR [29, 30]. Overstimulation of the RAS-RAF-MEK-MAPK pathway can promote cell proliferation [31]. Similarly, PLCγ has been identified as an essential regulator of tumor cell metastasis through angiogenic activity [32]. The abnormal activation of the JAK-STAT pathway has been linked with enhanced tumor invasion, metastasis, and immune evasion [33], and the PI3K-AKT-mTOR pathway can promote tumor growth and impede the natural process of apoptosis [34].

The mesenchymal-epithelial transition (MET) factor receptor is another significant HGFR subfamily receptor widely expressed over the surface. Upon interacting with the HGF ligand, it initiates various signaling cascades involved in several biological processes such as proliferation, migration, angiogenesis, and morphogenesis of epithelial cells. In cancer, amplification or mutation of HGFR can occur de novo or in response to therapy, causing aberrant overexpression of the MET receptor and implicating it in cellular proliferation, survival, and metastasis to distant organs [35, 36].

3 Mechanisms of Activation of Receptor Tyrosine Kinases and Downstream Signaling Pathways

Without a particular ligand, the activation loop maintains its closed conformation through cis-auto inhibition and limits the catalytic activity. In general, the activation process of RTKs usually begins with the binding of a specific monomer ligand (e.g., VEGF) or dimer ligand (e.g., KIT) with the ECD of the receptor [37, 38] and triggers receptor dimerization. However, in some RTK (such as EGFR), the dimerization can occur independently of ligand binding, which indicates the potential for receptor-mediated processes that do not necessitate ligand interaction with the extracellular domain (ECD). IR family members (IRR, IGFR1) have two subunits-α and β, and its constitutive form resides as a disulfide-linked dimer, even in the absence of its ligand [39]. In other RTKs, such as the FGFR family, the receptor's dimerization is restricted to ligand or receptor-induced, and the co-factors (such as heparin sulfate) are also involved [40].

The non-covalent interaction between the ligand and ECD is the pivotal event that induces dynamic conformational changes in the structure, and destabilizes the auto-inhibitory interaction. This leads to the exposure of the ATP binding site for the trans-auto phosphorylation of key tyrosine residues by the tyrosine kinase domain (TKD), assuming an active confirmation. The phosphotyrosines are created to serve as docking sites for a wide range of other downstream signaling proteins, such as the phosphotyrosine-binding (PTB) domain or Src homology 2 (SH2) domains [41, 42]. The TKD can also phosphorylate the tyrosine on related docking proteins, which offer additional possible binding sites but lack enzymatic activity [43]. These proteins serve as signaling platforms for the assembly of effector proteins and stimulate various other signaling pathways, such as RAF/MEK/ERK, PI3K/AKT/mTOR, and JAK/STAT. These docking proteins can also crosstalk with other activated RTK pathways [44, 45].

The regulatory C-terminal domains, which offer several phosphotyrosines and differ throughout RTK families, allow the cells to receive diverse downstream signals. An activated RTK is a nodal point in a complex network, transmitting extracellular signals to its interior. This transmission is essential for maintaining intricate cellular functions such as growth, proliferation, differentiation, and survival [13, 20].

4 Pathological Role of Receptor Tyrosine Kinases and Their Implications in Cancer

Under normal physiological conditions, the phosphorylation/dephosphorylation of RTKs is tightly regulated through the antagonizing effect of kinases or phosphatases. There are various mechanisms by which kinase may gain transforming activities and attain the constitutive activation of normal downstream pathways irrespective of ligand. These abnormal activations impede the other regulatory functions of cells, such as apoptosis and cell death and lead to the formation of immortal or cancerous cells (Fig. 2). Some plausible mechanisms for developing the transform ability of RTKs are outlined below:

4.1 Activating Mutation in Receptor Tyrosine Kinases

Oncogenic activation or gain of function of RTK can be acquired by nonsynonymous mutations (such as missense and In-Frame deletions). These alteration leads to changes in the structure and function of RTKs and acts as a driver mutation by providing a selective growth advantage to the proliferating cells. For instance, a mutation occurs in the TKD of EGFR, destabilizing the inactive conformation and opening the activation loop for phosphorylation. In lung cancer, the mutation was predominantly found in the encoded region of TKD, including in-frame deletions within exon 19. It is a specific mutation in the EGFR gene, where portions of the DNA coding sequence for this region are deleted without disrupting the gene's reading frame. These deletions typically result in the loss of 3–6 nucleotides, leading to the removal of a few amino acids from the protein product while retaining the overall protein reading frame, hence the term 'in-frame'. Other important mutations include G719X missense mutations (where the glycine residue at position 719 is substituted with another amino acid), in-frame insertions in exon 20, as well as point mutation (L858R) within exon 21 [46–48]. However, mutation is not only confined to the TKD but it can also be found in other additional domains, such as ECD, TMD, and JMD, which adopt an alternate mechanism to activate RTK. In glioblastoma multiforme (GBM), point mutations such as A289X, G598X, and R108K are frequently observed and positioned at the ECD of EGFR [49]. Similarly, serine substitution to cysteine at 249 positions in uterine cervix carcinoma was found in the ECD of FGFR3. These mutations lead to unpaired cysteine residues, which allow aberrant receptor dimerization via intermolecular disulfide bonding [50]. Mutations such as V659E and G660D were identified in the TMD, which acts as an oncogenic driver and stabilizes the dimerization of Her2 in lung adenocarcinomas [51]. In the case of gastrointestinal stromal tumors (GIST), V560G can be found within the JMD of the c-Kit receptor. In the MET receptor, the METΔEx14 variant in JMD increases signaling and is often associated with cancer. Mutation in the JMD domain releases autoinhibitory interaction and hyperactivates the RTKs [52, 53].

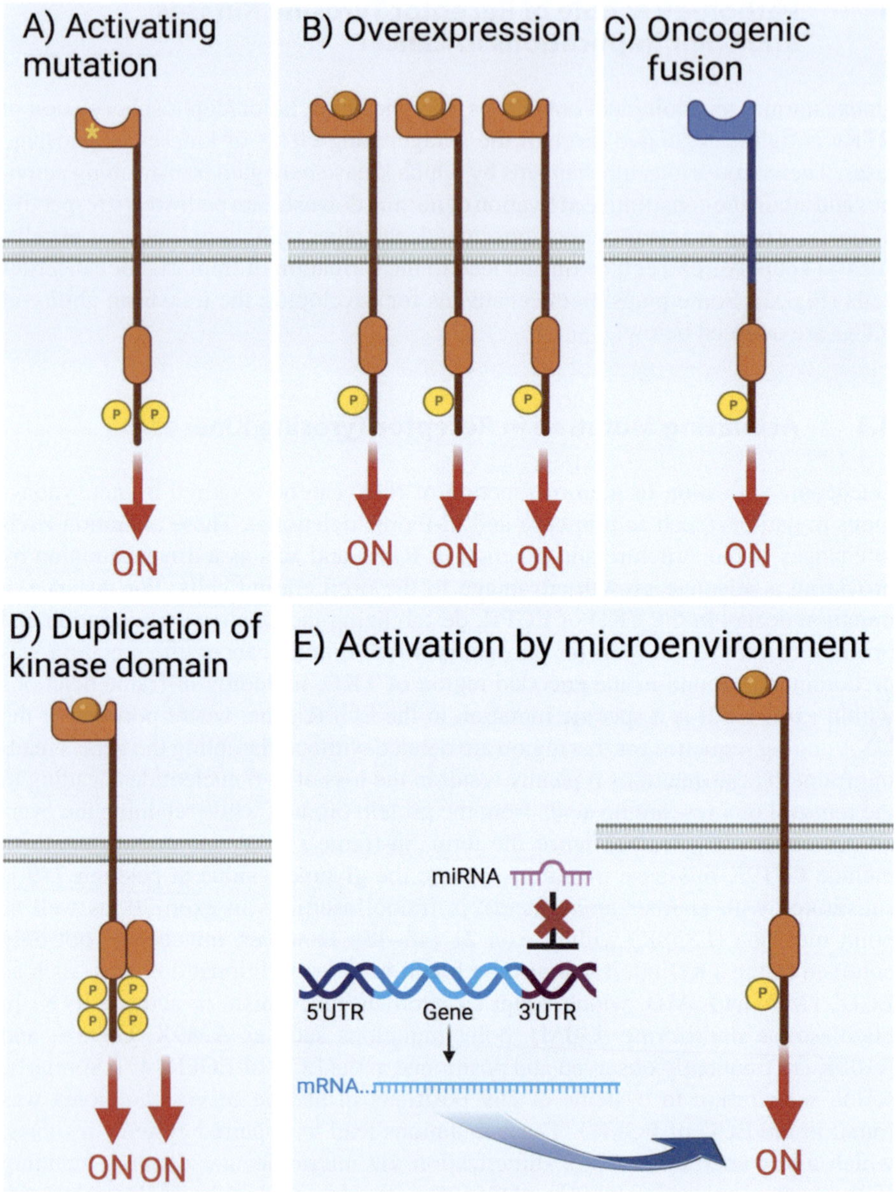

Fig. 2 Mechanisms of receptor tyrosine kinase (RTK) alterations in cancer pathology. The figure illustrates the various mechanisms through which receptor tyrosine kinases (RTKs) contribute to oncogenesis. (**a**) Activating mutations in RTKs in cancer: A schematic representation of common gain-of-function mutations found in RTK genes that drive aberrant signaling and promote tumorigenesis. (**b**) Overexpression of RTKs in cancer: the overabundance of RTKs due to genomic amplifications and other mechanisms, resulting in elevated signaling pathways linked to cancer progression. (**c**) Oncogenic fusion of RTKs in cancer: the formation of fusion proteins through

chromosomal rearrangements, showcasing how these novel oncoproteins result in constitutive activation of RTK signaling cascades. (**d**) Duplication of kinase domain of RTKs in cancer: displays instances of kinase domain duplications that lead to increased kinase activity and contribute to oncogenic signaling. (**e**) Activation of RTKs by microenvironment in cancer: the mechanism by which microRNAs (miRNAs) modulate receptor tyrosine kinase (RTK) activation in the tumor microenvironment

4.2 Overexpression of Receptor Tyrosine Kinases

Overexpression of RTK contributes significantly to the oncogenesis and is associated with the poor prognosis. Generally, it arises due to an increased copy number of specific exonic regions that encode RTK receptors. For instance, EGFR amplification was found in cfDNA of patients diagnosed with colorectal (16.3%), NSCLC (9%), and genitourinary cancers (8.1%) [54]. Similarly, overexpression of HER2 was predominant in various malignancies, including bladder (12.4%), breast (10.5%), and cervical (3.9%) cancers [55]. In NSCLC, MET overexpression was linked to a significantly higher risk of mortality (hazard ratio (HR): 1.52 [95% confidence interval (CI): 1.08–2.15]) [56] and resistance to anti-tumorigenic agents [57, 58]. Likewise, amplification of the FGFR family was observed in breast, lung, urothelial, and gastric cancer [59].

4.3 Chromosomal Rearrangement or Oncogenic Fusion of Receptor Tyrosine Kinases

As a result of chromosomal translocation, proteins fuse and form novel oncoproteins. These fusion proteins make the signaling pathway constitutively activated and raise the potential of tumorigenesis in normal cells. ABL-BCR was the first tyrosine kinase fusion protein identified in chronic myelogenous leukemia resulting from the translocation of chromosomes between 9 and 22 [60]. Another notable fusion oncoprotein is FLT3-ITD (Fms-like tyrosine kinase 3 internal tandem duplication), found in acute myeloid leukemia. This fusion suppresses the negative regulatory role of the JMD and causes a gain of function, which in turn promotes uncontrolled cell proliferation and survival of hematopoietic cells in acute myeloid leukemia (AML) [61].

4.4 Duplication of Kinase Domain in Receptor Tyrosine Kinases

The presence of an additional kinase domain in RTKs can induce allosteric effects that lead to persistent activation of the receptor. This phenomenon occurs because the additional kinase domain can cause conformational changes within the RTK structure, stabilizing an active conformation miming the state usually induced by ligand binding. For example, the EGFR and other RTKs have been observed to undergo conformational shifts that stabilize the active kinase state, promoting continuous signaling even in the absence of a ligand [62]. It may also provide additional binding sites for downstream signaling proteins (such as SH2 domain-containing proteins) and enhance downstream signal transduction [63].

4.5 Activation of Receptor Tyrosine Kinases by Altered Tumor Microenvironment

MicroRNAs (miRNAs), small non-coding RNA molecules, play a crucial role in modulating the tumor microenvironment through post-transcriptional dysregulation of gene expression. In general, miRNAs function to suppress the expression of protein by targeting its mRNA [64]. The unchecked expression of RTKs may result from the loss or downregulation of these miRNAs. miR-7 is known to target the 3′ untranslated region (UTR) of the EGFR mRNA. Downregulation of miR-7 can result in increased EGFR expression and activation [65].

In some cases, miRNAs can indirectly activate RTKs by targeting and downregulating negative regulators of RTK signaling pathways. For instance, miR-21 binds to the 3′ UTR of PTEN mRNA in colorectal cancer, leading to its degradation or translational repression. With reduced PTEN, the inhibitory effect on the PI-3K/Akt pathway is lifted, resulting in increased PI-3K activity and Akt phosphorylation [66].

5 Targeting Receptor Tyrosine Kinases for Therapeutic Intervention

Tyrosine kinases play a significant role in the molecular pathogenesis of cancer. Over the past two decades, RTKs have gained popularity as possible targets for anticancer treatments. As a result, many RTK inhibitors have been approved by the US FDA and other drug regulatory organizations for clinical use, and numerous additional inhibitors are currently in various phases of clinical trials [67]. With the advent of advancement in sequencing technology, the complexity and involvement of RTKs in disease progression increases, providing new avenues for drug discovery. Recent understanding of the pathophysiology of cancer marked the importance of RTKs in targeted therapy and revolutionized the course of disease management. Here we discussed some of the approved small molecule kinase inhibitors against the specific RTKs primarily involved in the development and progression of cancer.

5.1 EGFR-Tyrosine Kinase Inhibitors

The epidermal growth factor receptor (EGFR/HER1) is a member of the ErbB family, which comprises four structurally related RTKs: EGFR/HER1, HER2, HER3, and HER4 [68]. EGFR or members of the HER family are known to promote tumor growth by activating four key signaling cascades, including MAPK, PI3K/AKT/

mTOR, JAK/STAT, and PLCγ [69]. Mutations in EGFR are critical for the progression and spread of tumor cells, particularly non-small cell lung cancer (NSCLC) where 40–80% of patients display EGFR overexpression resulting from epigenetic aberrations and changes in gene copy number [19, 70]. Research on EGFR-TKIs has been relatively extensive with four generations being developed, and they have proven indispensable in the treatment of patients with EGFR-mutant non-small cell lung cancer (NSCLC) [71].

Reversible, first-generation EGFR-TKIs (gefitinib, erlotinib) were initially intended for patients with wild-type EGFR overexpression [72]. Nevertheless, individuals with advanced non-small cell lung cancer (NSCLC) who carry EGFR-activating mutations (exon 19 deletion and exon 21 L858R substitution) have benefited greatly from these TKIs in terms of survival advantages [73, 74]. Given that the combination of erlotinib with gemcitabine considerably prolongs overall survival (OS) of pancreatic cancer patients overexpressing EGFR compared to gemcitabine alone, the combination was also approved for use as first-line treatment for the disease [75]. The pan-HER inhibitors, dacomitinib, and afatinib, are irreversible second-generation EGFR TKIs that exhibit enhanced cellular efficacy toward EGFR oncogenic variants (EGFR-L858R/T790M) [76–78]. Both dacomitinib and afatinib, however, present higher toxicities and acquired resistance development (about 60% of these resistant cases contain the secondary T790M mutation), restricting their application in clinical settings [79–81]. This led to the development of third-generation EGFR TKIs (osimertinib, olmutinib and rociletinib) in an effort to manage T790M-directed resistance [82]. Osimertinib displays the potential to suppress both EGFR-sensitizing and T790M mutations and was first approved for the treatment of T790M positive advanced NSCLC [83]. Additionally, osimertinib has the potential to cross the blood-brain barrier [84]. The efficacy of osimertinib was further identified to be favorable in patients with central nervous system (CNS) metastases treated with or without EGFR-TKIs [85] or those harboring T790M EGFR mutation [83]. Subsequently, in the phase III FLAURA trial, osimertinib was found to be more effective than erlotinib or gefitinib as a first-line treatment for NSCLC with an EGFR mutation (NCT02296125) [86, 87]. A significant survival benefit was also observed with osimertinib as adjuvant therapy after complete tumor resection in stage IB–IIIA NSCLC patients carrying EGFRmutations (NCT02511106) [88, 89].

Another third-generation EGFR-TKI, almonertinib received approval from the National Medical Products Administration (NMPA) of China. Almonertinib was found to be efficacious with an objective response rate (ORR) of 68.9% along with CNS ORR of 60.9% in patients with prior therapy for EGFR T790M-positive NSCLC (NCT02981108) [90]. In advanced NSCLC with positive EGFR T790M mutation, several other third-generation EGFR-TKIs, such as furmonertinib (AST2818) [91, 92], rezivertinib (BPI-7711) [93, 94], lazertinib (YH25448) [95–97], and nazartinib (EGF816) [98], have also demonstrated acceptable safety and reasonable outcomes. Acquired resistance mutations frequently impair the long-term effects of third-generation EGFR-TKIs and the most common cause is the EGFR exon 20 C797S mutation [99]. To overcome osimertinib resistance facilitated by mutation mechanisms involving EGFR, fourth-generation EGFR-TKIs are currently being developed, for instance EAI045 against mutations affecting EGFR T790M and C797S [100], suppression of EGFR triple mutants (D746-750/T790M/C797S and L858R/T790M/C797S) with TQB3804 [101] and (19del/T790M/C797S and L858R/T790M/C797S) with LS-106 [102]. Furthermore, patients with the triple mutation of EGFR/T790M/cis-C797S have demonstrated preliminary effectiveness while administering cetuximab and brigatinib combination [103]. Though combination treatment shows efficacious results, research into the mechanisms underlying resistance to first-, second-, and third-generation EGFR TKIs is necessary to develop novel therapeutic approaches for NSCLC with EGFR-activating mutations.

5.2 VEGFR-Tyrosine Kinase Inhibitors

Vascular endothelial growth factor receptor (VEGFR) is an important cell membrane receptor that interacts with VEGF to stimulate angiogenesis, which is the formation of new blood vessels, and to maintain existing vasculature. This interaction is vital not only for normal physiological processes but also for pathological conditions, such as tumor growth. VEGFR-associated multi-targeted TKIs (e.g., sorafenib, sunitinib, pazopanib, axitinib, cabozantinib, regorafenib, lenvatinib, anlotinib, and others) have shown great promise as anti-tumor agents in the fight against various solid cancers in the last few years [104]. Sorafenib is the first approved anti-angiogenic receptor TKI as front-line treatment for the advanced hepatocellular carcinoma (HCC), targeting VEGFR-1/2/3, PDGFR-β, FGFR, and c-Kit receptors [105, 106]. In contrast to sorafenib, lenvatinib showed non-inferior median progression-free survival (mPFS) (7.4 vs 3.7 months; $p < 0.0001$) and median overall survival (mOS) (13.6 vs 12.3 months; HR 0.92). While Lenvatinib was associated with fewer cases of grade 3 palmar-plantar erythrodysesthesia (hand-foot syndrome), it was linked to a higher incidence of hypertension and proteinuria in unresectable HCC (NCT01761266) [107]. More recently, individuals with untreated or unresectable HCC demonstrated encouraging anti-tumor efficacy with mPFS of 8.6 months, and a mOS of 22.0 months when treated with a combination of lenvatinib and pembrolizumab, an anti-programmed cell death protein-1 (PD-1) antibody (NCT03006926) [108]. The U.S. FDA (United States, Food and Drug Administration) designated this combination as a breakthrough treatment. Sunitinib, was the first cancer medication to be licensed by the U.S. FDA for the concurrent treatment of gastrointestinal stromal tumors (GISTs) refractory to imatinib (NCT00075218) [109] and renal cell carcinoma (RCC) [110]. Later other TKIs like cabozantinib [111, 112], pazopanib [113], and axitinib [114] were also approved for the treatment of metastatic RCC. Anlotinib has shown promising results in the treatment of lung cancer. The National Medical Products Administration of China has approved it as a third-line or further-line therapeutic for both NSCLC [115] and relapsed small cell lung cancer (SCLC) [116]. Recently, a phase 2 study demonstrated that combining anlotinib with sintilimab, an anti-PD-1 monoclonal antibody, as a second or third-line treatment for extensive disease SCLC, yielded further encouraging anticancer activity with mild toxicities [117]. Additionally, patients with thyroid carcinoma [118, 119], soft tissue sarcoma (STS) [120], and other solid malignancies showed improved survival rates when treated with VEGFR-associated multi-targeted TKIs.

5.3 PDGFR-Tyrosine Kinase Inhibitors

Platelet-derived growth factor receptor (PDGFR) is discovered to play a critical function in the proliferation, metastasis, invasion, and angiogenesis of cancer cells via modulation of several regulatory pathways, such as PI3K/AKT, MAPK, and JAK/STAT signaling [25]. PDGFR inhibition is crucial, particularly in the therapy of gastrointestinal stromal tumors (GIST) as approximately 10% of GISTs involve PDGFR mutations [121]. Imatinib is the first-line treatment, an oral TKI for advanced GISTs which competitively inhibits KIT and PDGFRA's ATP binding sites [122, 123]. However, imatinib exhibits activity against non-D842V PDGFRA-mutated tumors but is found to be ineffective against tumors with the D842V mutation, with notable variations in the objective response rates between the two subtypes [124]. For advanced GIST, VEGFR multi-kinase TKIs sunitinib and regorafenib are recommended as second and third-line treatments, respectively [125, 126]. Other multi-kinase inhibitors targeting VEGFR (sunitinib, sorafenib, axitinib, lenvatinib, regorafenib, and pazopanib) have also been identified to target PDGFR to enhance anti-angiogenesis effects and suppress tumor growth in RCC [127–130] and HCC [131]. Avapritinib and ripretinib, two selective TKIs aimed at KIT and PDGFRA mutations, were recently approved as a fourth-line treatment for GIST. Avapritinib's phase I trial showed an 86% ORR in GIST individuals with a PDGFRA exon 18 mutation and a 22% ORR in individuals who had not responded to at least third-line treatment [132]. Likewise, in Chinese patients with GIST who had PDGFRA D842V mutations, avapritinib was found to be well tolerated, to exhibit significant anti-tumor activity, and to be notably effective as fourth- or later-line treatment (NCT04254939) [133]. Comparing ripretinib to a placebo, a phase III trial showed better mPFS (6.3 vs. 1.0 months; HR 0.15; $p < 0.0001$) and mOS (15.1 vs. 6.6 months; HR 0.36; $p = 0.0004$) in case of advanced GIST [134].

5.4 FGFR-Tyrosine Kinase Inhibitors

Fibroblast growth factor receptors (FGFRs) are important for stimulating cancer cell proliferation, differentiation, and migration [59]. Given the broad range of FGFR genetic alterations reported across various cancer types, it is not surprising that numerous FGFR-directed therapies are currently being explored in preclinical and clinical investigations. These therapies for targeting FGFR activation include small-molecule receptor TKIs, which can be categorized as either non-selective or selective.

5.4.1 Non-selective FGFR TKIs

The FGFR kinase domain shares significant similarities with other receptor tyrosine kinases, such as PDGFR and VEGFR. Non-selective tyrosine kinase inhibitors (TKIs), which compete with ATP for the ATP-binding site, are basically multi-target TKIs. Examples include dovitinib, ponatinib, nintedanib, lucitanib, derazantinib, anlotinib, and sunitinib. These inhibitors target FGFR but also have the potential to affect a variety of other tyrosine kinases. For instance, dovitinib (TKI258) has inhibitory activity against VEGFR1-3, PDGFR-β, FLT-3, FGFR1-3, KIT, TrkA, RET, and CSF-1. Despite its broad targeting capabilities, dovitinib's clinical application is limited by its toxicity [135], a common issue with many non-selective inhibitors. Nevertheless, non-selective FGFR inhibitors may have the potential to overcome therapy-related secondary alterations that lead to resistance. For instance, ponatinib, which is approved for treating Philadelphia chromosome-positive acute lymphoblastic leukemia, has shown effectiveness against the BCR–ABL T315I gatekeeper mutation which causes resistance to imatinib (NCT03589326) [136, 137]. FDA-approved nintedanib (BIBF1120) for treating idiopathic pulmonary fibrosis, a rare condition. However, it did not demonstrate sufficient activity as a monotherapy to justify a phase III trial compared to chemotherapy in patients with recurrent clear cell carcinoma of the ovary or endometrium [138]. Likewise, sunitinib did not show an antitumor effect in individuals with breast cancer who had FGFR1 or FGFR2 mutations (NCT02693535) [139]. Additionally, other non-selective TKIs such as Lucitanib [137], derazantinib [138], zotatifin [140], and anlotinib [141, 142] are currently being evaluated in clinical studies for tumors with FGFR mutations.

5.4.2 Selective FGFR TKIs

Selective inhibitors targeting FGFRs have been developed to improve on-target efficacy and reduce adverse effects associated with multi-target FGFR TKIs. The first generation of these selective inhibitors, known as pan-FGFR (1-4) non-covalent inhibitors, includes erdafitinib (JNJ-42756493), AZD4547, infigratinib (BGJ398), Debio1347, LY2874455, pemigatinib, E7090, and rogaratinib [143]. Due to structural differences, FGFR4 is less effectively targeted by most pan-FGFR inhibitors, with erdafitinib and LY287445 being notable exceptions [144, 145].

The U.S. FDA has approved three FGFR inhibitors on the basis of their efficacy for specific cancers, such as erdafitinib for urothelial carcinoma, pemigatinib, and infigratinib for cholangiocarcinoma. Additionally, alofanib, a selective FGFR2 inhibitor, blocks FGFR2 signaling through an allosteric mechanism [146]. Recent developments have focused on irreversible inhibitors, which form a covalent bond that prevents ATP from easily dislodging them, resulting in prolonged suppression [147]. Phase 1/2 clinical trials are currently evaluating futibatinib (TAS-120), a new pan-FGFR inhibitor that forms a covalent adduct with a cysteine side chain in the FGFR phosphate-binding loop [148, 149]. Similar irreversible interaction approaches have been employed by other covalent inhibitors, such as fisogatinib (BLU-554), FGF401 (roblitinib), PRN1371, and INCB062079, though they primarily target FGFR4.

5.5 HGFR/MET-Tyrosine Kinase Inhibitors

The interaction of hepatocyte growth factor receptor (HGFR), also known as mesenchymal-epithelial transition factor (MET), and its ligand HGF stimulates several molecular pathways crucial for cancer cell growth and development [150]. Recent advancements have identified potential responders to MET inhibitors through the detection of MET amplification or MET exon 14 skipping mutations (METex14).

Crizotinib (PF-02341066) is a multitargeted TKI that inhibits MET, ALK, and ROS1 by competing with ATP. It has shown significant efficacy against METex14 and MET amplification in non-small cell lung cancer (NSCLC) [151, 152]. Other multi-targeted TKIs with notable activity against MET include cabozantinib, merestinib, glesatinib, and sitravatinib [153].

Selective MET inhibitors also show promise for enhancing survival. Tepotinib, camaptinib, and savolitinib are examples. Following the failure of platinum-based therapy, the FDA designated temotinib as a breakthrough therapy for the treatment of metastatic or METex14-skipping NSCLC (NCT02864992) [154]. Capmatinib has been approved for the treatment of adult NSCLC patients with METex14 mutations (NCT05110196), regardless of prior medication history. However, further Phase III studies are needed to compare capmatinib's efficacy and tolerability with other established treatments, as well as real-world investigations to assess the drug's adverse effects in clinical settings [155, 156]. Savolitinib, also known as volitinib, has demonstrated encouraging anti-tumor effectiveness and acceptable toxicity in pulmonary sarcomatoid carcinoma (PSC) and other NSCLC types with METex14-mutation, achieving an overall response rate (ORR) of 47.5%. Notably, liver damage and hypersensitivity were the most common adverse events, occurring in 2.9% of patients, and accounted for 14.3% of discontinuation due to treatment-related adverse events (TRAEs) [157].

6 Conclusion and Future Perspective

Ongoing research over the past few decades on RTKs has substantially advanced our understanding of cellular signaling pathways. Recent breakthroughs in structural biology, molecular biology, and sequencing technology have deepened our comprehension of dysregulated RTK signaling pathways in various diseases, including cancer. This expanded knowledge lays the groundwork for precision treatments that could positively impact millions of cancers worldwide. RTK-targeted therapies, including monoclonal antibodies and small molecule inhibitors, have marked the advent of a transformative phase in cancer management, representing a significant advancement in precision medicine. The efficacy of RTK-targeted therapies lies in their ability to selectively interfere with aberrant signaling pathways that drive tumor growth and progression while sparing normal cells. This targeted approach minimizes systemic toxicity and offers patients a more tolerable treatment option compared to traditional chemotherapy. The remarkable success of RTK-targeted therapies has not only reshaped cancer treatment but also sparked interest in their potential applications for other diseases beyond oncology.

Despite their advancements, challenges remain, including drug resistance and the need for more personalized approaches. However, emerging technologies such as CRISPR-based gene editing and single-cell sequencing present new opportunities to identify promising therapeutic strategies. By leveraging computational algorithms and multi-omics data, we can gain a deeper understanding of RTK signaling networks and develop combination therapies targeting multiple nodes within these pathways [158–160]. This approach promises to enhance progression-free survival and refine cancer management. Ultimately, the integration of molecular research with clinical applications will pave the way for a new era in cancer management, characterized by personalized and efficient therapies.

Acknowledgements The figures are created with Biorender.com

References

1. Schlessinger J. Receptor tyrosine kinases: legacy of the first two decades. Cold Spring Harb Perspect Biol. 2014;6:a008912.
2. Nair A, et al. Conceptual evolution of cell signaling. Int J Mol Sci. 2019;20:3292.
3. Li E, Hristova K. Role of receptor tyrosine kinase transmembrane domains in cell signaling and human pathologies. Biochemistry. 2006;45:6241–51.
4. Sabarwal A, et al. A combination therapy using an mTOR inhibitor and Honokiol effectively induces autophagy through the modulation of AXL and Rubicon in renal cancer cells and restricts renal tumor growth following organ transplantation. Carcinogenesis. 2022;46:360–70.
5. Levi-Montalcini R, Cohen S. *In vitro* and *in vivo* effects of a nerve growth-stimulating agent isolated from snake venom. Proc Natl Acad Sci USA. 1956;42:695–9.
6. Cohen S, Levi-Montalcini R. Purification and properties of a nerve growth-promoting factor isolated from mouse sarcoma 180. Cancer Res. 1957;17:5–20.
7. Cohen S. Origins of growth factors: NGF and EGF. J Biol Chem. 2008;283:33793–7.
8. Gospodarowicz D, et al. The angiogenic activity of the fibroblast and epidermal growth factor. Exp Eye Res. 1979;28:501–14.
9. Gospodarowicz D, et al. Fibroblast growth factor: structural and biological properties. J Cell Physiol. 1987;133:15–26.
10. Antoniades HN, Hunkapiller MW. Human platelet-derived growth factor (PDGF): amino-terminal amino acid sequence. Science. 1983;220:963–5.
11. Heldin C-H, Westermark B. Mechanism of action and *in vivo* role of platelet-derived growth factor. Physiol Rev. 1999;79:1283.
12. Ullrich A, et al. Human epidermal growth factor receptor cDNA sequence and aberrant expression of the amplified gene in A431 epidermoid carcinoma cells. Nature. 1984;309:418–25.
13. Lemmon MA, Schlessinger J. Cell signaling by receptor tyrosine kinases. Cell. 2010;141:1117–34.
14. Diwanji D, et al. More than the sum of the parts: toward full-length receptor tyrosine kinase structures. IUBMB Life. 2019;71:706–20.
15. Regad T. Targeting RTK signaling pathways in cancer. Cancers. 2015;7:1758–84.
16. Trenker R, Jura N. Receptor tyrosine kinase activation: from the ligand perspective. Curr Opin Cell Biol. 2020;63:174–85.

17. Robinson DR, et al. The protein tyrosine kinase family of the human genome. Oncogene. 2000;19:5548–57.
18. Ségaliny AI, et al. Receptor tyrosine kinases: characterisation, mechanism of action and therapeutic interests for bone cancers. J Bone Oncol. 2015;4:1–12.
19. Yamaoka T, et al. Receptor tyrosine kinase-targeted cancer therapy. Int J Mol Sci. 2018;19:3491.
20. Zhang N, Li Y. Receptor tyrosine kinases: biological functions and anticancer targeted therapy. MedComm. 2023;4:e446.
21. Sigismund S, et al. Emerging functions of the EGFR in cancer. Mol Oncol. 2018;12:3–20.
22. Thomas R, Zhang W. Rethink of EGFR in cancer with its kinase independent function on board. Front Oncol. 2019;9:800.
23. Farooqi AA, Siddik ZH. Platelet-derived growth factor (PDGF) signalling in cancer: rapidly emerging signalling landscape. Cell Biochem Funct. 2015;33:257–65.
24. Pandey P, et al. New insights about the PDGF/PDGFR signaling pathway as a promising target to develop cancer therapeutic strategies. Biomed Pharmacother. 2023;161:114491.
25. Zou X, et al. Targeting the PDGF/PDGFR signaling pathway for cancer therapy: a review. Int J Biol Macromol. 2022;202:539–57.
26. Apte RS, et al. VEGF in signaling and disease: beyond discovery and development. Cell. 2019;176:248–1264.
27. Liu Y, et al. Recent progress on vascular endothelial growth factor receptor inhibitors with dual targeting capabilities for tumor therapy. J Hematol Oncol. 2022;15:89.
28. Patel SA, et al. Molecular mechanisms and future implications of VEGF/VEGFR in cancer therapy. Clin Cancer Res. 2023;29:30–9.
29. Tiong KH, et al. Functional roles of fibroblast growth factor receptors (FGFRs) signaling in human cancers. Apoptosis. 2013;18:1447–68.
30. Du S, et al. Current progress in cancer treatment by targeting FGFR signaling. Cancer Biol Med. 2023;20:490.
31. Bahar ME, et al. Targeting the RAS/RAF/MAPK pathway for cancer therapy: from mechanism to clinical studies. Signal Transduct Target Ther. 2023;8:455.
32. Sjöberg E, et al. Endothelial VEGFR2-PLCγ signaling regulates vascular permeability and antitumor immunity through eNOS/Src. J Clin Invest. 2023;133:e161366.
33. Hu X, et al. The JAK/STAT signaling pathway: from bench to clinic. Signal Transduct Target Ther. 2021;6:402.
34. He Y, et al. Targeting PI3K/Akt signal transduction for cancer therapy. Signal Transduct Target Ther. 2021;6:425.
35. Lee CC, et al. Epithelial cell adhesion molecule (EpCAM) regulates HGFR signaling to promote colon cancer progression and metastasis. J Transl Med. 2023;21:530.
36. Sattler M, et al. The role of the c-Met pathway in lung cancer and the potential for targeted therapy. Ther Adv Med Oncol. 2011;3:171–84.
37. Leppänen VM, et al. Structural and mechanistic insights into VEGF receptor 3 ligand binding and activation. Proc Natl Acad Sci USA. 2013;110:12960–5.
38. Yuzawa S, et al. Structural basis for activation of the receptor tyrosine kinase KIT by stem cell factor. Cell. 2007;130:323–34.
39. Ward CW, et al. The insulin and EGF receptor structures: new insights into ligand-induced receptor activation. Trends Biochem Sci. 2007;32:129–37.
40. Yayon A, et al. Cell surface, heparin-like molecules are required for binding of basic fibroblast growth factor to its high affinity receptor. Cell. 1991;64:841–8.
41. Schlessinger J, Lemmon MA. SH2 and PTB domains in tyrosine kinase signalling. Sci STKE. 2003;2003:re12.
42. Hubbard SR, Miller WT. Receptor tyrosine kinases: mechanisms of activation and signaling. Curr Opin Cell Biol. 2007;19:117–23.
43. Wintheiser GA, Silberstein P. Physiology, tyrosine kinase receptors. In: StatPearls [Internet]. Treasure Island (FL): StatPearls Publishing; 2019.

44. Cordover E, et al. Signaling pathways downstream to receptor tyrosine kinases: targets for cancer treatment. J Cancer Metastasis Treat. 2020;6:45.
45. Montor WR, et al. Receptor tyrosine kinases and downstream pathways as druggable targets for cancer treatment: the current arsenal of inhibitors. Mol Cancer. 2018;17:1–18.
46. Jänne PA, et al. Epidermal growth factor receptor mutations in non–small-cell lung cancer: implications for treatment and tumor biology. J Clin Oncol. 2005;23:3227–34.
47. Sharma SV, et al. Epidermal growth factor receptor mutations in lung cancer. Nat Rev Cancer. 2007;7:169–81.
48. Sondka Z, et al. The COSMIC Cancer Gene Census: describing genetic dysfunction across all human cancers. Nat Rev Cancer. 2018;18:696–705.
49. Lee JC, et al. Epidermal growth factor receptor activation in glioblastoma through novel missense mutations in the extracellular domain. PLoS Med. 2006;3:e485.
50. Wu R, et al. Somatic mutations of fibroblast growth factor receptor 3 (FGFR3) are uncommon in carcinomas of the uterine cervix. Oncogene. 2000;19:5543–6.
51. Ou SHI, et al. HER2 transmembrane domain (TMD) mutations (V659/G660) that stabilize homo-and heterodimerization are rare oncogenic drivers in lung adenocarcinoma that respond to afatinib. J Thorac Oncol. 2017;12:446–57.
52. Hirota S, et al. Gain-of-function mutations of c-kit in human gastrointestinal stromal tumors. Science. 1998;279:577–80.
53. Estevam GO, et al. Conserved regulatory motifs in the juxtamembrane domain and kinase N-lobe revealed through deep mutational scanning of the MET receptor tyrosine kinase domain. bioRxiv. 2024;12:RP91619.
54. Kato S, et al. Revisiting epidermal growth factor receptor (EGFR) amplification as a target for anti-EGFR therapy: analysis of cell-free circulating tumor DNA in patients with advanced malignancies. JCO Precis Oncol. 2019;3:1–14.
55. Yan M, et al. HER2 expression status in diverse cancers: review of results from 37,992 patients. Cancer Metastasis Rev. 2015;34:157–64.
56. Guo B, et al. Prognostic value of MET gene copy number and protein expression in patients with surgically resected non-small cell lung cancer: a meta-analysis of published literatures. PLoS One. 2014;9:e99399.
57. Sabarwal A, et al. A novel combination treatment with honokiol and rapamycin effectively restricts c-Met-induced growth of renal cancer cells, and also inhibits the expression of tumor cell PD-L1 involved in immune escape. Cancers. 2020;12:1782.
58. Rawat L, et al. A novel combination therapy with Cabozantinib and Honokiol effectively inhibits c-Met-Nrf2-induced renal tumor growth through increased oxidative stress. Redox Biol. 2023;68:102945.
59. Krook MA, et al. Fibroblast growth factor receptors in cancer: genetic alterations, diagnostics, therapeutic targets and mechanisms of resistance. Br J Cancer. 2021;124:880–92.
60. Melo JV, et al. The ABL-BCR fusion gene is expressed in chronic myeloid leukemia. Blood. 1993;81:158–65.
61. Ahn JS, Kim HJ. FLT3 mutations in acute myeloid leukemia: a review focusing on clinically applicable drugs. Blood Res. 2022;57:32–6.
62. Du Z, Lovly CM. Mechanisms of receptor tyrosine kinase activation in cancer. Mol Cancer. 2018;17:1–13.
63. Liu BA, et al. RC Homology 2 Domain Binding Sites in Insulin, IGF-1 and FGF receptor mediated signaling networks reveal an extensive potential interactome. Cell Commun Signal. 2012;10:1–23.
64. MacFarlane LA, Murphy PR. MicroRNA: biogenesis, function and role in cancer. Curr Genomics. 2010;11:537–61.
65. Webster RJ, et al. Regulation of epidermal growth factor receptor signaling in human cancer cells by microRNA-7. J Biol Chem. 2009;284:5731–41.
66. Xiong B, et al. MiR-21 regulates biological behavior through the PTEN/PI-3 K/Akt signaling pathway in human colorectal cancer cells. Int J Oncol. 2013;42:219–28.

67. Kumar R, et al. Recent developments in receptor tyrosine kinase inhibitors: a promising mainstay in targeted cancer therapy. Med Drug Discov. 2024;23:100195.
68. Yarden Y, Pines G. The ERBB network: at last, cancer therapy meets systems biology. Nat Rev Cancer. 2012;12:553–63.
69. Ayati A, et al. A review on progression of epidermal growth factor receptor (EGFR) inhibitors as an efficient approach in cancer targeted therapy. Bioorg Chem. 2020;99:103811.
70. Tan AC, Tan DSW. Targeted therapies for lung cancer patients with oncogenic driver molecular alterations. J Clin Oncol. 2022;40:611–25.
71. Zhang C, et al. Emerging therapies for non-small cell lung cancer. J Hematol Oncol. 2019;12:45.
72. Cappuzzo F, et al. Gefitinib in pretreated non-small-cell lung cancer (NSCLC): analysis of efficacy and correlation with HER2 and epidermal growth factor receptor expression in locally advanced or metastatic NSCLC. J Clin Oncol. 2003;21:2658–63.
73. Sim EH, et al. Gefitinib for advanced non-small cell lung cancer. Cochrane Database Syst Rev. 2018;1:CD006847.
74. Landi L, Cappuzzo F. Experience with erlotinib in the treatment of non-small cell lung cancer. Ther Adv Respir Dis. 2015;9:146–63.
75. Moore MJ, et al. Erlotinib plus gemcitabine compared with gemcitabine alone in patients with advanced pancreatic cancer: a phase III trial of the National Cancer Institute of Canada Clinical Trials Group. J Clin Oncol. 2007;25:1960–6.
76. Wirth SM. Afatinib in non-small cell lung cancer. J Adv Pract Oncol. 2015;6:448–55.
77. Brzezniak C, et al. Dacomitinib, a new therapy for the treatment of non-small cell lung cancer. Expert Opin Pharmacother. 2013;14:247–53.
78. Jain P, et al. Afatinib and lung cancer. Expert Rev Anticancer Ther. 2014;14:1391–406.
79. Denis MG, et al. EGFR T790M resistance mutation in non small-cell lung carcinoma. Clin Chim Acta. 2015;444:81–5.
80. Tomasello C, et al. Resistance to EGFR inhibitors in non-small cell lung cancer: clinical management and future perspectives. Crit Rev Oncol Hematol. 2018;123:149–61.
81. Lim SM, et al. Acquired resistance to EGFR targeted therapy in non-small cell lung cancer: mechanisms and therapeutic strategies. Cancer Treat Rev. 2018;65:1–10.
82. Cheng Z, et al. The advance of the third-generation EGFR-TKI in the treatment of non-small cell lung cancer (Review). Oncol Rep. 2024;51:1–13.
83. Mok TS, et al. Osimertinib or platinum–pemetrexed in EGFR T790M–positive lung cancer. N Engl J Med. 2017;376:629–40.
84. Cross DAE, et al. AZD9291, an irreversible EGFR TKI, overcomes T790M-mediated resistance to EGFR inhibitors in lung cancer. Cancer Discov. 2014;4:1046–61.
85. Reungwetwattana T, et al. CNS response to osimertinib versus standard epidermal growth factor receptor tyrosine kinase inhibitors in patients with untreated EGFR-mutated advanced non-small-cell lung cancer. J Clin Oncol. 2018:JCO2018783118. https://doi.org/10.1200/JCO.2018.78.3118.
86. Soria JC, et al. Osimertinib in untreated EGFR-mutated advanced non–small-cell lung cancer. N Engl J Med. 2018;378:113–25.
87. Ramalingam SS, et al. Overall survival with osimertinib in untreated, EGFR-mutated advanced NSCLC. N Engl J Med. 2020;382:41–50.
88. Wu YL, et al. Postoperative chemotherapy use and outcomes from ADAURA: osimertinib as adjuvant therapy for resected EGFR-mutated NSCLC. J Thorac Oncol. 2022;17:423–33.
89. Herbst RS, et al. Osimertinib as adjuvant therapy in patients (pts) with stage IB–IIIA EGFR mutation positive (EGFRm) NSCLC after complete tumor resection: ADAURA. J Clin Oncol. 2020;38:LBA5.
90. Lu S, et al. Efficacy of aumolertinib (HS-10296) in patients with advanced EGFR T790M+ NSCLC: updated post-national medical products administration approval results from the APOLLO Registrational Trial. J Thorac Oncol. 2022;17:411–22.

91. Zhang X, et al. 48P Efficacy analysis and mechanism exploration of furmonertinib for advanced NSCLC with EGFR exon 20 insertion mutation. ESMO Open. Abstract. 2023;8:Supplement 2100906.
92. Ding J, et al. Furmonertinib for EGFR-mutant advanced non-small cell lung cancer: a glittering diamond in the rough of EGFR-TKI. Front Pharmacol. 2024;15:1357913.
93. Shi Y, et al. A phase I study to evaluate safety and antitumor activity of BPI-7711 in EGFRM+/T790M+ advanced or recurrent NSCLC patients. J Thorac Oncol. 2019;14:S1126–7.
94. Wu S, et al. Efficacy and safety of rezivertinib (BPI-7711) in patients with locally advanced or metastatic/recurrent EGFR T790M-mutated NSCLC: a phase IIb, multicenter, single-arm, open-label study. J Clin Oncol. 2022;40:9098.
95. Hong MH, et al. Phase II trial of lazertinib in patients with epidermal growth factor receptor (EGFR) mutation-positive (M+), metastatic non-small cell lung cancer (NSCLC) with asymptomatic or mild symptomatic brain metastases after failure of EGFR tyrosine kinase inhibitor (TKI) (KCSG LU20-15). J Clin Oncol. 2023;41:9054.
96. Cho BC, et al. Lazertinib as a frontline treatment in patients with EGFR-mutated advanced non-small cell lung cancer: long-term follow-up results from LASER201. Lung Cancer. 2024;190:107509.
97. Cho BC, et al. Lazertinib versus Gefitinib as first-line treatment in patients with EGFR-mutated advanced non–small-cell lung cancer: results from LASER301. J Clin Oncol. 2023;41:4208–17.
98. Tan DSW, et al. Nazartinib for treatment-naive EGFR-mutant non-small cell lung cancer: results of a phase 2, single-arm, open-label study. Eur J Cancer. 1990;2022(172):276–86.
99. Le T, Gerber DE. Newer-generation EGFR inhibitors in lung cancer: how are they best used? Cancers. 2019;11:366.
100. Wang S, et al. EAI045: the fourth-generation EGFR inhibitor overcoming T790M and C797S resistance. Cancer Lett. 2017;385:51–4.
101. Liu X, et al. Abstract 1320: preclinical evaluation of TQB3804, a potent EGFR C797S inhibitor. Cancer Res. 2019;79(13_Supplement):1320.
102. Liu Y, et al. LS-106, a novel EGFR inhibitor targeting C797S, exhibits antitumor activities both *in vitro* and *in vivo*. Cancer Sci. 2022;113:709–20.
103. Wang Y, et al. Effective treatment of lung adenocarcinoma harboring EGFR-activating mutation, T790M, and cis-C797S triple mutations by brigatinib and cetuximab combination therapy. J Thorac Oncol. 2020;15:1369–75.
104. Qin S, et al. Recent advances on anti-angiogenesis receptor tyrosine kinase inhibitors in cancer therapy. J Hematol Oncol. 2019;12:27.
105. Llovet JM, et al. Sorafenib in advanced hepatocellular carcinoma. N Engl J Med. 2008;359:378–90.
106. Cheng AL, et al. Efficacy and safety of sorafenib in patients in the Asia-Pacific region with advanced hepatocellular carcinoma: a phase III randomised, double-blind, placebo-controlled trial. Lancet Oncol. 2009;10:25–34.
107. Kudo M, et al. Lenvatinib versus sorafenib in first-line treatment of patients with unresectable hepatocellular carcinoma: a randomised phase 3 non-inferiority trial. Lancet. 2018;391:1163–73.
108. Finn RS, et al. Phase Ib study of lenvatinib plus pembrolizumab in patients with unresectable hepatocellular carcinoma. J Clin Oncol. 2020;38:2960–70.
109. Demetri GD, et al. Efficacy and safety of sunitinib in patients with advanced gastrointestinal stromal tumour after failure of imatinib: a randomised controlled trial. Lancet. 2006;368:1329–38.
110. Motzer RJ, et al. Overall survival and updated results for sunitinib compared with interferon alfa in patients with metastatic renal cell carcinoma. J Clin Oncol. 2009;27:3584–90.
111. Choueiri TK, et al. Cabozantinib versus everolimus in advanced renal-cell carcinoma. N Engl J Med. 2015;373:1814–23.

112. Choueiri TK, et al. Cabozantinib versus everolimus in advanced renal cell carcinoma (METEOR): final results from a randomised, open-label, phase 3 trial. Lancet Oncol. 2016;17:917–27.
113. Sternberg CN, et al. Pazopanib in locally advanced or metastatic renal cell carcinoma: results of a randomized phase III trial. J Clin Oncol. 2010;28:1061–8.
114. Zarrabi K, et al. New treatment options for metastatic renal cell carcinoma with prior anti-angiogenesis therapy. J Hematol Oncol. 2017;10:38.
115. Han B, et al. Effect of anlotinib as a third-line or further treatment on overall survival of patients with advanced non–small cell lung cancer. JAMA Oncol. 2018;4:1569–75.
116. Cheng Y, et al. Anlotinib vs placebo as third- or further-line treatment for patients with small cell lung cancer: a randomised, double-blind, placebo-controlled Phase 2 study. Br J Cancer. 2021;125:366–71.
117. Ma S, et al. Sintilimab plus anlotinib as second or further-line therapy for extensive disease small cell lung cancer: a phase 2 investigator-initiated non-randomized controlled trial. EClinicalMedicine. 2024;70:102543.
118. Elisei R, et al. Cabozantinib in progressive medullary thyroid cancer. J Clin Oncol. 2013;31:3639–46.
119. Brose MS, et al. Cabozantinib for radioiodine-refractory differentiated thyroid cancer (COSMIC-311): a randomised, double-blind, placebo-controlled, phase 3 trial. Lancet Oncol. 2021;22:1126–38.
120. Poddubskaya EV, et al. Personalized prescription of tyrosine kinase inhibitors in unresectable metastatic cholangiocarcinoma. Exp Hematol Oncol. 2018;7:21.
121. Schaefer IM, et al. The GIST of advances in treatment of advanced gastrointestinal stromal tumor. Am Soc Clin Oncol Educ Book. 2022;42:885–99.
122. Reichardt P, et al. Safety of imatinib mesylate in a multicenter expanded access program in adult patients with gastrointestinal stromal tumors in the adjuvant setting. Oncol Res Treat. 2019;42:629–35.
123. Awatsu T, et al. A retrospective analysis of the efficacy and safety of imatinib in elderly patients with advanced gastrointestinal stromal tumor. J Clin Oncol. 2024;42(suppl):730.
124. Yoo C, et al. Efficacy of imatinib in patients with platelet-derived growth factor receptor alpha-mutated gastrointestinal stromal tumors. Cancer Res Treat. 2016;48:546–52.
125. Goodman VL, et al. Approval summary: sunitinib for the treatment of imatinib refractory or intolerant gastrointestinal stromal tumors and advanced renal cell carcinoma. Clin Cancer Res. 2007;13:1367–73.
126. Demetri GD, et al. Efficacy and safety of regorafenib for advanced gastrointestinal stromal tumours after failure of imatinib and sunitinib (GRID): an international, multicentre, randomised, placebo-controlled, phase 3 trial. Lancet. 2013;381:295–302.
127. Yang L, et al. Efficacy and safety of sorafenib in advanced renal cell carcinoma patients: results from a long-term study. Oncol Lett. 2012;3:935–9.
128. Wang B, et al. First-line pazopanib treatment in metastatic renal cell carcinoma: real-world data from a Single Chinese Center. Front Pharmacol. 2020;11:517672.
129. Rini BI, et al. Axitinib (AG-013736; AG) in patients (pts) with metastatic renal cell cancer (RCC) refractory to sorafenib. J Clin Oncol. 2007;25(suppl):5032.
130. Van der Wijngaart H, et al. Candidate biomarkers for treatment benefit from sunitinib in patients with advanced renal cell carcinoma using mass spectrometry-based (phospho)proteomics. Clin Proteomics. 2023;20:49.
131. Vogel A, et al. Hepatocellular carcinoma: ESMO Clinical Practice Guidelines for diagnosis, treatment and follow-up. Ann Oncol. 2018;29(Suppl):iv238–55.
132. Heinrich MC, et al. Clinical activity of avapritinib in ≥ fourth-line (4L+) and PDGFRA Exon 18 gastrointestinal stromal tumors (GIST). J Clin Oncol. 2019;37(suppl):11022.
133. Li J, et al. Efficacy and safety of avapritinib in treating unresectable or metastatic gastrointestinal stromal tumors: a phase I/II, open-label, multicenter study. Oncologist. 2022;28:187–e114.

134. Blay JY, et al. Ripretinib in patients with advanced gastrointestinal stromal tumours (INVICTUS): a double-blind, randomised, placebo-controlled, phase 3 trial. Lancet Oncol. 2020;21:923–34.
135. Sharma M, et al. Phase II study of Dovitinib in recurrent glioblastoma. J Neuro-Oncol. 2019;144:359–68.
136. Jabbour E, et al. First report of PhALLCON: a phase 3 study comparing ponatinib (pon) vs imatinib (im) in newly diagnosed patients with Philadelphia chromosome-positive acute lymphoblastic leukemia (Ph+ALL). J Clin Oncol. 2023;41(suppl):398868.
137. Jabbour E, et al. Dose modification dynamics of ponatinib in patients with chronic-phase chronic myeloid leukemia (CP-CML) from the PACE and OPTIC trials. Leukemia. 2024;38:475–81.
138. Glasspool R, et al. 596 A randomised phase II study of nintedanib (BIBF1120) compared to chemotherapy in patients with recurrent clear cell carcinoma of the ovary or endometrium. Int J Gynecol Cancer. 2020;30(Suppl 4). https://doi.org/10.1136/ijgc-2020-ESGO.223.
139. Calfa CJ, et al. Sunitinib in patients with breast cancer with FGFR1 or FGFR2 amplifications or mutations: results from the targeted agent and profiling utilization registry study. JCO Precis Oncol. 2024;8:e2300513.
140. Zhao N, et al. Targeting eIF4A triggers an interferon response to synergize with chemotherapy and suppress triple-negative breast cancer. J Clin Invest. 2023;133(24):e172503.
141. Meng Z, et al. *In vitro* and *in vivo* investigations of anlotinib in bladder cancer treatment. J Clin Oncol. 2022;40(suppl):e16505.
142. Sun Y, et al. Safety, pharmacokinetics, and antitumor properties of anlotinib, an oral multi-target tyrosine kinase inhibitor, in patients with advanced refractory solid tumors. J Hematol Oncol. 2016;9(1):105.
143. Zheng J, et al. Signaling pathway and small-molecule drug discovery of FGFR: a comprehensive review. Front Chem. 2022;10:860985.
144. Wu D, et al. Crystal structure of the FGFR4/LY2874455 complex reveals insights into the pan-FGFR selectivity of LY2874455. PLoS One. 2016;11:e0162491.
145. Perera TPS, et al. Discovery and pharmacological characterization of JNJ-42756493 (erdafitinib), a functionally selective small-molecule FGFR family inhibitor. Mol Cancer Ther. 2017;16:1010–20.
146. Tsimafeyeu I, et al. Targeting FGFR2 with alofanib (RPT835) shows potent activity in tumour models. Eur J Cancer. 1990;2016(61):20–8.
147. Fumarola C, et al. Expanding the arsenal of FGFR inhibitors: a novel chloroacetamide derivative as a new irreversible agent with anti-proliferative activity against FGFR1-amplified lung cancer cell lines. Front Oncol. 2019;9:179.
148. Kalyukina M, et al. TAS-120 cancer target binding: defining reactivity and revealing the first fibroblast growth factor receptor 1 (FGFR1) irreversible structure. ChemMedChem. 2019;14:494–500.
149. Goyal L, et al. Futibatinib for FGFR2-rearranged intrahepatic cholangiocarcinoma. N Engl J Med. 2023;388:228–39.
150. Crepaldi T, et al. The MET oncogene: thirty years of insights into molecular mechanisms driving malignancy. Pharmaceuticals. 2024;17:448.
151. Camidge DR, et al. Crizotinib in patients with MET-amplified NSCLC. J Thorac Oncol. 2021;16:1017–29.
152. Drilon A, et al. Antitumor activity of crizotinib in lung cancers harboring a MET exon 14 alteration. Nat Med. 2020;26:47–51.
153. Spagnolo CC, et al. Targeting MET in non-small cell lung cancer (NSCLC): a new old story? Int J Mol Sci. 2023;24(12):10119.
154. Mazieres J, et al. Tepotinib treatment in patients with MET exon 14-skipping non-small cell lung cancer: long-term follow-up of the VISION phase 2 nonrandomized clinical trial. JAMA Oncol. 2023;9:1260–6.
155. Hsu R, et al. The development and role of capmatinib in the treatment of MET-dysregulated non-small cell lung cancer—a narrative review. Cancers. 2023;15:3561.

156. Groen HJM, et al. Capmatinib in patients with METex14-mutated or high-level MET-amplified advanced non–small-cell lung cancer (NSCLC): results from cohort 6 of the phase 2 GEOMETRY mono-1 study. J Clin Oncol. 2020;38(suppl):9520.
157. Lu S, et al. Phase II study of savolitinib in patients (pts) with pulmonary sarcomatoid carcinoma (PSC) and other types of non-small cell lung cancer (NSCLC) harboring MET exon 14 skipping mutations (METex14+). J Clin Oncol. 2020;38(suppl):9519.
158. Sliwoski G, et al. Computational methods in drug discovery. Pharmacol Rev. 2014;66:334–95.
159. Kumar R, et al. Computational drug discovery and development along with their applications in the treatment of women-associated cancers. Milton Park, Abingdon-on-Thames, UK: CRC Press; 2022. p. 51–63.
160. You Y, et al. Artificial intelligence in cancer target identification and drug discovery. Signal Transduct Target Ther. 2022;7:156.

Cell Death Pathways in Cancer and Potential Therapeutic Targets

Reetika Ghosh, Sanchayeta Dey, Anowar Hussain, Gagan Deep, and Anand Shankar Ramteke

1 Introduction

Cancer has been one of the leading causes of death globally, with over 20 million new cases diagnosed in 2022 and 9.7 million deaths. Current trends of increasing cases predict 35 million new cases in 2050, a sharp rise of 77% as compared to 2022, along with 19 million deaths. Additionally, it has been predicted that one in five people will develop cancer in their lifetime, and one in nine men and one in twelve women will die from the disease, highlighting the need for a better understanding of disease mechanisms and finding new generation therapeutics as well as preventive strategies [1].

The development of cancer is a multistep process involving inter- and cross-play between genetic/epigenetic alterations and environmental factors that disrupt the normal functioning of a cell's biochemical and molecular cargos. The transformation process of a normal cell into a transformed/malignant one initiates with the accumulation of mutations in oncogenes, tumor suppressor genes, DNA repair

Reetika Ghosh and Sanchayeta Dey contributed equally as co-first authors.

R. Ghosh · S. Dey · A. S. Ramteke (✉)
Department of Molecular Biology and Biotechnology, Tezpur University, Tezpur, Sonitpur, Assam, India
e-mail: anand@tezu.ernet.in

A. Hussain (✉)
Department of Biophysics, Pub Kamrup College, Baihata Chariali, Kamrup, Assam, India

G. Deep
Department of Internal Medicine-Gerontology and Geriatric Medicine, Wake Forest University School of Medicine, Winston-Salem, NC, USA

Department of Cancer Biology, Wake Forest University School of Medicine, Winston-Salem, NC, USA

A. S. Ramteke et al. (eds.), *Molecular Targets in Cancer Therapy*,
https://doi.org/10.1007/978-981-96-7336-0_2

genes, etc., and is further fueled by additional mutations, microenvironmental changes, etc. The transformed or malignant cells develop certain characteristics, known as hallmarks of cancer, that differentiate them from their normal counterparts. These hallmarks include sustaining proliferative signals, evading growth suppressors, enabling replicative immortality, resisting cell death, inducing angiogenesis, activating invasion and metastasis, metabolic reprogramming, and immune evasion [2, 3]. A precise understanding of the above hallmarks provides a basis for developing targeted therapies for better clinical outcomes.

Among the hallmarks, resisting cell death plays one of the major roles in malignant transformation and tumor development. In normal physiology, tissue homeostasis is maintained through the programmed death of damaged cells, but transformed cells acquire the ability to evade these death pathways, contributing to uncontrolled proliferation and disease advancement (Fig. 1) [4, 5]. The programmed cell deaths (PCDs) are regulated by intrinsic and extrinsic mechanisms such as apoptosis, necroptosis, ferroptosis, and autophagy-dependent cell death. Each of these mechanisms is regulated by distinct molecular regulators, such as the intrinsic pathway by the BCL-2protein family, the extrinsic pathway by death receptors (Fas and TRAIL), necroptosis by receptor-interacting protein kinases (RIPK1 and RIPK3), and ferroptosis by lipid peroxidation and glutathione metabolism [5]. Inactivation or impairment in one or more mechanisms involved in cell death, contributes to progression, metastasis, immune invasion, and therapeutic resistance.

As many cancer cells develop resistance to conventional therapies such as chemotherapy and radiotherapy, which exert their effects by inducing cell death,

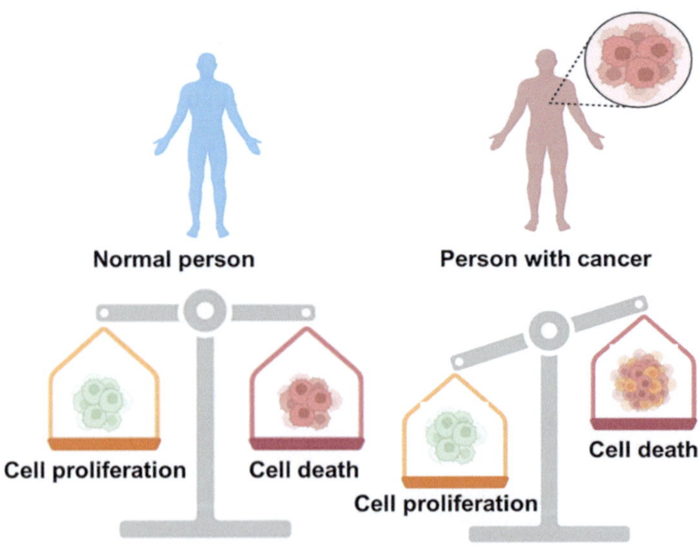

Fig. 1 Diagram illustrates the comparison between cell proliferation and cell death in normal person and cancer patient. (Created in BioRender. Hussain, A. (2025) https://BioRender.com/fvis87c)

restoration of apoptosis, enhancing ferroptosis, or activating necroptosis have emerged as promising approaches to sensitize the cancer cells to cytotoxic therapeutics. Therefore, understanding the role of these pathways is important for the development of novel and precise therapeutic strategies for effectively inducing the death of cancer cells without causing harm to normal cells. This chapter explores and describes different mechanisms of cell deaths pertaining to cancers, emphasizing their usage as therapeutic targets along with insights from experimental research, clinical translations, and challenges associated with them. This comprehensive review will pave the way to critically analyze the cell death pathways in order to develop newer and personalized therapeutic strategies to improve patient outcomes.

2 Cell Death Mechanisms and Pathways

Normal cells adopt various modalities of cell death to eliminate the damaged or unwanted cells to maintain tissue homeostasis, which have been found to be impaired in cancer cells. These impaired mechanisms not only play a role in the advancement of cancer but also in the development of therapeutic resistances thereby reducing patient outcomes (Fig. 2). Some of the cell death modalities are discussed below.

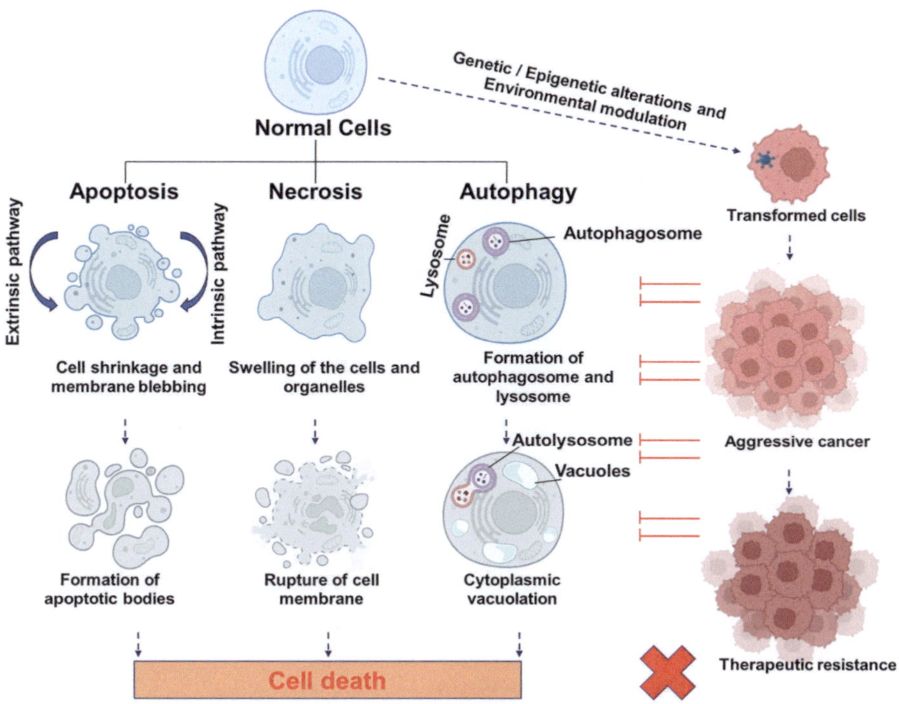

Fig. 2 Pathways of cell death and their role in cancer progression and development of therapeutic resistance. The red inhibitory marks hold promise as therapeutic targets. (Created in BioRender. Hussain, A. (2025) https://BioRender.com/317zhy4)

2.1 Programmed Cell Death Mechanisms

2.1.1 Apoptosis

Apoptosis is a highly controlled form of PCD that maintains cellular homeostasis and removes damaged cells from our body. Though it occurs through two distinct pathways: the extrinsic pathway (death receptor-mediated) and the intrinsic (mitochondrial-dependent) pathway, both ultimately converge on activation of caspase, resulting in a regulated form of cellular disassembly. The dysregulation of the apoptosis pathway allows cancer cells to avoid cell death, promoting uncontrolled growth of tumors and resistance to therapy [6].

The extrinsic pathway is initiated when death ligands such as Fas Ligand (FasL) and Tumor Necrosis Factor-alpha (TNF-α) bind to their respective death receptors (Fas/CD95, TNFR), which then recruit adaptor proteins such as Fas-associated death domain (FADD) and TNF receptor-associated death domain (TRADD), leading to the formation of the death-inducing signalling complex (DISC). DISC facilitates the activation of caspase-8 and initiates the execution phase by activating downstream effector caspases, such as caspase-3 and caspase-7. However, this pathway is regulated by inhibitory molecules such as c-FLIP, which inhibits caspase-8 activation and hence prevents the process of apoptosis [7]. Additionally, cytotoxic T lymphocytes (CTLs) and natural killer (NK) cells use the perforin/granzyme pathway to bypass the death receptor requirement by releasing granzyme B, a serine protease that either cleaves Bid or directly activates caspase-3 to activate the mitochondrial apoptotic pathway and eliminate target cells [6, 8].

The proteins belonging to the Bcl-2 family regulate the intrinsic apoptotic pathway, and this maintains a balance between pro-apoptotic (Bax, Bak) and anti-apoptotic (Bcl-2, Bcl-xl, Mcl-1) signals. Cellular stressors such as DNA damage, oxidative stress, or chemotherapy induce apoptosis by activating BH3-only proteins (Puma, Noxa, and Bim), which inhibit Bcl-2 and promote oligomerization of Bax/Bak. This results in mitochondrial outer membrane permeabilization (MOMP) followed by the release of cytochrome c, apoptosome formation, and activation of caspase-9, which ultimately triggers executioner caspases [6].

By overexpressing anti-apoptotic proteins, mutating pro-apoptotic genes, or impairing caspase function, cancer cells evade apoptosis. While death receptor agonists increase extrinsic pathway activation, therapeutic approaches like BH3 mimetics (e.g., Venetoclax) aim to restore apoptosis by blocking Bcl-2. Designing targeted cancer therapies with minimal toxicity requires an understanding of how apoptosis is regulated [6].

2.1.2 Mitochondrial Dysfunction-Driven Cell Death Mechanisms

Cell survival and death are mainly controlled by mitochondria. Mitochondrial dysfunction can lead to activation of apoptosis through two main mechanisms: mitochondrial permeability transition pore (mPTP) opening and mitochondrial outer membrane permeabilization (MOMP). These mechanisms cause pathological conditions like cancer and cognitive impairment, disturb mitochondrial homeostasis and trigger caspase activation.

mPTP and Cell Death

The non-selective ion channel called mitochondrial permeability transition pore (mPTP), present on the inner membrane, opens in response to oxidative stress, depletion of ATP, and calcium overload. Prolonged opening of this pore causes loss of membrane potential, swelling of mitochondria, breach of the outer membrane, and release of apoptogenic proteins, which triggers apoptosis or necrosis, whereas the transient opening of these pores maintains mitochondrial homeostasis [9, 10].

The exact molecular composition of mPTP is still controversial. However, the main regulators of mPTP are cyclophilin D (CypD), adenine nucleotide translocase (ANT), and F_1F_0-ATP synthase. CypD promotes pore formation and is blocked by cyclosporin A (CsA), whereas ANT is a Ca^{2+}-dependent channel, regulated by the inhibitor bongkrekate and activator atractyloside. Studies also indicate that F_1F_0-ATP synthase forms channels with mPTP-like properties, further linking it to mitochondrial dysfunction [9, 10]. The pro-apoptotic proteins, Bax and Bak, influence the mPTP opening and this links intrinsic apoptosis to mitochondrial stress [11].

Caspase-Dependent and Independent Cell Death

Mitochondrial outer membrane permeabilization (MOMP) generally results in caspase-dependent apoptosis. However, when caspases are inhibited, caspase-independent cell death (CICD) takes place through mitochondrial proteins such as AIF and EndoG, which migrate to the nucleus and then cause fragmentation of DNA [12]. Although complete MOMP ensures apoptosis, partial MOMP (iMOMP) facilitates survival, thereby enhancing therapy resistance and promoting oncogenic transformation through the mechanism of inflammation and genomic instability [13, 14]. Targeting Bcl-2 family proteins and mPTP regulators represents a promising strategy in cancer therapy.

2.1.3 Parthanatos

Parthanatos is a major form of regulated cell death. It is a multistep pathway driven by excessive activation of poly-ADP-ribose polymerase-1 (PARP-1); a key enzyme involved in DNA repair. PARP-1 finds and repairs damaged DNA by facilitating the synthesis of poly-ADP-ribose (PAR) chains under normal conditions. However, in the case of acute DNA damage, PARP-1 becomes hyperactivated, leading to an overaccumulation of PAR chains. As a result, contrary to promoting DNA repair, this excessive PAR synthesis triggers a toxic response, ultimately leading to cell death [15].

Parthanatos is evident by the mitochondrial release of apoptosis-inducing factor (AIF). Excessive stimulation of PARP-1 causes the accumulation of PAR chains, which physically interacts with AIF, causing its translocation from mitochondria to the nucleus. On entering inside the nucleus, AIF influences large-scale DNA fragmentation and chromatin condensation, indicating the final steps of parthanatos [15]. This activity is forwarded by the depletion of cellular NAD+ and ATP, which disrupts energy metabolism and accelerates cell death. Neurodegenerative disorders, cardiovascular diseases, and cancer are some of the pathological areas where parthanatos plays a crucial role. It is also one of the potential therapeutic targets, as it is evident that tumors with extensive DNA damage and persistent PARP-1 activation can paradoxically promote cancer cell death.

2.1.4 Immunogenic Cell Death (ICD)

ICD, a specialized form of PCD activates cell death through the immune system; it is different from apoptosis, which often fails to elicit an immune response. ICD is distinguished by cellular changes on the cell surface and the release of soluble mediators that trigger an immune response. In this process dendritic cells promote exposure of antigen to T cells, thus stimulating an adaptive immune response [16].

Key triggers of ICD include chemotherapeutic agents, oncolytic viruses, and physical or chemical therapies. These triggers lead to the release of damage-associated molecular patterns (DAMPs), which are recognized by pattern recognition receptors (PRRs) like Toll-like receptors (TLRs) and NOD-like receptors (NLRs). In addition to producing tumor-specific immune responses, the interaction between DAMPs and TLRs/NRLs aids in the development of long-term immunological memory, which increases the efficacy of anticancer therapies [5, 17]. Molecular changes preceding ICD include DNA damage responses, endoplasmic reticulum stress, and apoptotic signaling. Tumor cells undergoing therapy express surface danger signals and release immunostimulatory factors, activating the innate immune system. The interplay between these molecular changes determines whether the immune system will mount a tumor-specific response [16, 17].

2.2 Necrosis and Related Pathways

These pathways have relatively less control than the PCDs and are often resulted from the pathological consequences.

2.2.1 Necroptosis

Necroptosis is a variant of necrotic cell death controlled by RIPK1, RIPK3, and MLKL. When apoptosis is obstructed, it serves as an alternative pathway for cell death, especially through the inhibition of caspase-8. In contrast to apoptosis, necroptosis results in the rupture of the cell membrane and release of the intracellular components, which induce inflammation. Necrostatin-1 (Nec-1) is a well-defined inhibitor that specifically reduces the activity of RIPK1, consequently averting the initiation of necroptosis [18].

The process begins with the binding of TNF to TNFR1, which leads to the formation of Complex I that comprises RIPK1, TRADD, cIAP1, cIAP2, TRAF2, and TRAF5. When RIPK1 undergoes polyubiquitination via cIAP1/2, the NF-κB pathway becomes activated, thereby promoting cell survival. However, the deubiquitination by CYLD disrupts this survival signal and facilitates the formation of Complex II, which includes RIPK1, TRADD, FADD, and caspase-8. The active caspase-8 cleaves both RIPK1 and RIPK3, setting in apoptosis. In situations when caspase-8 is inhibited, the formation of a necrosome takes place, wherein RIPK3 phosphorylates MLKL, triggering its oligomerization and translocation to the membrane. This process results in the permeabilization of the plasma membrane, damage to the organelles, and necrotic cell death along with inflammation [18].

Necroptosis plays a dual role in cancer. It can inhibit tumor growth by eliminating malignant cells; however, it also encourages tumor progression through the release of DAMPs and inflammatory signals. Increased levels of expression of RIPK1 and RIPK3 in pancreatic ductal adenocarcinoma indicate potential involvement of tumorigenesis [19]. In the models of lung metastasis, deficiency of endothelial caspase-8 led to an increase in necroptosis, thereby facilitating the extravasation of tumor cells. In contrast, deletion of RIPK3 or MLKL and treatment with Nec-1 resulted in decreased levels of metastasis driven by necroptosis [19].

2.2.2 Pyroptosis

The activation of inflammatory caspases, especially caspase-1, which cleaves gasdermin family proteins, drives pyroptosis. Pyroptosis leads to cell swelling, rupture, and release of inflammatory cytokines such as IL-1β and IL-18 into the tumor microenvironment. The permeabilization of the plasma membrane results in osmotic imbalance, ultimately causing cell lysis and immunogenic cell death [20].

Pyroptosis plays a contradictory role in cancer. On one hand, the discharge of damage-associated molecular patterns and cytokines can alert the immune cells for anti-tumoral activity against the tumors, promoting tumor clearance. On the other hand, severe inflammation can give rise to a tumor-supportive environment, fostering immune evasion, and tumor progression [20, 21]. The molecular pathways controlling pyroptosis that are being investigated for possible therapeutic approaches are inflammasome activation and gasdermin cleavage. Pharmacological medications or gene editing to induce pyroptosis, especially in tumor cells, may be promising cancer treatment strategies.

2.2.3 Ferroptosis

Ferroptosis is an intracellular iron-dependent form of cell death that is different from apoptosis, necrosis, and autophagy. Extensive studies demonstrated that ferroptosis plays a pivotal role in tumor suppression, thus providing a new avenue for cancer therapy. It is characterized by the accumulation of iron and reactive oxygen species (ROS) and eventually results in the death of cells [7, 22]. In ferroptosis, the generation of reactive oxygen species during oxidative phosphorylation is substantially offered by mitochondria. The increase in ROS production, damaging lipids, proteins, and DNA, is due to the disproportion between electron transport and oxygen reduction. All this leads to mitochondrial shrinkage, increased membrane density, and ultimately ferroptotic cell death. Ferroptosis is distinct by enhanced membrane density and rupture, forming vesicles not similar to apoptosis, which does not disrupt the cell membrane. Ferroptosis is evident by mitochondrial atrophy and loss of cristae while preserving nuclear integrity, aside from chromatin condensation [23, 24] making it unique from necroptosis or pyroptosis, which involve cytoplasmic swelling and DNA fragmentation.

2.2.4 Lysosomal Cell Death (LCD)

LCD occurs due to malfunction of the lysosomes in the cells in which lysosomal membrane permeabilization (LMP) is a characteristic event. LMP can be triggered by lysosomotropic agents or reactive oxygen species (ROS) during oxidative stress. In the case of cancer cells, the lysosomal membrane is damaged and the enzymes, like cathepsins, leak out to the cytosol. This process has two contrasting scenarios in cancer. One is that the enzymes will leak out of the organelle, causing damage to cell parts; the other is that it can trigger inflammatory responses that promote tumor progression [25, 26].

Cleaving pro-apoptotic proteins and suppressing anti-apoptotic proteins like Bcl-2, Bcl-xl, and Mcl-1 is done by cathepsin-mediated apoptosis; also, they amplify the caspase activation by degrading inhibitors such as XIAP. Cystatins, natural cathepsin inhibitors, prevent cathepsins from being overstimulated and act as a shield from dying prematurely by binding to the active site of the cathepsins. Sometimes cystatins support cancer cells, allowing them to survive and spread easily and resist cathepsin-targeted chemotherapy [27]. So, it acts as a balance between cell survival and cell death and acts as a regulator of LCD as it controls the extent of cathepsin activation.

2.3 Autophagy-Dependent Cell Death

It has been well-proposed that autophagy has a close link to helping cancer cells to survive; that is, autophagy is associated with promoting tumor cell survival [28]. If the tumor is somehow established and starts developing, autophagy is said to be induced due to stress, hypoxia, and nutrient deficiency. Some tumor cells use autophagy to withstand the therapeutic attempts by using chemotherapy, hormone, or radiation therapy. Now, there are even many reports that portray cancers (like pancreatic cancer) as "addicted" to autophagy in advanced stages [29]. Autophagy is also reported to have been induced by signaling pathways like that of transforming growth factor-β (TGF-β) [30]. Autophagy was found to have a role in the regulation of apoptosis and the progression of cancer, like prostate cancer [28, 31].

So here, we can conclude that autophagy suppresses tumors by lowering or inhibiting the formation of reactive oxygen species (ROS), controlling and minimizing the chances for the development of mitochondrial abnormalities, and preventing the degradation of the genomic stability. Autophagy, contrastingly, supports tumor growth in the later stages when, under certain circumstances, like failure of autophagy or loss of function of some factors, the tumor continues growing and gets established. In such cases, proliferation is promoted and supported by the mechanism of autophagy and inhibits apoptosis or cell death, that is, supports the survival of the proliferating tumor cells. Not only does autophagy support the cells to survive the therapies but it also increases drug resistance in the tumor cells, thus ensuring the survival of the cells [32].

2.4 Cell Cycle Related Cell Death

2.4.1 Cellular Senescence

In cellular senescence, the cells stop dividing but remain metabolically active. It is an irreversible growth arrest. Earlier it was considered to be a protective mechanism against tumorigenesis, but later studies suggested that the accumulation of senescent cells promoted tumor progression. This duality is governed by the senescence-associated secretory phenotype (SASP), cytokines, chemokines, and proteases, which modify the tissue microenvironment [33].

The G1-to-S phase transition and cyclin-dependent kinases (CDKs) are blocked by key tumor suppressors like p21 and p16INK4a, which triggers senescence-associated proliferation arrest (SAPA). The main role of the p53 signaling pathway is the activation of downstream regulators like pRb, p107, and p130, which suppresses genes needed for DNA replication. Cell cycle arrest is reinforced by p16INK4a, a biomarker of senescence, which inhibits CDK4/6 [34, 35]. Increased senescence-associated beta-galactosidase (SA-β-gal) activity, growth arrest, enlarged morphology, nuclear prominence, and cytoplasmic granularity are some distinct characteristics of senescent cells [36].

DNA damage is one of the primary inducers of cellular senescence that activates the p53 and p21 pathways, a hallmark of senescence [36]. Repeated cell divisions cause the telomere to lose its protection and shorten in length, thus inducing replicative senescence. Damaged DNA gets accumulated, and the p53/p16 pathway is activated when oncogene-induced senescence occurs. This results when Ras-driven MAPK signaling (for example) triggers proliferative stress.

In cancer, senescence plays a complex role. SASP components can promote angiogenesis, immunosuppression, epithelial-to-mesenchymal transition (EMT), drug resistance, and cancer stemness, ultimately facilitating tumor metastasis, even though tumor progression is prevented by initial growth arrest. On the other hand, senescence-associated growth arrest can inhibit proliferation while the immune response by CD4+ T cells, macrophages, and natural killer cells helps in eliminating senescent tumor cells [36].

2.4.2 Mitotic Catastrophe

Mitotic catastrophe is a mechanism of cell death occurring during mitosis; it happens when the cells enter mitosis prematurely. The main cause of mitotic catastrophe is DNA damage or malfunction in cell cycle progressions. Mitotic catastrophe is accompanied by chromatin condensation, mitochondrial release of pro-apoptotic proteins (in particular Cyt c and AIF), caspase activation, and DNA degradation. This indicates that mitotic catastrophe is accompanied by the key molecular events defining apoptosis, namely, caspase activation and MMP [37]. Mitotic catastrophe involves (1) cell death during or near metaphase in a p53-independent manner, as observed in Chk2-inhibited heterokaryon, and (2) cell death following failed mitosis, activated by the polyploidy checkpoint, which is partially p53-dependent. It is primarily induced due to included defects in DNA structure checkpoints, which arrest cells at the G2/M transition, and spindle assembly checkpoints, which prevent anaphase until proper chromosome attachment is achieved [37, 38]. Genetic suppression of G2 checkpoint regulators such as ATM, ATR, Chk1, Chk2, Plk1-3, Pin1, Mlh1, and 14-3-3-σ, as well as microtubule-targeting agents like taxanes, vinca alkaloids, and colchicine, are the additional inducers [39].

3 Current Status of Targeting Cell Death Pathways in Cancer Therapy

Scientists have been striving to develop effective therapeutics to cure cancers targeting the hallmarks and attained success to certain degrees. However, variations among and within cancer cells and stages, along with genetic and microenvironmental influences and the development of resistances, reduce the effectiveness of many therapeutics [40]. In this chapter, we have discussed the intricate role of cell death pathways in cancer progression and the window they provide to target. Among them, apoptosis is widely recognized for tumor-suppressing properties, and others, like autophagy-dependent cell death, paradoxically promote the survival of cancer cells, complicating the treatment strategies. Activation of autophagy-dependent cell death supports the survival of cancer cells under stress conditions and also triggers cell death depending on context. Cancer cells often use this mechanism to evade apoptosis and survive radiation and chemotherapeutic insults, making them resistant [41, 42].

To date, several chemical compounds and phytochemicals have been tested in various preclinical cancer models targeting key proteins or cellular pathways involved in cancer progression. Some of these molecules demonstrated excellent results and progressed to advanced or clinical trials, and some already received FDA approval and are currently being used in clinics. These molecules/phytochemicals act by selectively modulating specific targets involved in growth and proliferation, cell death, angiogenesis, etc. In the following section, we discuss some of the experimental molecules, promising candidates, and FDA-approved drugs that have been examined for their ability to modulate cell death pathways.

3.1 Experimental Targets in Preclinical Research

Various natural and synthetic compounds have been tested using in vitro (cell culture) and in vivo (animal) models to examine their potential for tumor inhibition through the induction of cell death. Some of the most promising candidates under experimental evaluation, highlighting their mechanism and therapeutic relevance, are discussed below:

3.1.1 Lupeol

A pentacyclic triterpenoid found in medicinal plants exerts cytotoxic activity in breast cancer cells through modulation of apoptotic pathways. Computational studies confirmed MAPK and Akt as its major targets making them ideal candidates for anticancer therapy [43, 44].

3.1.2 Berberine

It is an isoquinoline alkaloid found in various medicinal plants that inhibits cell proliferation and induces apoptosis in breast, liver, lung, and colorectal cancers by targeting molecules like nuclear factor-kappa B (NF-κB), mitogen-activated protein kinases (MAPKs), and phosphoinositide 3-kinase (PI3K)/Akt. It was also observed that berberine could help in the expression of tumor suppressor genes [45].

3.1.3 Cordycepin and Its Derivatives

Epidermal Growth Factor Receptor (EGFR) tyrosine kinases are known to be overexpressed in breast cancers. Studies have demonstrated that cordycepin and its derivatives emerged as potential anticancer therapies as they inhibit the high expression of EGFR tyrosine kinases and induce apoptosis to suppress the growth of tumors [46].

3.1.4 Callyspongia aerizusa

The extract of *Callyspongia aerizusa*, a marine sponge, reduces malignancy, metastasis, and invasion in the A549 lung cancer cells. Downregulation of STAT-3, which is responsible for the survival and proliferation of the cell and VEGF, a key driver of angiogenesis, also supports the above observation of the anti-cancer effects of this extract [47]

3.1.5 Mastoparan-M

This peptide-based toxin extracted from the venom of a wasp induces the death of liver, colorectal, and breast cancer cells, giving a promising aspect for cancer therapy [48].

3.1.6 Trimethylamine N-oxide (TMAO)

TMAO, a metabolite commonly produced by gut bacteria has the ability to induce pyroptosis mediated by GSDME. When TMAO is combined with PD-1, it enhances the anti-tumor effects of anti-PD-1, and now it is in the pre-clinical stage of testing and has been experimented in breast cancer cells [49]

3.1.7 Cyclo-phosphamide

It is an alkylating drug that indirectly induces autophagy in B-cell lymphoma. It mainly works by inducing cellular stress by damaging the DNA [50]

3.1.8 Fe(III)-Shikonin Supramolecular Nanomedicine (FSSN)

FSSN prepared by coordinating naturally occurring shikonin with iron ions (Fe (III)) induce necroptosis and ferroptosis along with improved delivery and much-reduced toxicity. It was tested in Colon cancer and now has reached the preclinical stage [42, 51].

3.1.9 Myricetin-Loaded Solid Lipid Nanoparticles (MYC-SLNs)

MYC-SLNs are made of solid lipids that encapsulate Myricetin, a plant compound with anti-tumor properties. It enhanced necroptosis in A549 lung cancer cells by inducing the expression of RIPK3 and MLKL. Encapsulating myricetin in SLNs also enhances its delivery and absorption compared to administering it alone. It is now under preclinical testing [42, 52].

Preclinical studies help in the identification of molecules having the potential to modulate pathways involved in cancer, based on which researchers may proceed with advanced clinical studies. However, before advancing into advanced studies, understanding molecular mechanisms is crucial for refining targeted therapeutic strategies.

3.2 Candidates in Clinical Research

Based on the findings of preclinical studies, several phytochemicals and molecules with promising anticancer activities are currently being evaluated in clinical trials to assess their efficacy and therapeutic potential along with safety assessment.

3.2.1 SLAMF7

Modulation of Tumor Microenvironment (TME) by targeting Signaling Lymphocyte Activation Molecule Family member 7 (SLAMF7) emerged as a novel therapeutic target in Hepatocellular carcinoma (HCC). SLAMF7, or CD319, originally identified in natural killer (NK) cells, plays a key role in modulating TME. Clinical studies also suggest that the efficacy of immunotherapies can be enhanced by targeting SLAMF7 in HCC patients who do not respond to the current treatments [53, 54].

3.2.2 Chloroquine and Hydroxychloroquine

Chloroquine (CQ) and its derivative hydroxychloroquine (HCQ) have been investigated in several cancers, including glioblastoma and pancreatic cancers. They have proven autophagy inhibitory activity specifically by blocking the fusion of lysosomes and auto-phagosomes. Although the FDA has approved them as antimalarial drugs, their use as anticancer drugs is currently under clinical trials for potential repurposing [42, 55, 56].

3.2.3 Rapamycin (Sirolimus)

Rapamycin targets the cell cycle and slows down the progression of the tumor. They do so by inhibiting the mechanistic target of rapamycin (mTOR), a key regulator of cell growth and proliferation. This compound is currently under clinical trials (phase I) as a combination therapy with hydroxychloroquine for various cancers [57, 58].

3.2.4 Autophinib

Autophinib is a selective inhibitor of ULK1, which blocks the initiation of autophagy. It has been tested in vitro for anticancer activity against various cancers and found to initiate autophagy, decrease stemness of cancer cells, and prevent renewal of cancer cell populations [42, 59].

3.2.5 Veliperib

Veliparib induces parthanatos by inhibiting the PARP enzyme, which is crucial for repair of DNA. This effectively causes the accumulation of damaged DNA in the cancer cells, thus leading to cell death. It was tested in various cancer cell lines and is currently under Phase II/III clinical trials [60, 61].

3.2.6 Resveratrol

It is commonly found in grapes in nature and was found to induce autophagy through the activation of SIRT1 and inhibition of mTOR. It was studied in various cancers and is currently under preclinical and clinical trials [62].

3.2.7 Curcumin

It is one of the most commonly used phytochemicals found in turmeric. Curcumins induce autophagy via multiple pathways, like inhibiting the activity of transcription factors and many more. It was demonstrated in breast cancer and colon cancer cells and is currently under preclinical and clinical trials [63, 64].

3.2.8 Erastin

This molecule induces ferroptosis in cancer cells by binding to the voltage-dependent anion channels (VDACs). This molecule inhibits system xCT, which impairs the cysteine and glutamine antiport; thus, a reduction in cysteine levels disrupts GSH synthesis. It has been tested and experimented in various cancer cells and is now in the preclinical status of testing [65, 66].

For viable cancer treatment, the translation of preclinical laboratory discovery into clinical research lab is critical, and the molecules entering the clinical evaluation phase hold promise for elevating patient outcomes, especially for aggressive malignancies and drug-resistant cases.

3.3 FDA-Approved Drugs Targeting Cell Death Pathways

Several molecules demonstrated efficacy in preclinical and clinical research and progressed for full regulatory approval by Food and Drug Administration (FDA). Some of such molecules are discussed below:

3.3.1 Arsenic Trioxide (ATO)

It induces autophagy via oxidative stress and activates the pathways of apoptosis, disrupting the mitochondrial membrane potential. The FDA has approved this drug for acute promyelocytic leukemia, while it is in investigational status for other types of cancer [67, 68].

3.3.2 Cisplatin (DDP)

This chemotherapeutic drug cross-links DNA by binding to purine bases, guanine and adenine, and prevents repairing, further dividing, and finally causing cell death. This drug has also been found to induce pyroptosis in cancer cells, offering an alternative approach to the treatment of cancer [69, 70].

3.3.3 Olaparib (Lynparza)

Olaparib, a potent PARP inhibitor used in targeted therapy, inhibits the enzymatic function of PARP-1/2, which leads to the accumulation of damaged DNA. The effects of this drug were studied in BRCA-mutated ovarian; breast, pancreatic, and prostate cancers and FDA have approved its use for the treatment of cancer [42, 71, 72].

3.3.4 Rucaparib

Rucaprib, is also a PARP inhibitor that gained the approval of FDA and works by binding to the active site of the enzyme and thereby stopping the formation of PAR-polymers. It was tested in BRCA-mutated ovarian cancer and prostate cancer [73, 74].

3.3.5 Niraparib

Another PARP-inhibitor that prevents repair of damaged DNA and even may form PARP-DNA complexes, inducing parthanatos and apoptosis. It has gained FDA approval, and its mechanism was mainly studied in ovarian cancers [75].

3.3.6 Talazoparib

This PARP inhibitor also induces parthenogenesis and has received FDA approval for use as an anti-cancer medication. As compared to other PARP inhibitors, Talazoparib has the strongest PARP-trapping ability [76].

Significant efficacy of these drugs in preclinical, clinical studies has paved the way for further research to develop more effective treatment strategies. The combination of these molecules or drugs needs to be evaluated in a more comprehensive way to develop more refined, personalized therapeutic strategies to enhance patient outcomes.

4 Limitations of Current Research

Although significant progress has been made so far to target the cell death pathways to effectively treat cancer, several challenges limit their clinical application. One reason is the variability in tumor responses due to genetic and epigenetic alterations, making them more or less sensitive to substances that induce cell death, resulting in variable treatment responses [42]. Development of therapy-induced resistance is another obstacle in clinical applications. Cancer cells invade the therapeutic insults by activating alternate survival pathways such as autophagy and necroptosis and by decreasing the expression of anti-apoptotic proteins (e.g., Bcl-2, Mcl-1, etc.). Other common reasons are the side effects and multi-organ toxicity [42, 77]. Drugs that target these pathways cause cytotoxicity in non-cancerous tissues; for example, autophagy inhibitors like chloroquine disrupt the immune balance, while PARP inhibitors may induce hematological toxicities in patients with dysfunctional DNA repair. The lack of reliable biomarkers to estimate patient responses to the known therapies limits their precise clinical application. Current biomarkers such as caspase activation and mitochondrial depolarization offer limited understanding, thus highlighting the requirement for more personalized biomarkers [78, 79]. Addressing these limitations requires a better understanding of the molecular interplay between cell-death pathways, advanced drug-delivery systems, and the integration of personalized therapeutic approaches, which would enhance efficiency with minimal toxicity. This warrants the scientist and researcher to find next-generation therapeutic molecules and to develop combinatorial and personalized treatment protocols along with biomarkers to overcome the resistances.

5 Conclusion and Future Scope

Cell death pathways play a major role in the maintenance of tissue homeostasis through the removal of damaged or unwanted cells. But transformed or malignant cells either take advantage of defects in the pathways involved in cell death or acquire the ability to evade these pathways, contributing to uncontrolled proliferation and therapeutic resistance. Modulation of cell death pathways has gained impetus as an emerging therapeutic approach. Here, in this review, we have summarized different cell death mechanisms such as apoptosis, mitochondrial dysfunction-driven cell death, parthanatos, immunogenic cell death, necroptosis, pyroptosis, ferroptosis, etc., pertaining to the targets that can effectively be exploited for therapeutic outcomes. Some of these targets have been effectively exploited, and various drugs/pro-drugs have been developed, such as SLAMF7, Olaparib, and arsenic trioxide. However, the complexity of tumor heterogeneity, therapy-induced resistance, and systemic toxicity are the key factors that hinder clinical success.

To overcome the obstacles of these drugs, multitargeted approaches combining one or more targets may become more viable. The development of effective biomarkers for the identification of target populations and real-time monitoring of the prognosis will also benefit the enhancement of patient outcomes. Moreover, the development of a nanotechnology-based drug delivery system for site-specific and sustained delivery holds promise in targeting cancer cells with minimal or no off-target effects. Further, the integration of multi-omics approaches along with artificial intelligence in mapping and selecting dysregulated cell death pathways and developing therapeutic protocols will optimize treatment regimens.

Acknowledgements We are sincerely grateful to the Department of Molecular Biology and Biotechnology, Tezpur University, Tezpur, Assam, India and Department of Biophysics, Pub Kamrup College, Baihata Chariali, Kamrup, Assam, India for giving the opportunity and resources to compile this review. The authors gratefully acknowledge BioRender (https://BioRender.com) for all support in creating the Figures used in the chapter. The support from SERB-DST, DBT, and ICMR, New Delhi is also acknowledged.

Conflict of Interest GD is the founder of LiBiCo LLC, which has no influence or contribution to the work presented in this manuscript. Other authors have no conflict of interest.

References

1. Ferlay J, et al. Global cancer observatory: cancer today. Lyon: Int. Agency Res. Cancer; 2024. Available from: https://gco.iarc.who.int/today. Accessed on 6 Feb 2025.
2. Hanahan D, Weinberg RA. Hallmarks of cancer: the next generation. Cell. 2011;144(5):646–74. https://doi.org/10.1016/j.cell.2011.02.013.
3. Hanahan D. Hallmarks of cancer: new dimensions. Cancer Discov. 2022;12(1):31–46. https://doi.org/10.1158/2159-8290.CD-21-1059.
4. Yuan J, Ofengeim D. A guide to cell death pathways. Nat Rev Mol Cell Biol. 2024;25(5):379–95. https://doi.org/10.1038/s41580-023-00689-6.
5. Tang D, et al. The molecular machinery of regulated cell death. Cell Res. 2019;29(5):347–64. https://doi.org/10.1038/s41422-019-0164-5.
6. Elmore S. Apoptosis: a review of programmed cell death. Toxicol Pathol. 2007;35(4):495–516. https://doi.org/10.1080/01926230701320337.
7. Jan R, Chaudhry GE. Understanding apoptosis and apoptotic pathways targeted cancer therapeutics. Adv Pharm Bull. 2019;9(2):205–18. https://doi.org/10.15171/apb.2019.024.
8. Hay ZLZ, Slansky JE. Granzymes: the molecular executors of immune-mediated cytotoxicity. Int J Mol Sci. 2022;23(3):1833. https://doi.org/10.3390/ijms23031833.
9. Endlicher R, et al. The mitochondrial permeability transition pore-current knowledge of its structure, function, and regulation, and optimized methods for evaluating its functional state. Cells. 2023;12(9):1273. https://doi.org/10.3390/cells12091273.
10. Bernardi P, et al. Identity, structure, and function of the mitochondrial permeability transition pore: controversies, consensus, recent advances, and future directions. Cell Death Differ. 2023;30(8):1869–85. https://doi.org/10.1038/s41418-023-01187-0.
11. Wolf P, et al. Pro-apoptotic complexes of BAX and BAK on the outer mitochondrial membrane. Biochim Biophys Acta, Mol Cell Res. 2022;1869(10):119317. https://doi.org/10.1016/j.bbamcr.2022.119317.
12. Bock FJ, Tait SWG. Mitochondria as multifaceted regulators of cell death. Nat Rev Mol Cell Biol. 2020;21(2):85–100. https://doi.org/10.1038/s41580-019-0173-8.
13. Ichim G, et al. Limited mitochondrial permeabilization causes DNA damage and genomic instability in the absence of cell death. Mol Cell. 2015;57(5):860–72. https://doi.org/10.1016/j.molcel.2015.01.018.
14. Kalkavan H, Green D. MOMP, cell suicide as a BCL-2family business. Cell Death Differ. 2018;25(1):46–55. https://doi.org/10.1038/cdd.2017.179.
15. Huang P, et al. Molecular mechanisms of parthanatos and its role in diverse diseases. Int J Mol Sci. 2022;23(13):7292. https://doi.org/10.3390/ijms23137292.
16. Kroemer G, et al. Immunogenic cell death in cancer therapy. Annu Rev Immunol. 2013;31:51–72. https://doi.org/10.1146/annurev-immunol-032712-100008.
17. Galluzzi L, et al. Immunogenic cell death in cancer: concept and therapeutic implications. J Transl Med. 2023;21(1):162. https://doi.org/10.1186/s12967-023-04017-6.
18. Gong Y, et al. The role of necroptosis in cancer biology and therapy. Mol Cancer. 2019;18(1):100. https://doi.org/10.1186/s12943-019-1029-8.
19. Zhu F, et al. Complex roles of necroptosis in cancer. J Zhejiang Univ Sci B. 2019;20(5):399–413. https://doi.org/10.1631/jzus.B1900160.
20. Van Opdenbosch N, Lamkanfi M. Caspases in cell death, inflammation, and disease. Immunity. 2019;50(6):1352–64. https://doi.org/10.1016/j.immuni.2019.05.020.

21. Lu L, et al. Emerging mechanisms of pyroptosis and its therapeutic strategy in cancer. Cell Death Discov. 2022;8(1):338. https://doi.org/10.1038/s41420-022-01101-6.
22. Zhang C, et al. Ferroptosis in cancer therapy: a novel approach to reversing drug resistance. Mol Cancer. 2022;21(1):47. https://doi.org/10.1186/s12943-022-01530-y.
23. Li J, et al. Ferroptosis: past, present and future. Cell Death Dis. 2020;11(1):88. https://doi.org/10.1038/s41419-020-2298-2.
24. Chen Z, et al. Ferroptosis as a potential target for cancer therapy. Cell Death Dis. 2023;14(7):460. https://doi.org/10.1038/s41419-023-05930-w.
25. Repnik U, et al. Lysosomes and lysosomal cathepsins in cell death. Biochim Biophys Acta. 2012;1824(1):22–33. https://doi.org/10.1016/j.bbapap.2011.08.016.
26. Appelqvist H, et al. Lysosome-mediated apoptosis is associated with cathepsin D-specific processing of BID at Phe24, Trp48, and Phe183. Ann Clin Lab Sci. 2012;42(3):231–42.
27. Xie Z, et al. Cathepsin B in programmed cell death machinery: mechanisms of execution and regulatory pathways. Cell Death Dis. 2023;14(4):255. https://doi.org/10.1038/s41419-023-05786-0.
28. Ahmadi-Dehlaghi F, et al. Autophagy: a challengeable paradox in cancer treatment. Cancer Med. 2023;12(42):11542–69. https://doi.org/10.1002/cam4.5577.
29. Hashimoto D, et al. Autophagy is needed for the growth of pancreatic adenocarcinoma and has a cytoprotective effect against anticancer drugs. Eur J Cancer. 2014;50(7):1382–90. https://doi.org/10.1016/j.ejca.2014.01.011.
30. Kiyono K, et al. Autophagy is activated by TGF-beta and potentiates TGF-beta-mediated growth inhibition in human hepatocellular carcinoma cells. Cancer Res. 2009;69(23):8844–52. https://doi.org/10.1158/0008-5472.CAN-08-4401.
31. Loizzo D, et al. Novel insight into autophagy and prostate cancer: a comprehensive review. Int J Mol Sci. 2022;23(7):3826. https://doi.org/10.3390/ijms23073826.
32. Lim SM, et al. Is targeting autophagy mechanism in cancer a good approach? The possible double-edge sword effect. Cell Biosci. 2021;11(1):56. https://doi.org/10.1186/s13578-021-00570-z.
33. Hernandez-Segura A, et al. Hallmarks of cellular senescence. Trends Cell Biol. 2018;28(6):436–53. https://doi.org/10.1016/j.tcb.2018.02.001.
34. Schmitt CA, et al. Senescence and cancer—role and therapeutic opportunities. Nat Rev Clin Oncol. 2022;19(10):619–36. https://doi.org/10.1038/s41571-022-00668-4.
35. Wang L, et al. Exploiting senescence for the treatment of cancer. Nat Rev Cancer. 2022;22(6):340–55. https://doi.org/10.1038/s41568-022-00450-9.
36. Yang J, et al. The paradoxical role of cellular senescence in cancer. Front Cell Dev Biol. 2021;9:722205. https://doi.org/10.3389/fcell.2021.722205.
37. Sazonova EV, et al. A link between mitotic defects and mitotic catastrophe: detection and cell fate. Biol Direct. 2021;16(1):25. https://doi.org/10.1186/s13062-021-00313-7.
38. Bai Z, et al. Perspectives and mechanisms for targeting mitotic catastrophe in cancer treatment. Biochim Biophys Acta Rev Cancer. 2023;1878(5):188965. https://doi.org/10.1016/j.bbcan.2023.188965.
39. Castedo M, et al. Cell death by mitotic catastrophe: a molecular definition. Oncogene. 2004;23(16):2825–37. https://doi.org/10.1038/sj.onc.1207528.
40. Anand U, et al. Cancer chemotherapy and beyond: current status, drug candidates, associated risks and progress in targeted therapeutics. Genes Dis. 2022;10(4):1367–401. https://doi.org/10.1016/j.gendis.2022.02.007.
41. Tong X, et al. Targeting cell death pathways for cancer therapy: recent developments in necroptosis, pyroptosis, ferroptosis, and cuproptosis research. J Hematol Oncol. 2022;15(1):174. https://doi.org/10.1186/s13045-022-01392-3.
42. Saxena R, et al. Targeting regulated cell death pathways in cancers for effective treatment: a comprehensive review. Front Cell Dev Biol. 2024;12:1462339. https://doi.org/10.3389/fcell.2024.1462339.

43. Luo X, et al. Exploring the therapeutic potential of lupeol: a review of its mechanisms, clinical applications, and advances in bioavailability enhancement. Food Chem Toxicol. 2025;196:115193. https://doi.org/10.1016/j.fct.2024.115193.
44. Liu K, et al. Lupeol and its derivatives as anticancer and anti-inflammatory agents: molecular mechanisms and therapeutic efficacy. Pharmacol Res. 2021;164:105373. https://doi.org/10.1016/j.phrs.2020.105373.
45. Haque S, et al. Prospects of phytonutrient berberine as an anticancer agent. J Biochem Mol Toxicol. 2025;39(1):e70073. https://doi.org/10.1002/jbt.70073.
46. Bibi S, et al. Cordycepin and its structural derivatives suppress EGFR tyrosine kinase in breast carcinomas: a computational approach. Curr Med Chem. 2025. https://doi.org/10.2174/0109298673333884241119063458.
47. Hadisaputri YE, et al. STAT3 suppression in A549 lung carcinoma cells by marine sponge *Callyspongia aerizusa*. J Exp Pharmacol. 2025;17:15–25. https://doi.org/10.2147/JEP.S494158.
48. Zhao H, et al. Phytosomes loaded with mastoparan-M for breast cancer treatment. Int J Nanomedicine. 2025;20:109–24. https://doi.org/10.2147/IJN.S481871.
49. Wang H, et al. The microbial metabolite trimethylamine N-oxide promotes antitumor immunity in triple-negative breast cancer. Cell Metab. 2022;34(4):581–594.e8. https://doi.org/10.1016/j.cmet.2022.02.010.
50. Amaravadi RK, et al. Autophagy inhibition enhances therapy-induced apoptosis in lymphoma. J Clin Invest. 2007;117(2):326–36. https://doi.org/10.1172/JCI28833.
51. Feng W, et al. Fe(III)-Shikonin nanomedicine for tumor therapy via ferroptosis and necroptosis. Adv Healthc Mater. 2022;11(2):e2101926. https://doi.org/10.1002/adhm.202101926.
52. Alidadi H, et al. Myricetin in solid lipid nanoparticles induces apoptosis in HT-29 colorectal cancer cells. Mol Biol Rep. 2022;49(9):8537–45. https://doi.org/10.1007/s11033-022-07683-9.
53. Bruneau A, Hammerich L. SLAMF7 reprogramming of macrophages in hepatocellular carcinoma. Transl Cancer Res. 2024;13(12):6995–7001. https://doi.org/10.21037/tcr-24-876.
54. Weng J, et al. SLAMF7-regulated CCL2 signaling sensitizes hepatocellular carcinoma to immunotherapy. Cancer Res. 2024;84(11):1817–33. https://doi.org/10.1158/0008-5472.CAN-23-3106.
55. Varisli L, Cen O, Vlahopoulos S. Chloroquine's pharmacological effects in cancer treatment: interference with inflammatory signaling. Immunology. 2020;159(3):257–78. https://doi.org/10.1111/imm.13160.
56. Agalakova NI. Chloroquine and chemotherapeutic compounds in experimental cancer treatment. Int J Mol Sci. 2024;25(2):945. https://doi.org/10.3390/ijms25020945.
57. Blagosklonny MV. Cancer prevention with rapamycin. Oncotarget. 2023;14:342–50. https://doi.org/10.18632/oncotarget.28410.
58. Li Q, et al. Inetetamab with sirolimus and chemotherapy for HER2-positive metastatic breast cancer. Cancer Innov. 2024;3(5):e145. https://doi.org/10.1002/cai2.145.
59. Aleksandrova KV, Suvorova II. Effectiveness of autophagy inhibitors in A549 cancer stem cells. Acta Nat. 2023;15(1):19–25. https://doi.org/10.32607/actanaturae.11891.
60. Tuli R, et al. Veliparib with gemcitabine and radiotherapy in locally advanced pancreatic cancer. EBioMedicine. 2019;40:375–81. https://doi.org/10.1016/j.ebiom.2018.12.060.
61. Zeng Y, et al. PARP inhibitors: a review of the pharmacology, pharmacokinetics, and pharmacogenetics. Semin Oncol. 2024;51(1–2):19–24. https://doi.org/10.1053/j.seminoncol.2023.09.005.
62. Song B, et al. Inhibitory potential of resveratrol in cancer metastasis: from biology to therapy. Cancers (Basel). 2023;15(10):2758. https://doi.org/10.3390/cancers15102758.
63. Shakeri A, et al. Curcumin: a naturally occurring autophagy modulator. J Cell Physiol. 2019;234(5):5643–54. https://doi.org/10.1002/jcp.27404.
64. Wang W, et al. Curcumin in cancer therapy: exploring molecular mechanisms and overcoming clinical challenges. Cancer Lett. 2023;570:216332. https://doi.org/10.1016/j.canlet.2023.216332.

65. Sinha BK, et al. Mechanisms of cell death induced by erastin in human ovarian tumor cells. Int J Mol Sci. 2024;25(16):8666. https://doi.org/10.3390/ijms25168666.
66. Wu F, et al. Identification of ferroptosis-related genes and pathways in prostate cancer cells under erastin exposure. BMC Urol. 2024;24(1):78. https://doi.org/10.1186/s12894-024-01472-1.
67. Jiang Y, et al. An overview of arsenic trioxide-involved combined treatment algorithms for leukemia: basic concepts and clinical implications. Cell Death Discov. 2023;9(1):266. https://doi.org/10.1038/s41420-023-01558-z.
68. Gill H. Chemotherapy-free approaches to newly diagnosed acute promyelocytic leukaemia: is oral-arsenic trioxide/all-trans retinoic acid/ascorbic acid the answer? Expert Rev Hematol. 2024;17(10):661–7. https://doi.org/10.1080/17474086.2024.2391098.
69. Yan H, et al. Cisplatin induces pyroptosis via activation of MEG3/NLRP3/caspase-1/GSDMD pathway in triple-negative breast cancer. Int J Biol Sci. 2021;17(10):2606–21. https://doi.org/10.7150/ijbs.60292.
70. Xuzhang W, et al. Cisplatin-induced pyroptosis enhances the efficacy of PD-L1 inhibitor in small-cell lung cancer via GSDME/IL12/CD4Tem axis. Int J Biol Sci. 2024;20(2):537–53. https://doi.org/10.7150/ijbs.89080.
71. Paik J. Olaparib: a review as first-line maintenance therapy in advanced ovarian cancer. Target Oncol. 2021;16(6):847–56. https://doi.org/10.1007/s11523-021-00842-1.
72. Joris S, et al. Efficacy of olaparib in advanced cancers with germline or somatic mutations in BRCA1, BRCA2, CHEK2, and ATM: a Belgian Precision tumor-agnostic phase II study. ESMO Open. 2023;8(6):102041. https://doi.org/10.1016/j.esmoop.2023.102041.
73. Monk BJ, et al. A randomized, phase III trial to evaluate rucaparib monotherapy as maintenance treatment in patients with newly diagnosed ovarian cancer (ATHENA-MONO/GOG-3020/ENGOT-ov45). J Clin Oncol. 2022;40(34):3952–64. https://doi.org/10.1200/JCO.22.01003.
74. Sayyid RK, et al. Rucaparib monotherapy in the heavily pre-treated metastatic castrate-resistant prostate cancer setting: practical considerations and alternate treatment approaches. Transl Androl Urol. 2024;13(5):884–8. https://doi.org/10.21037/tau-23-671.
75. Wu X, et al. Niraparib maintenance therapy using an individualized starting dose in patients with platinum-sensitive recurrent ovarian cancer (NORA): final overall survival analysis of a phase 3 randomized, placebo-controlled trial. EClinicalMedicine. 2024;72:102629. https://doi.org/10.1016/j.eclinm.2024.102629.
76. Heiss BL, et al. US Food and Drug Administration approval summary: Talazoparib in combination with enzalutamide for treatment of patients with homologous recombination repair gene-mutated metastatic castration-resistant prostate cancer. J Clin Oncol. 2024;42(15):1851–60. https://doi.org/10.1200/JCO.23.02182.
77. D'Amico M, De Amicis F. Challenges of regulated cell death: implications for therapy resistance in cancer. Cells. 2024;13(13):1083. https://doi.org/10.3390/cells13131083.
78. Peng F, et al. Regulated cell death (RCD) in cancer: key pathways and targeted therapies. Signal Transduct Target Ther. 2022;7(1):286. https://doi.org/10.1038/s41392-022-01110-y.
79. Murthy P, Muggia F. PARP inhibitors: clinical development, emerging differences, and the current therapeutic issues. Cancer Drug Resist. 2019;2(3):665–79. https://doi.org/10.20517/cdr.2019.002.

Telomerase as a Target for Cancer Therapy

Suresh Kumar Ramakrishnan, Ram Raj Prasad, Deepak Kalyanasundaram, Akhil Varshney, and Pramod K. Yadava

1 Introduction

Cancer cells proliferate independently of the normal regulatory mechanisms by acquiring a few characteristics presented as six hallmarks [1]; the same is updated now as ten hallmarks and four enabling characters [2]. One of the attributes is limitless replicative potential. The limitless replicative potential ensures uncontrolled cell division due to enhanced expression of telomerase in cancer cells, while it is barely detectable in terminally differentiated cells. Telomerase is the ribonucleoprotein that synthesizes species-specific repeat sequences called telomeres at chromosome termini. Telomerase is an enzyme (TERT or telomerase reverse transcriptase, in humans hTERT) associated with an RNA molecule (TR or hTR) that acts as a template for telomere synthesis. Telomerase is over-expressed in stem cell cancer cells except sarcomas and is

S. K. Ramakrishnan (✉)
ICMR-National Institute of Cancer Prevention and Research,
Noida, Uttar Pradesh, India
e-mail: suresh.kr73@gov.in

R. R. Prasad
Cardiovascular Pulmonary Research Laboratories, Anschutz Medical Campus, University of Colorado, Aurora, Denver, CO, USA

D. Kalyanasundaram
School of Life Sciences, Bharathidasan University, Palkalaiperur Campus,
Tiruchirappalli, Tamil Nadu, India

A. Varshney
Centre for Advanced Vision Science, Department of Ophthalmology, University of Virgina,
Charlottesville, VA, USA

P. K. Yadava
School of Life Sciences, Jawaharlal Nehru University, New Delhi, India

A. S. Ramteke et al. (eds.), *Molecular Targets in Cancer Therapy*,
https://doi.org/10.1007/978-981-96-7336-0_3

present at very low or undetectable levels in normal cells. This attribute of exclusiveness in cancer is believed to have a role in cell cycle regulation and must be directly or indirectly influencing control over cell division.

The telomerase enzyme was discovered by Elizabeth Blackburn and Carol Grieder in 1984 [3], who showed that there is a specialized enzyme called telomerase that synthesizes the telomere sequence. They have also shown that the telomerase enzyme is complexed with an RNA component (hTR), which acts as a template in telomere synthesis. [4]. Primary cell cultures undergo a limited number of doublings, typically around 52–60 divisions for human cells referred to as the Hayflick limit or Phase III phenomenon [5]. Subsequent research proved that human telomerase is a ribonucleoprotein that synthesizes TTAGGG repeats and telomerase expression is associated with increased telomere length which results in an extension of Hayflick limit for the cell population [6, 7]. Ectopic expression of telomerase in normal cells have proven that telomerase extends telomere length and life span at the cellular level and correspondingly extends the lifespan of the organism. The first association of life span came up from studies of Jerry Shay, who found that telomerase presence provides proliferative advantage to the cells [8]. Later on, it was demonstrated that telomerase has an important role in all cancers and places itself as one of the hallmarks and an attractive target for cancer therapy.

Every cell in an organism has machineries to safeguard cellular processes and mechanisms to protect genome integrity. Any aberration in the process or machinery will lead to the omission of cells from re-entry to the cell cycle and lead to apoptosis. Such strictly controlled mechanism sometimes goes awry and cells acquire all the adaptive molecular processes to continue uninterrupted cell proliferation. When cell doubling is not defined or limited, there is always a likelihood of acquiring new aberrations either at the molecular or cellular level that leads to dysfunctional cellular regulatory mechanisms. To avoid such abnormality, cellular processes are wired through a strict regimen of control mechanisms and number of divisions, genome size, chromosome number, and lifetime. Telomere lengthening is maintained by telomerase, in some exceptional circumstances alternative mechanism called Alternative lengthening of telomere (ALT) is employed in cancers of muscle origin [9]. Telomeres of human chromosomes consist of arrays of TTAGGG sequences. Telomeres provide genomic integrity and by assuming a special structure they are protected from inter-chromosomal fusion by capping chromosome ends. Telomere length actually defines the number of divisions a cell can undergo.

It was demonstrated that in primary cultures, untransformed cells can undergo a certain limited number of doublings, (Hayflick limit) or [5] when normal cells encounter various physiological stresses they enter into senescence. Senescence is a kind of growth arrest of cells that arises when cells are propagated for a longer period of time and is termed as replicative senescence. Telomere length reduces at every cell division and after a particular critical length, it induces cell death signalling.

2 Telomere Crisis and Tumorigenesis

Double knockout of telomerase RNA and P53 (mTR−/−, P53−/−) in a mouse model results in enhanced transformation [10]. Telomere recession causes a crisis that creates genetic instability resulting in cell death or transformation depending on the genetic background and activation of other adaptive mechanisms. In human cells, the final cell division precedes replicative senescence, which is considered as an early crisis. Senescence is manifested by the slowdown of growth and altered gene expression profiles. It is hypothesized that telomere attrition prevents or promotes tumorigenesis based on the p53 and pRB status [11]. Early evidence showed that only a few telomere ends get shorter and they are perceived as DNA damage signals that result in growth arrest or replicative senescence or Mortality Stage 1 (M1) or STASIS. This is dependent on p53 and pRB/p16 status. In the absence of these genetic backgrounds, due to mutation, the M1 barrier is breached and continues to divide with further reduction in telomere length and reaching a second barrier called M2 or CRISIS (Fig. 1) [12]. This stage is characterized by uncapped chromosome ends, inter-chromosomal fusions, and a larger number of apoptotic cells. However, a few rare cells emerge in M2 due to activation of telomerase, and continue to divide indefinitely and propagate further [13]. Recent evidence has also shown that breaching of barrier M2 reaches M3 in the presence of telomerase and in the absence of autophagy and c-GAS-STING pathway, which is a component of the innate immune system [14].

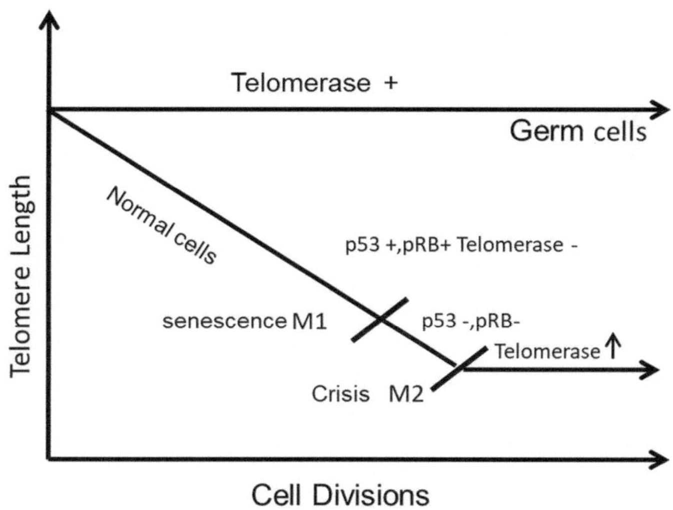

Fig. 1 Tumorigenesis process has to breach two barriers in the absence of telomerase. Telomere attrition leads to M1 stage senescence and cell death which are p53- and pRb-dependant. In the absence of p53 and pRb, the cells breach the barrier and reach crisis stage M2 marked by mass cell death. However, a few cells get telomerase activated and escape from cell death by extension of telomere length

3 Emergence of Telomerase as a Probable Target for Therapeutic Intervention

Differential and exclusive expression of telomerase only in cancer cells has primarily made them a target of choice. The components associated with telomerase and end replication problems that are implicated in attrition of telomere length and their relevance in cell doubling have been utilized as an attractive target for therapeutic intervention.

3.1 End Replication Problem

Russian scientists Olovnikov and James Watson in 1971,1972 respectively [15, 16] pointed to the problem of extension of the lagging strand beyond the terminal primer while the 3′ end of the leading strand could be extended to the very end of the template. They speculated that the ends of chromosomes would require some special mechanism for lengthening and maintenance. An observation from later studies showed that the 3′ end of the chromosome had a single-stranded overhang and 5′ recessed end. This was realized that normal DNA polymerase may not copy the very end of a chromosome, creating unequal lengths at the end, i.e., failure of normal DNA polymerase to complete the lagging strand replication, which is synthesized by short extension from RNA primers called Okazaki fragments. This has been referred to as the end replication problem [17].

The double-stranded DNA has two linear anti-parallel (5′-3′ and 3′-5′) complementary DNA strands. During replication, the replication machinery unwinds the two strands and the leading strand synthesis takes place by continuous synthesis in 5′ to 3′ directions without interruption, whereas the opposite strand, i.e., lagging strand is synthesized in repeated primer synthesis and extension, removal of primer and repair of the gaps. Due to a failure of conventional DNA polymerase to synthesize DNA in 3′ to 5′ direction, the short patches of synthesis transiently result in Okazaki fragments. Subsequently, the RNA primers are removed by RNase H, gaps are filled by DNA polymerase and ends sealed by ligases. Removal of primer at the very end of the lagging strand leaves a gap and generates the overhang of complementary strand sequences (Fig. 2). So, the 3′ end is found to have an overhang of 150 nucleotides due to the fact that 5′ end gets further recessed by erosion of C-strand-specific exonucleases. This generates an overhang of the 3′ end of the template′, and recessed 5′ end of the copy strand. In this recessed strand in consequent cycles of DNA synthesis, the terminal length gets reduced to critical lengths that lead to DNA damage signalling. Recent evidence proves that the end replication problem is not only for lagging strands but it is also for the leading strand that generates blunt ends at the ends. However, we are not discussing new evidence of the leading strand problem here.

The reduction in telomere length is due to the inability of conventional polymerase to maintain the chromosome ends. The progressive telomere shortening leads to replicative senescence and is believed to enter senescence or cell death.

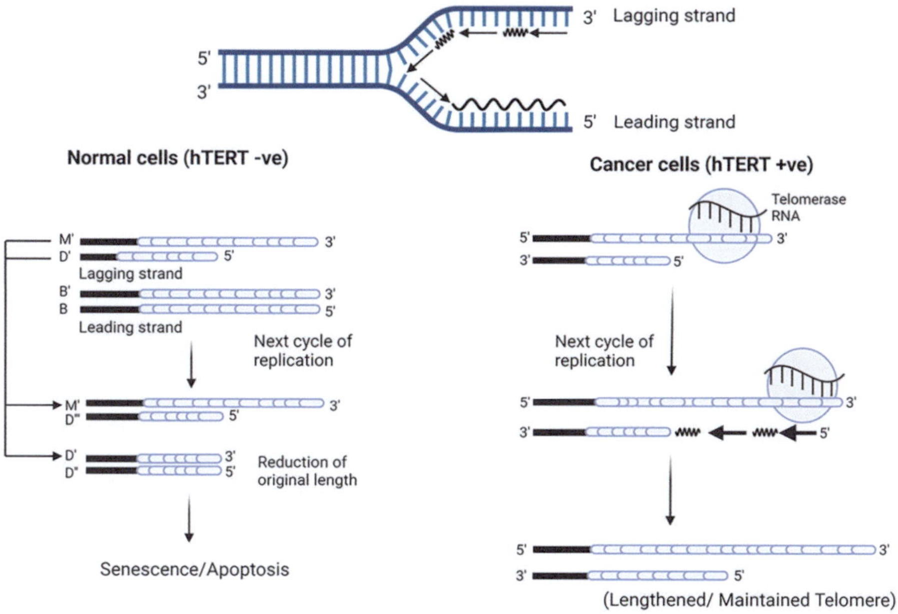

Fig. 2 End replication problem in the presence and absence of telomerase and telomere attrition. M—Mother, D—Daughter strand, B—Mother strand of Leading strand, D′, D″—Daughter I, II, respectively. When recessed D′ act as a template the daughter strand gets the same reduced length

Elizabeth Blackburn in the early days of her career observed that there is a special array of repeat sequences present at Tetrahymena and subsequent experiments with her graduate student Carol Grieder unraveled the mechanism and presence of a special enzyme called telomerase that synthesize telomeres [3, 4]. Subsequently, they also found that telomerase bound with an RNA molecule which acts as a template in telomere synthesis [4]. Following the same, Morin (1989) reported that human terminal transferase is a ribonucleoprotein and synthesizes TTAGGG [10]. The expression of telomerase was probed in cancer tissues and found that telomerase expression was found exclusively in cancer cells, germ cells (normal ovaries, sperms cells), immortal cells, and strongly repressed in normal somatic cells [7]. All these novel observations made telomerase an important molecule in many diseases, particularly in cancers. This discovery gave a major impediment in understanding of regulation of cell division, the role of telomeres in normal and abnormal cell divisions.

3.2 Telomere Structure

Telomeres are specialized repeat sequence structures present at every chromosome end. Telomere length varies from species to species and in humans it is approximately 4–15 kb in length and overhang of about 150–200 bases in length. Due to the

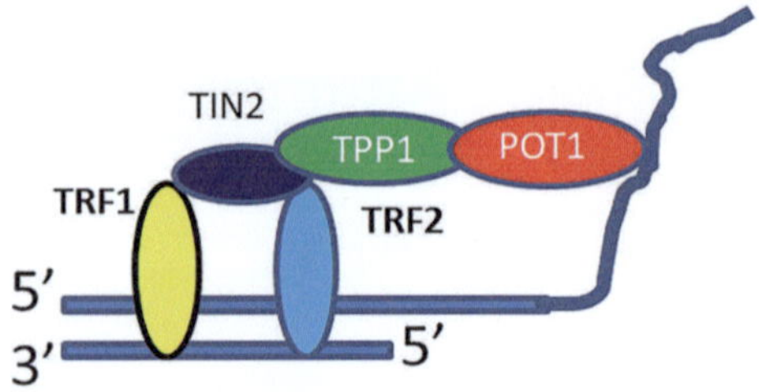

Fig. 3 Telomere binding Shelterin proteins TRF1, TRF2, TIN2 TPP1, and POT1. All components act as targets. TIN2 (TRF1 interacting protein 2) is recruited by TRF1, and TIN2 bridges the TRF1 and TRF2

prevailing end replication problem, telomeres get shortened and after a certain limit of telomere shortening it is considered as DNA damage and starts sending the DNA damage signalling. However, the telomeric-specific proteins particularly 6 protein complexes called Shelterin (TRF1, TRF2, POT1, RAP1, TIN2, and TPP1) binds to TTAGGG sequence (Fig. 3) and protects the telomere end. Due to the Shelterin binding, it assumes the T loop structure, protects telomeres from recognizing it as damaged DNA and suppresses the DNA damage signaling. Some of the factors like nucleases, and helicases act as accessory proteins for this Shelterin function [18].

Telomeric repeat binding factor 1, 2 (TRF1, TRF2) binds to the duplex telomeric sequences, whereas the protection of telomeres protein 1(POT1) binds with the single-stranded overhang. TRF1 recruits TIN2 (TRF1 interacting protein 2) and TRF2 recruits Rap1 (Repressor activator protein). TIN2 (TRF1 interacting protein 2) is recruited by TRF1, and TIN2 bridges the TRF1 and TRF2. As telomere sequence 5′-TTAGGG-3′ gets reduced at every cell division, longer telomeres get stuck into to T loop and shorter loops fail to do so. The absence of such protection leads to telomere shortening in subsequent cell divisions and ends up with senescence or apoptosis. Shortened telomeres fail to get protected and induce replicative senescence. In addition, the double-stranded DNA of telomeres due to its richness in G nucleotides, assumes the G quadruplex structure which is a 4-stranded guanine-rich non-canonical structure and has also been found in *invitro* and in vivo. Such structures are present in human telomeres, certain promoter regions of oncogenes, and UTR region of certain genes [19].

3.3 Telomerase Structure

Human telomerase (hTERT) is found in chromosome 5 (5p15.33), with 40 kb with 16 exons. Human telomerase (hTERT) found as a dimer along with dimeric RNAs in in vivo is known for its catalytic function. TERT contains four domains—telomerase essential N terminal domain (TEN), TERRA RNA binding domain, catalytic reverse transcriptase (RT) domain, and C terminal domain (CTE). Sequence alignment of telomerase from various species reveals that TERT is about 1132 amino acids. Telomerase N terminal domain (TEN) has TPP1/TIN2 protein of telomere and single-stranded telomeric DNA binding affinity. TEN is linked to telomerases RNA binding domain (TRBD) by linker (Fig. 4).

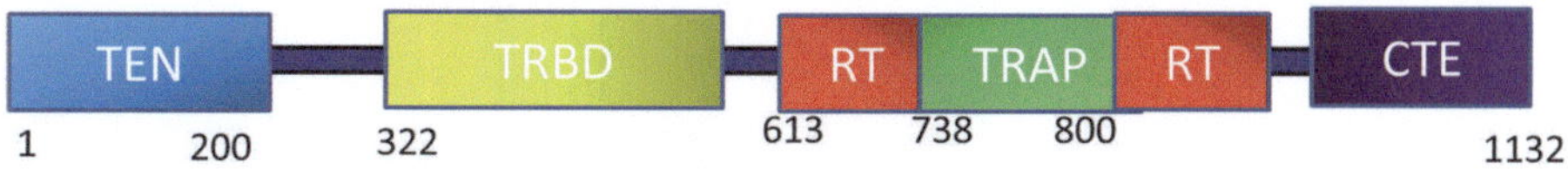

Fig. 4 Telomerase structure: TEN: N terminal Domain, TRBD: Telomerase RNA binding domain, RT: Revers transcriptase, TRAP: CTE: C terminal Element

TRBD interacts with hTR's conserved region of CR4/5. The central RT domain shares homology with retroviral reverse transcriptase and has TRBD at N terminus and CTE at C terminus. The reverse transcriptase domain has an overlapped domain called TRAP, which has been recently reported using multiple cryo-electron microscopy [20].

Telomerase RNA (TR) varies in length across species and human telomerase RNA (hTR) contains 451 bases assuming conserved RNA secondary structure. Aligning them across different species, it was found to have 4 conserved secondary structures, that include single-stranded large loop containing a template region that bears a template for telomere synthesis, template boundary element, pseudoknot region, a helix with loop and stem terminus element (STE) (Fig. 5). From many mutational studies, phylogenetic aligning and conservation of secondary structure reveals that it is the structure rather than the sequence that plays a role in telomerase RNA and telomerase function [21].

By assessing the homology alignment of 35 vertebrate telomerase RNA keeping the conserved region as a reference, hTR was designated with CR1 to CR89 (Conserved region) and a paired region (P). The first conserved region (CR1) represents template region CR1, and CR4, and CR5 represents the longest region. There are 10 conserved helical regions, which are universal, i.e., present in all vertebrate telomerase RNAs. These helices represent four important domains (1) Pseudoknot domain, (2) CR4-CR5 domain, (3) H/ACA domain, (4) CR7 domain. The helices, pseudoknot region and CR4-CR5 domain are bound by telomerase protein, and the template region of hTR is present at 5′ to the pseudoknot region. These regions, either the template region or single-stranded regions of these important domains can be potential targets for telomere inhibition.

3.4 Targeting Telomerase

The revelation of DNA replication by telomerase and the presence of an End replication problem in inducing cell cycle arrest gave a great window of opportunity for therapeutic intervention. Most of the earlier studies targeted telomerase replication targets and its exclusivity of expression only in cancers made to use telomerase promoters in certain gene therapy applications. Many efforts have been undertaken through nuclei acid-based, antisense, DNA-based DNA zyme, RNA-based ribozyme, short interfering RNA, etc. and peptide-based vaccine strategies and other inhibitors (Fig. 6). There were inevitable challenges in using telomerase as a therapeutic target drug as most of the effects are mediated through telomerase inhibition

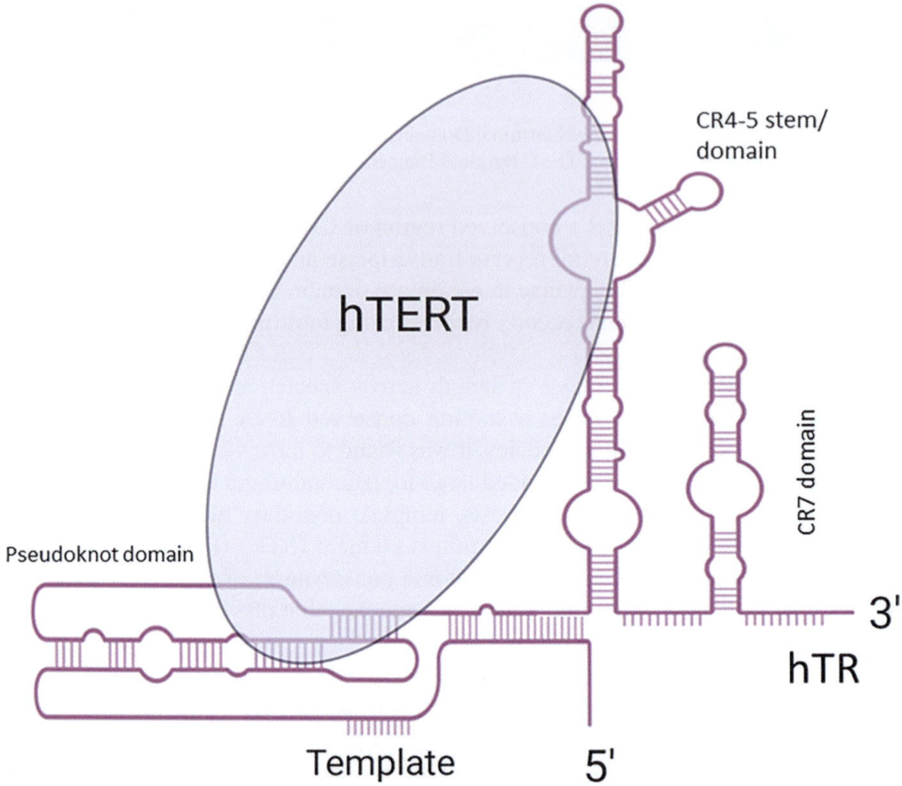

Fig. 5 Telomerase RNA structure: showing template region, Pseudoknot domain, CR4-CR5 domain, and CR7 domain

mediated telomere attrition process and attrition takes a longer time to induce DNA damage signal to get activated to pronounce cell death. Other challenge was the non-availability of crystal structures limited to design structure-based design of active inhibitors against telomerase. However, a recent development in technology has brought recent advances in telomerase structure that enable updated structural inhibitors. Particularly recent near atomic structural details generated using multiple high resolution Cryogenic electron microscopy (Cryo-EM) revealed a new component within a particular domain (reverse transcriptase) of telomerase. This also revealed structural details of how telomerase protein binds with RNA and how it is recruited and bound with telomeres. Such new detailed structural revealing may pave for the development of new inhibitors [22].

Telomerase inhibition by means of using antisense, ribozyme, siRNA, shRNA, and phytochemical-based methods has shown that it can elicit telomere length attrition and induction of senescence or apoptosis in p53 dependant and independent manner.

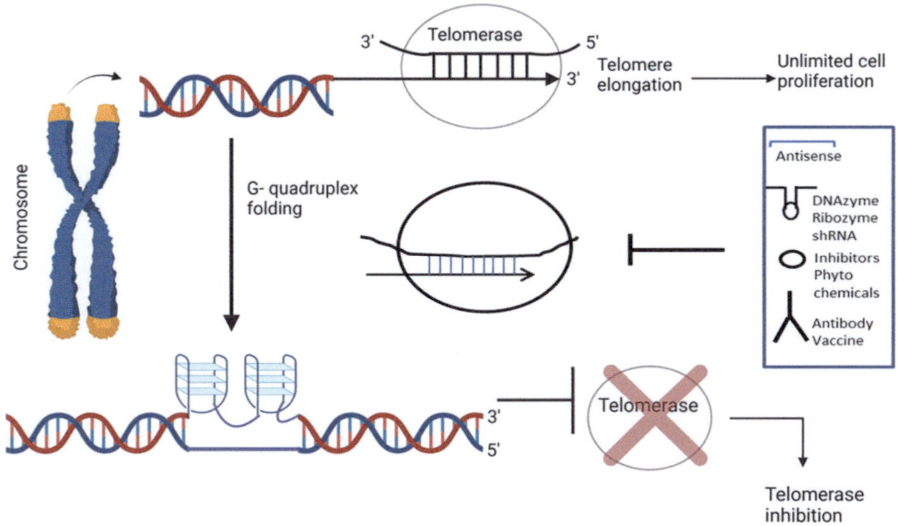

Fig. 6 Telomerase as targets by various means, antisense, ribozyme, DNA zyme, shRNA, small molecule inhibitors, G- quadruplex inhibitors, antibody/vaccine-based approaches used in therapeutic trials

Short-term treatment with telomerase inhibitors causes cell death in cancer cells, which show telomerase might have other functions apart from telomere maintenance. Many laboratories started working on the extra telomeric function of telomerase and found telomerase indeed has non-telomeric function, i.e. having a role in other cellular physiology like cellular growth, cell survival, and metabolism of the cells. It was known that telomerase does have a function related to NF-kb signaling, WNT/b catenin signaling pathways.

Here we have tried to represent most of the therapeutic efforts by nucleic acid-based (DNA, RNA), peptide-based, and natural phytochemical-based inhibitors of telomerase and components that have been reported. Though it is not very comprehensive, we have tried to represent the concepts and similar effects that could be elicited by other agents.

4 Targeting Telomerase and Components

Approaches that focus on inhibiting telomerase activity or inhibiting expression of its components, especially hTERT, may stop the proliferation of cancer cells and lead to cell death or senescence. Natural telomerase inhibitors, often derived from plants, include secondary metabolites such as polyphenols, alkaloids, terpenoids, xanthones, and other plant-based compounds, offering a natural approach to counteracting telomerase activity in cancer cells. These compounds function through various mechanisms, primarily by inhibiting telomerase activity, interfering with transcriptional, post-transcriptional, post-translational modifications, nuclear localization, and stabilization of G-quadruplex.

4.1 Nucleic Acid-Based Therapeutics Against Telomerase, Telomerase RNA and Telomeres

4.1.1 Antisense Oligos/Molecules

Antisense molecules are either DNA or RNA molecules that bind complementarily with the target sequence and induce RNase H activity against duplex region and degrade the targets. Antisense against hTR/hTERT has been studied with modification in oligonucleotides that includes 2′5′-oligodenylate(2-5A) linkages, 2-O methyl RNA, Phosphorothioate modified oligo-nucleotide. Imetelstat is an antisense phosphorothioate-modified 13-mer oligonucleotide initially known as GRN163, and is the only therapeutic molecule approved by the FDA to utilize as therapeutics, targeting against telomerase [23].

4.1.2 Ribozyme and SiRNAs

Ribozymes are RNA molecules that have the capability to break and form covalent bonds. In the same way, DNA having such capability is called as DNA zyme. These molecules require metal ions for its activity mainly Mg ions. Ribozymes were discovered from pre-rRNA of *Tetrahymena thermophilia* by Thomas Cech and RNA component of bacterial RNAse P from *E. coli* by Syndney Altman and colleagues [24, 25]. There are different types of ribozymes based on its origin and shapes. Hammerhead ribozyme was the simplest RNA molecule having a catalytic core and flanking region that specifies the target, making it easier, so that it can be designed against any target.

Hammerhead Ribozyme against hTR template region was shown to reduce telomerase activity in two human melanoma cell lines (JR8 and M14) [26]. Stable expression of targeted ribozyme against telomerase RNA, inhibits telomerase activity, reduces telomerase RNA level, and induces differentiation in a P53negative background [27]. Targeted ribozyme against hTERT mRNA expressed in MCF-7 breast cancer cell line and HBL immortalized breast cell line diminished hTERT RNA abundance and reduced the telomerase activity. And it also sensitizes the cells to topoisomerase inhibitor [28]. Similarly, siRNA could also induce telomerase inhibition, telomere attrition and cell death (Fig. 7).

4.1.3 DNAzyme

DNAzyme is an RNA cleaving single-stranded DNA molecule, that is 15–40 nt in length, which catalyzes the cleaving reaction. These molecules don't exist in nature but are developed through invitro procedure with defined nucleotides which is an essential core for catalysis, and flanking arms to impose specificity against target sequence. DNA zyme targeted against hTERT has been shown to have its maximal activity with differing Mg^{2+} ions and in cell lines, the DNAzyme which was experimented in low Mg^{2+} worked better in inhibiting telomerase [29].

4.1.4 Aptamers

Aptamers are nucleic acid molecules (RNA or DNA) that were selected against the target molecule by Systematic Evolution of Ligands by Exponential Enrichment

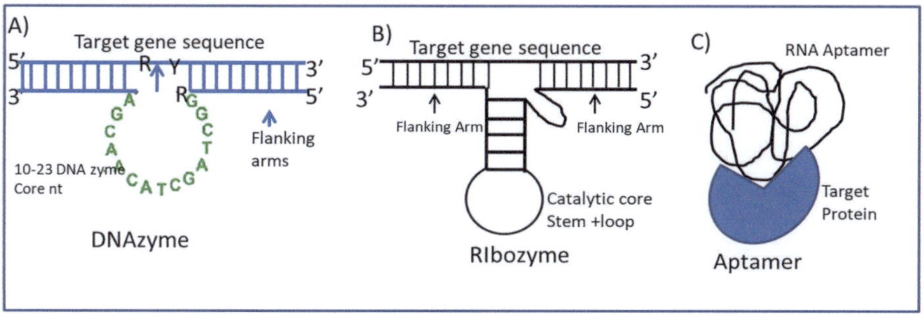

Fig. 7 Nucleic acid-based therapeutics': (**a**) DNA zyme with its 10–23 model core sequence with target. (**b**) Hammerhead ribozyme. (**c**) Structured RNA aptamer selected against the protein

(SELEX). Aptamers are used to block a molecule or used as a diagnostic tool [30]. RNA aptamers against synthetic peptides of RID2 domain of telomerase were selected by in vitro selection protocol SELEX. The selected RNAs could bind to the free peptide, and expressing this aptamer in MCF7 cells showed a reduction in telomerase activity as estimated by TRAP assay [31].

4.1.5 Small Molecule Inhibitors

Small molecule inhibitors can be designed to inhibit target proteins by binding in any pocket of the protein. This can bind to any extracellular and intracellular targets owing to its smaller size. A small molecule can also be taken orally.

4.1.6 Quadruplex-Based Telomerase Inhibitors

G-quadruplex structures, found in telomeres and oncogene promoters, have emerged as promising targets for anticancer therapy. The use of ligands that stabilize or induce G-quadruplex structures is a valid approach for cancer treatment. These ligands can operate through three main mechanisms: by disrupting telomere maintenance, reducing oncogene expression, and increasing genome instability [32]. G4s can form within the telomeric DNA, particularly in the G-rich overhang at the 3′ end. These structures can hinder telomerase access to the telomere, effectively blocking its activity. Stabilization of G4s by specific ligands can further enhance this inhibitory effect, making G4-stabilizing ligands potential telomerase inhibitors (Table 1).

4.2 Targeting by Vaccine

Vaccines are generally considered for infectious viral diseases and have historically been used for preventive and protective measures. Therapeutic vaccines were not successful due to its poor clinical efficacy. However recent developments have made it possible for development of multiple therapeutic cancer vaccines. As telomerase expression is found in 90% of cancers, it could be utilized as a vaccine molecule to

Table 1 Nucleic acid-based inhibitors against Telomerase, Telomerase RNA, and Telomeres

Compound	Target	Mechanism of action	References
Antisense			
2′- O-methyl-RNA	hTR	The compound binds to the template region of hTR to prevent catalysis of telomere repeat addition	[33]
N3′ → P5′ thio- phosphoramidate			[34]
Imetelstat		Lipid-conjugated phosphorothioate 13-mer antisense oligonucleotide that binds to hTR to inhibit telomerase	[35, 36]
Antisense oligonucleotides (AS- ODNs)	hTERT	Reduction in hTERT and telomere length	[37]
Hammerhead ribozyme	hTERT	Reduction in hTERT and telomere length	[28, 38]
Hammerhead ribozyme	hTR	Reduction in telomerase activity, telomere length, induction of differentiation	[27]
siRNA	hTERT	Reduction in hTERT and telomere length	[37]
Aptamers	hTERT	Binds to RID domain of TERT	[31]
Other inhibitors			
BIBR1532	hTERT	Binds to non-catalytic site on hTERT to inhibit translocation	[39]
AZT	Telomeres (nucleoside analogues)	Upon incorporation in the telomere, it abolishes telomere elongation due to absence of a 3′-OH group in its nucleotide	[40]
5-MeCITP (non-natural nucleotide)		Docks in the catalytic site and prevent telomere extension	[41]
Acyclic nucleoside phosphonates, e.g., PMEGpp		After getting incorporated into telomere, it inhibits telomere extension due to the absence of OH group	[42]
Telomestatin	G-quadruplex stabilizers	Prevent telomere extension by Stabilizing G-quadruplexes and shelterin binding, creating to replication stress	[43]
RHPS4			[44]

generate immunity. Two types of vaccination strategies have been developed, one is against tumour-associated antigens (TAA) and the other is against neo-antigens. Neo-antigens are present as unique individual-specific neo-antigens. Neo-antigens are the peptides generated by an individual's own cancer cells due to the mutations in their DNA that lead to the generation of neoantigens. These neo-antigens are individual-specific and vary in expression from person to person. The TAA vaccine has been successful when they were used along with immune checkpoint inhibitors. Most of the vaccines generated against telomerase are TAA-based vaccines.

Recent developments in immunotherapy include adoptive immune cell therapy that relies on genetic modification of T cells using T-cell receptor or chimeric antigen receptor (CAR) as a guiding factor for targeting tumour. CAR-specific receptor antigens particularly in B cell receptor CD19 molecules have been used in cancer treatments.

Various clinical strategies have been explored, including the use of DNA, RNA-based vaccines, MHC class I or II-restricted hTERT peptides, autologous antigen-presenting cells loaded with hTERT peptides, or cells transduced with stabilized hTERT mRNA for transient antigen expression. Most of the efforts were successful with peptide vaccines that were attempted against telomerase has been presented in Table 2.

Very early efforts have shown that telomerase is processed naturally into peptides and can be presented on the cell surface through major histocompatibility complexes I,II (MHCI and MHC II). Upon muscular injection of peptides, dendritic cells present to T cells-CD4 + T cells migrate to tumours and activate CD+8 T cells

Table 2 Therapeutic vaccines against telomerase and their clinical trials

Name of the vaccine	Adjuvant	Cancer types tested	Clinical phase trail	References
DNA based				
pDUV5 Canine specific(cTERT)	Chemokine ligand 21 (CCL21)	Canine induced cancer	Phase I	[45]
phTERT	IL 12 plasmid	HPV16 bearing tumour model of mice and Monkeys	NIL	[46]
INVAC1	hTERT Fused with ubiquitin	Solid tumors	Phase II	[47]
INO5401	NO		Phase I	[48]
RNA based				
mRNA vaccine	GM-CSF	Advanced renal cancer	Phase I/II	[48]
Peptide based				
hTERT:540–548 peptide	Montanide	Metastatic cancer	Phase I	[49]
GV1001	GM-CSF or TLR-7,	Pancreatic cancer, melanoma	Phase II/ III	[50]
Vx-001(two 9 amino acid peptides)	Montanide ISA51VG	Advanced non-small cell lung cancer (NSCLC)	Phase II	[51]
GX301(Four peptide of hTERT)	Montanide	Advanced prostate cancer	Phase II	[52]
UCPVax (universal cancer peptide-based vaccine)	Montanide	Non-small cell lung cancer (NSCLC)	Phase II	[53]
UV1	GM-CSF	Prostate cancer, NON small cell lung cancer	Phase II	[54]

against hTERT expressing cells and kill them by releasing cytokines or tumour necrosis factor (TNF) related apoptosis-inducing ligand (TRAIL) ways. GV1001, is a 16 amino acid-based peptide that significantly extends the survival of patients with pancreatic cancer. DNA vaccines (plasmid-based) have shown good immune response, and peptide vaccines have shown good clinical efficacy, such as UCPVax, UV1, and Vx-001 in certain cancer types. Recent advancements in vaccine design, including multiple peptides and adjuvants, have enhanced immune responses. However, challenges remain in achieving consistent and durable anti-tumor immunity. The table on vaccine strategy though not comprehensive, represents the available records on vaccine strategy against Telomerase. An extensive review of cancer vaccines against telomerase may be referred from Vahidi and Zabeti Touchaei (2024) [55].

4.3 Targeting Telomerase and Components by Phytochemicals

Approaches that focus on inhibiting telomerase activity or inhibiting expression of its components, especially hTERT, may stop the proliferation of cancer cells and lead to cell death or senescence. Natural telomerase inhibitors, often derived from plants, include secondary metabolites such as polyphenols, alkaloids, terpenoids, xanthones, and other plant-based compounds, offer a natural approach to counteract telomerase activity in cancer cells. These compounds function through various mechanisms, primarily by inhibiting telomerase activity, interfering with transcriptional, post-transcriptional, post-translational modifications, nuclear localization and stabilization of G-quadruplex.

4.3.1 Polyphenols/Flavonoids

Polyphenols are a diverse group of phytochemicals found in fruits, vegetables, tea, and herbs, showing promise as telomerase inhibitors. Recent studies suggest that polyphenols possess significant anticancer properties by influencing pathways related to hTERT expression activity.

Curcumin Curcumin, derived from turmeric (*Curcuma longa* L.), possesses anticancer properties. Curcumin inhibits telomerase activity in MCF-7 breast cancer cells via down-regulation of hTERT and c-myc mRNA [56]. It also affects the nuclear localization of telomerase, thereby inhibiting its activity. Furthermore, curcumin promotes the production of reactive oxygen species (ROS) in A-549 cells, leading to reduced hTERT levels [57]. Curcumin inhibits cell proliferation and telomerase activity in various cancer cell lines, such as Bel7402 and HL60 (Cui). Similar findings were also reported, highlighting curcumin's role in reducing cell viability and promoting apoptosis in leukemia cell lines [58].

Quercetin Quercetin is a flavonoid found in various fruits (apples, berries, grapes, and citrus), vegetables (broccoli, onion, tomato, and fennel), and green tea.

Quercetin possesses anti-carcinogenic and pro-apoptotic activities [59, 60]. Quercetin has demonstrated significant potential in cancer treatment and management by blocking telomerase activity and inducing cell death [61].

Resveratrol It is a phenolic compound found in fruits, such as grapes and berries. Pterostilbene, a natural analogue of resveratrol, has been shown to significantly decrease the expression and activity of telomerase in lung cancer cell lines [62] and in MCF and NCI-H460 cells [62, 63]. It alters transcription by altering promoter activity in colorectal cancer cell lines, human epidermoid carcinoma cells in a dose-dependent manner [64–67].

Epigallocatechin-3-gallate (EGCG) EGCG is a polyphenol, found in tea, fruits, vegetables, and nuts. EGCG impairs telomerase activity and leads to cell cycle arrest or apoptosis [68] and suppresses C myc [69]. Chronic treatment with non-cytotoxic doses of EGCG caused telomere shortening, subsequent senescence, and telomere-dependent genomic damage [70]. EGCG treatment enhances DNA damage, causes cell cycle arrest in G2, M, S phases, and induces apoptosis by reducing hTERT and Bcl-2, in ovarian cancer cells [71].

Apigenin Apigenin is a flavonoid found in fruits and vegetables. It is recognized for its diverse biological properties, including anti-inflammatory, anti-proliferative, antioxidant, and anti-carcinogenic effects [72]. Apigenin downregulates telomerase activity and induces apoptosis by inhibiting c-Myc-mediated hTERT expression in leukemia cells [73].

Luteolin Luteolin is a flavonoid found in carrots, celery, and herbs. It is known for its pharmacological benefits, including hepatoprotection, neuroprotection, and anti-cancer properties, such as lung, prostate, and breast cancer. In MDA-MB-231 cells, luteolin causes S-phase cell cycle arrest and apoptosis by inhibiting hTERT expression and reducing telomerase levels [74]. In-silico studies also indicate that structural modifications of luteolin analogs could lead to promising telomerase inhibitors [75].

Genistein Genistein, a natural isoflavone in soy products, has been shown to down-regulate telomerase expression through altering chromatin structure, and activity, by increasing trimethyl-H3K9 and decreasing dimethyl-H3K4 at the hTERT promoter, in time, dose-dependent manner in breast cancer cell line [76], head and neck, prostate, neuroblastoma, breast cancers [77, 78], and inhibits translocation of telomerase to the nucleus [78].

Silibinin Silibinin is a flavonoid derived from *Silybum marianum* [79]. Treatment with silibinin on a human leukemia cell line K562, LNCap prostate cancer cells by lowering hTERT mRNA resulted in a notable reduction in telomerase activity [80, 81]. The treatment of silibinin combined with another drug doxorubicin [82], another polyphenol, curcumin, shows more effectiveness in inhibiting telomerase in T47D breast cancer cells [83].

Sulforaphane Sulforaphane recognized for its anti-cancer properties, inhibits cell viability and telomerase activity by downregulating telomerase reverse transcriptase (hTERT) [84].

Other polyphenols, including gossypol, pectenotoxin-2 (PTX-2), sulforaphane alterperylenol, rubromycins, and purpuromycins, are also known to inhibit telomerase activity or expression in cancer cells. Gossypol, a compound derived from cotton seeds, causes direct cytotoxicity and inhibits telomerase activity by downregulating c Myc, hTERT expression, and nuclear translocation [85, 86]. Pectenotoxin-2 (PTX-2), first identified in marine sponges, reduces cell viability and inhibits telomerase activity in human leukemia cells by downregulating hTERT expression [87]. Chromatin immunoprecipitation assays demonstrate that PTX-2 decreases c-Myc and Sp1 expressions along with their DNA binding activity. It also reduces Akt phosphorylation, resulting in lower hTERT phosphorylation and nuclear translocation [88]. Alterperylenol is an orange-red antifungal pigment sourced from *Alternaria* sp., known to inhibit telomerase. Rubromycins, derived from *Streptomyces collinus*, and purpuromycins from *Actinoplanes ianthinogenes*, both demonstrate significant anti-telomerase effects [89, 90].

4.3.2 Alkaloids

Cephalotaxus alkaloids Cephalotaxus alkaloids, derived from *cephalotaxus* trees, include inactive cephalotaxine and potent compounds such as homoharringtonine and deoxyharringtonine [91] demonstrated that these compounds decrease the telomerase activity by altering transcription and trigger apoptosis [92].

Boldine Boldine, a natural alkaloid extracted from the *Boldo* and *Lindera* tree possesses cytotoxic and anti-tumor effects in hepatic, bladder, and brain cancer cell lines by lowering telomerase activity and hTERT expression [93–95].

Berberine Berberine is an alkaloid found in several plants, including *Berberis vulgaris* (barberry) and *Coptis chinensis* (Chinese goldthread) [96]. Berberine chloride exhibits anti-cancer activity with moderate telomerase inhibition. It interacts with G-quadruplexes (G4), enhancing their stability [97]. This stabilization impedes both telomerase activity and the growth of cancer cells. It was observed that berberine induces apoptosis in HL-60 and U937 human leukemia cells by down-regulating nucleophosmin/B23 mRNA and telomerase activity [98, 99].

Berbamine Berbamine, a dimeric tetrahydroisoquinoline alkaloid derived from *Berberis vulgaris*. Berbamine has demonstrated anti-telomerase activity in HL-60 leukemia cells, leading to a reduction in telomerase activity [100].

Other alkaloids, including cryptolepine derived from the roots of *Cryptolepis triangularis* and *C. sanguinolenta*, interact with G qudruplex moderate anti-telomerase activity [101]. Dictyodendrins and its members (A-E) isolated from the Japanese marine sponge *Dictyodendrilla verongiformis* were discovered to inhibit telomerase. Meridine, derived from the ascidian *Amphicarpa meridian*, shares structural similarities with ascididemin, which is isolated from the Okinawan tunicate [102, 103], meridine and ascididemin effectively stabilize G-quadruplexes and suppress telomerase activity [104].

4.3.3 Terpenoids

Terpenoids are derivatives of terpenes that contain various functional groups found in flowers, roots, plant materials, and essential oils [105], and have shown anticancer properties, via cell cycle arrest, impairing cancer cell differentiation, and promoting apoptosis [106].

Crocin Crocin, the main pigment found in *Crocus sativus*, shows antiproliferative effects on HepG2 liver cancer cells by lowering telomerase activity [107].

Tanshinone I Tanshinone is derived from the dried roots and rhizomes of *Salvia miltiorrhiza*. Tanshinone I demonstrates potential as an anti-cancer agent, reducing telomerase in various types of cancer, including breast, cervical, and prostate cancers [108].

Helenalin Helenalin, a sesquiterpene lactone derived from *Arnica montana*, exhibits anti-inflammatory, anti-tumor, and anti-microbial properties [109–111]. Helenalin and its derivatives (costunolide and parthenin) inhibit NF-κB and have shown potent anti-telomerase activity in hematopoietic cancer cells Jurkat and HL-60, selectively targeting telomerase activity [112, 113].

Ginsenoside Rk1 Ginsenoside Rk1 is isolated from heat-processed *Panax ginseng* [114]. Ginsenoside Rk1 treatment inhibited the growth of the HepG2 human hepatoma cell line by decreasing c-Myc and hTERT and suppressing telomerase activity [115] (Table 3).

Table 3 Natural phytochemicals having antitelomerase activity

Cass of compound	Source	Compound	Mechanism of action	References
Polyphenols/ flavonoids	*Curcuma longa*	Curcumin	Down-regulation of hTERT at mRNA and protein level	[56, 57]
	Phyllanthus urinaria	Quercetin	Reduction of hTERT expression and telomerase activity.	[61]
	Arachis hypogeal, *cassia* sp., *vitis vinifera*, *veratrum grandiflorum*	Resveratrol	Down-regulation hTERT expression, nuclear localization, and telomerase activity	[63, 66, 67]
	Camellia sinensis	Epigallocatechin-3-gallate (egcg)	Down-regulation hTERT expression and inhibition of telomerase activity by competitively at the active site.	[69, 71]
	Fruits and vegetables	Apigenin	Inhibition of binding of c-Myc at hTERT promoter, down-regulation hTERT expression and telomerase activity.	[73]
	Melissa officinalis	Luteolin	Reduction of hTERT expression and telomerase activity	[74]
	Plant secondary metabolites, glycine max	Genistein	Down-regulation hTERT expression, phosphorylation and nuclear localization, telomerase activity and stabilization of G4	[76–78]
	Silybum marianum	Silibinin	Reducing hTERT expression and its activity with reduction of c-Myc transcript.	[81, 83]
	Cottonseed	Gossypol	Down-regulation hTERT expression, phosphorylation and nuclear localization, telomerase activity and post-translational modification hTERT.	[85, 86]
	Dinophysis fortii	Pectenotoxin-2 (ptx-2)	Inhibition of binding of c-Myc at hTERT promoter, down-regulation hTERT expression and telomerase activity.	[87]
	Broccoli and cauliflower	Sulforaphane	Inhibition of post-translational modification of hTERT, telomerase activity	[84]
	Alternaria sp.	Alterperylenol	Inhibition of telomerase activity	[89]
	Streptomyces collinus	Rubromycins	Inhibition of telomerase activity	[90]
	Actinoplanes ianthinogenes	Purpuromycins	Inhibition of telomerase activity	[116]

Cass of compound	Source	Compound	Mechanism of action	References
Alkaloids	*Cephalotaxus sp*	Cephalotaxus alkaloids	Reduction of hTERT expression, and its activity	[92]
	Peumus boldus	Boldine	Down-regulation of hTERT expression	[93–95]
	Berberis vulgaris chinensis	Berberine	Inhibition of telomerase activity and stabilization of G4	[98, 99]
	Berberis vulgaris (barberry)	Berbamine	Inhibition of telomerase activity	[100]
	Cryptolepis triangularis	Cryptolepine	Inhibition of telomerase activity and stabilization of G4	[101]
	Dictyodendrilla verongiformis	Dictyodendrin	Inhibition of telomerase activity	[117]
	Amphicarpa meridian	Meridine	Stabilization of G4 Quadruplex and inhibiting in Telomerase	[104]
	Okinawan tunicate didenum sp.	Ascididemin	Inhibition of telomerase activity and stabilization of G4	[104]
Terpenoids	*Crocus sativus l.*	Crocin	Down-regulation of the telomerase activity and htert expression	[107]
	Salvia miltiorrhiza	Tanshinone I	Reduction of hTERT expression and activity	[118]
	Arnica montana	Helenalin	Inhibits NF Kb, and Reduce hTERT transcription	[112]
	Sun ginseng	Ginsenoside rk1	Reduces c Myc and hTERT transcript levels	[115]
	Tripterygium wilfordii	Triptolide	Down-regulation hTERT transcription through down-regulation of transcription factor SP1	[119]
	Nigella sativa	Thymoquinone	Down-regulation of the telomerase activity and htert expression	[120]
Triterpenoid	*Celosia argentea, maytenus ilicifolia, salacia reticulata, quassia species*	Pristimerin	Down-regulation of hTERT transcription through inhibition of nuclear factor kappa beta (Nf-κb), Inhibition of telomerase activity	[121]
	Woody angiosperms	Oleanane	Down-regulation hTERT expression	[122]

(continued)

Table 3 (continued)

Cass of compound	Source	Compound	Mechanism of action	References
Xanthone	*Garcinia hurburyi*	Gambogic acid	Abolishes hTERT activity and inhibits serine/threonine-protein kinase (Akt)	[123]
Macrocyclic	*Streptomyces anulatus*	Telomestatin	Interference of G-quadruplex stabilization, dissociation of TRF and POT1 telomere-specific binding proteins	[43, 124, 125]
Fatty acids	Oils, nuts	Oleic acid and other long-chain fatty acids	Competitively binding at the telomerase active site	[126]
Liposaccharide	*Axinella infundibula*	Axinelloside	Inhibition of telomerase activity	[127]
3-alkyl indoles	*Brassica oleracea*	Indole-3-carbinol	Reduces hTERT expression and its activity	[128]
Azaphilones	*Talaromyces flavus*	Diazaphilonic acid	Inhibition of telomerase activity	[129]
Organosulfur compounds	*Allium sativum*	Allicin and ajoene	Reduction of hTERT and telomerase activity	[130]

Additional terpenoids such as triptolide, sesquiterpene lactone, and thymoquinone are noted for their ability to inhibit telomerase activity. Triptolide, derived from *Tripterygium wilfordii*, lowers the levels of human telomerase reverse transcriptase (hTERT) mRNA, which in turn diminishes telomerase activity in Epstein-Barr virus (EBV)-positive B lymphoma cells [119]. Sesquiterpene lactones are well-known as strong inhibitors of human telomerase, highlighting their anticancer potential [131]. Thymoquinone, the active ingredient in *Nigella sativa*, also inhibits telomerase activity, leading to telomere shortening, DNA damage, and cell death in glioblastoma cancer cells [120].

4.3.4 Triterpenoid

Pristimerin Pristimerin, a triterpenoid quinone methide derived from plants in the Celastraceae and Hippocrateaceae families, exhibits strong anticancer effects [132]. It inhibits telomerase activity by inhibiting NF-kb-mediated phosphorylation of hTERT [121].

Oleanane Oleanane, a triterpenoid derivative of oleanolic acid, suppresses telomerase activity, hTERT gene expression, and regulatory proteins in pancreatic and prostate cancers, making hTERT a key target for its potential therapeutic applications [122].

Gambogic acid Gambogic acid, a xanthone from *Garcinia hurburyi* resin, is traditional Chinese medicine, that demonstrated potent anti-cancer effects against leukemia, gastric, lung, breast, liver, and pancreatic cancers, it inhibits telomerase by suppressing c-myc and blocking Akt phosphorylation, reducing hTERT activity [123, 133]

Telomestatin Telomestatin is a macrocyclic natural product, isolated from *Streptomyces analatus* [134], selectively binds to G-quadruplex structures, inhibits telomerase, disrupts telomere maintenance and induces apoptosis [43, 135].

Analogs of telomestatin, including macrocyclic polyoxazoles and cyclo[n]pyrroles, show strong inhibitory effects on telomerase. These analogs exhibit strong binding to telomeric DNA, with some showing in vivo activity against human cancer xenografts [136, 137]. Other Natural metabolites, metabolites such as fatty acids (including oleic acid and other long-chain fatty acids) [126], polysaccharides (like axinelloside A) [127], 3-alkyl indoles (such as indole-3-carbinol) [128], azaphilones (including diazaphilonic acid) [129], and organosulfur compounds (such as allicin and ajoene) [130], which are derived from plants, microbes, or marine sources, show potential for telomerase inhibition, making them valuable for cancer therapy.

5 Challenges

Due to the exclusivity of expression in cancer cells, telomerase emerged as an important target for therapeutic intervention. Most of the earlier efforts were to artificially induce end replication problems in cancer cells and render the cell devoid of telomerase and reduce telomere length, so cancer cells can induce cell death. The time it takes to induce the reduction in telomere length beyond the critical limit takes longer time to induce apoptotic signalling. The lag period may act positively or negatively, as cells may adopt any alternative signalling that may negate the telomere dependant apoptotic signalling. Moreover, the apoptosis programmes are P53 and PRb dependent which put forth the major challenge in P53 pRB negative tumours. However, recent evidence emerged that telomerase have non-canonical function and inhibition induces immediate cell death in certain cancer cells in invitro, which brings great promise for utilizing telomerase as a therapeutic target.

6 Conclusion

In spite of many challenges, telomerase still stays as good target to be utilized either by RNA or DNA or phytochemical-based agents to inhibit its function. Some phytochemical agents have multiple targets that can be used in inhibiting many of the pathways necessary for the initiation of cancer as telomerase is essential to breach barriers necessary for tumorigenesis. Recent efforts on vaccines against telomerase have produced mixed responses by using other novel adjuvants or using multiple peptides. However, Imtelstat (Rytelo) is the only drug that has got FDA approval for the treatment of myelodysplastic syndrome, a group of blood cancer disorders, and has shown promising results. Many telomerase-based vaccines still are in clinical trials and may bring promises for clinical response against cancers. Though many challenges are put forth, telomerase field may bring a great promise in the treatment of cancers in the foreseeable future.

Acknowledgement The author thanks ICMR-NICPR for all the support to bring out this work. Few diagrams were drawn with the help of Bio Render and MS Office PowerPoint.

References

1. Hanahan D, Weinberg RA. The hallmarks of cancer. Cell. 2000;100:57–70.
2. Hanahan D. Hallmarks of cancer: new dimensions. Cancer Discov. 2022;12(1):31–46.
3. Greider CW, Blackburn EH. Identification of a specific telomere terminal transferase activity in Tetrahymena extracts. Cell. 1985;43:405–13.
4. Greider CW, Blackburn EH. A telomeric sequence in the RNA of Tetrahymena telomerase required for telomere repeat synthesis. Nature. 1989;337:331–7.
5. Hayflick L, Moorhead PS. The serial cultivation of human diploid cell strains. Exp Cell Res. 1961;25:585–621.

6. Morin GB. The human telomere terminal transferase enzyme is a ribonucleoprotein that synthesizes TTAGGG repeats. Cell. 1989;59:521–9.
7. Kim NW, et al. Specific association of human telomerase activity with immortal cells and cancer. Science. 1994;266:2011–5.
8. Bodnar AG, et al. Extension of life-span by introduction of telomerase into normal human cells. Science. 1998;279:349–52.
9. Queisser A, et al. Inhibition of telomerase induces alternative lengthening of telomeres during human esophageal carcinogenesis. Cancer Genet. 2013;206(11):374–86.
10. Blasco MA, et al. Telomere shortening and tumor formation by mouse cells lacking telomerase RNA. Cell. 1997;91(1):25–34.
11. Maciejowski J, de Lange T. Telomeres in cancer: tumour suppression and genome instability. Nat Rev Mol Cell Biol. 2017;3:175–86.
12. Wright WE, Shay JW. The two-stage mechanism controlling cellular senescence and immortalization. Exp Gerontol. 1992;27:383–9.
13. Shay JW, Wright WE. Senescence and immortalization: role of telomeres and telomerase. Carcinogenesis. 2005;26(5):867–74.
14. Nassour J, et al. Telomeres and cancer: resolving the paradox. Annu Rev Cancer Biol. 2021;1:59–77.
15. Olovnikov AM. Printsip marginotomii v matrichnom sinteze polinukleotidov. Dokl Akad Nauk SSSR. 1971;201:1496–9.
16. Watson JD. Origin of concatemeric T7 DNA. Nat New Biol. 1972;239:197–201.
17. Wynford-Thomas D, Kipling D. The end-replication problem. Nature. 1997;389:551.
18. Palm W, de Lange T. How shelterin protects mammalian telomeres. Annu Rev Genet. 2008;42:301–34.
19. Xu Y, Komiyama M. G-quadruplexes in human telomere: structures, properties, and applications. Molecules. 2023;29(1):174.
20. Nguyen THD. Structural biology of human telomerase: progress and prospects. Biochem Soc Trans. 2021;49(5):1927–39.
21. Chen JL, Greider CW. Telomerase RNA structure and function: implications for dyskeratosis congenital. Trends Biochem Sci. 2004;29(4):183–92.
22. Welfer GA, Freudenthal BD. Recent advancements in the structural biology of human telomerase and their implications for improved design of cancer therapeutics. NAR Cancer. 2023;5(1):zcad010.
23. Shay J, Keith W. Targeting telomerase for cancer therapeutics. Br J Cancer. 2008;98:677–83.
24. Cech TR, et al. *In vitro* splicing of the ribosomal RNA precursor of tetrahymena: involvement of a guanosine nucleotides in the excision of the intervening sequence. Cell. 1981;27:487–96.
25. Cecelia GT, et al. The RNA moiety of ribonuclease P is the catalytic subunit of the enzyme. Cell. 1983;35(3):849–57.
26. Folini M, et al. Inhibition of telomerase activity by a hammerhead ribozyme targeting the RNA component of telomerase in human melanoma cells. J Invest Dermatol. 2000;114(2):259–67.
27. Ramakrishnan SK, et al. Expression of targeted ribozyme against telomerase RNA causes altered expression of several other genes in tumor cells. Tumour Biol. 2014;35(6):5539–50.
28. Ludwig A, et al. Ribozyme cleavage of telomerase mRNA sensitizes breast epithelial cells to inhibitors of topoisomerase. Cancer Res. 2001;61(7):3053–61.
29. Yuan BF, et al. Two DNAzymes targeting the telomerase mRNA with large difference in Mg2+ concentration for maximal catalytic activity. Int J Biochem Cell Biol. 2007;39(6):1119–29.
30. Nimjee SM, et al. Aptamers as therapeutics. Annu Rev Pharmacol Toxicol. 2017;57:61–79.
31. Varshney A, et al. Identification of an RNA aptamer binding hTERT-derived peptide and inhibiting telomerase activity in MCF7 cells. Mol Cell Biochem. 2017;427(1–2):157–67.
32. Kosiol N, et al. G-quadruplexes: a promising target for cancer therapy. Mol Cancer. 2021;20:40.
33. Pitts AE, Corey DR. Inhibition of human telomerase by 2′-O-methyl-RNA. Proc Natl Acad Sci USA. 1998;95(20):11549–54.

34. Gryaznov SM. Oligonucleotide n3′→p5′ phosphoramidates and thio-phoshoramidates as potential therapeutic agents. Chem Biodivers. 2010;7(3):477–93.
35. Schrank Z, et al. Oligonucleotides targeting telomeres and telomerase in cancer. Molecules. 2018;23(9):2267.
36. Marian CO, et al. The telomerase antagonist, imetelstat, efficiently targets glioblastoma tumor-initiating cells leading to decreased proliferation and tumor growth. Clin Cancer Res. 2010;16(1):154–63.
37. Kraemer K, et al. Microarray analyses in bladder cancer cells: inhibition of hTERT expression down-regulates EGFR. Int J Cancer. 2006;119(6):1276–84.
38. Saretzki G, et al. Ribozyme-mediated telomerase inhibition induces immediate cell loss but not telomere shortening in ovarian cancer cells. Cancer Gene Ther. 2001;8(10):827–34.
39. Pascolo E, et al. Mechanism of human telomerase inhibition by BIBR1532, a synthetic, non-nucleosidic drug candidate. J Biol Chem. 2002;277(18):15566–72.
40. Gomez DE, et al. AZT as a telomerase inhibitor. Front Oncol. 2012;2:113.
41. Hernandez-Sanchez W, et al. A non-natural nucleotide uses a specific pocket to selectively inhibit telomerase activity. PLoS Biol. 2019;17(4):e3000204.
42. Hájek M, et al. Inhibition of human telomerase by diphosphates of acyclic nucleoside phosphonates. Biochem Pharmacol. 2005;70(6):894–900.
43. Kim MY, et al. Telomestatin, a potent telomerase inhibitor that interacts quite specifically with the human telomeric intramolecular g-quadruplex. J Am Chem Soc. 2002;124(10):2098–9.
44. Salvati E, et al. Telomere damage induced by the G-quadruplex ligand RHPS4 has an antitumor effect. J Clin Invest. 2007;117(11):3236–47.
45. Thalmensi J, et al. A DNA telomerase vaccine for canine cancer immunotherapy. Oncotarget. 2019;10(36):3361–72.
46. Yan J, et al. 417. Induction of potent cytotoxic and Antitumor Activity by a highly optimized hTERT DNA vaccine. Mol Ther. 2015;23:S165.
47. Thalmensi J, et al. Anticancer DNA vaccine base on human telomerase reverse transcriptase generates strong and specific T cell immune response. Onco Targets Ther. 2015;5:e1083670.
48. Ellingsen EB, et al. Telomerase as a target for therapeutic cancer vaccines and considerations for optimizing their clinical potential. Front Immunol. 2021;12:682492.
49. Parkhurst MR, et al. Immunization of patients with the hTERT:540-548 peptide induces peptide-reactive T lymphocytes that do not recognize tumors endogenously expressing telomerase. Clin Cancer Res. 2004;10(14):4688–98.
50. Inderberg-Suso EM, et al. Widespread CD4+ T-cell reactivity to novel hTERT epitopes following vaccination of cancer patients with a single hTERT peptideGV1001. Onco Targets Ther. 2012;1:670–86.
51. Gridelli C, et al. Clinical activity of a htert (vx-001) cancer vaccine as post-chemotherapy maintenance immunotherapy in patients with stage IV non-small cell lung cancer: final results of a randomised phase 2 clinical trial. Br J Cancer. 2020;122:1461–6.
52. Fenoglio D, et al. Immunogenicity of GX301 cancer vaccine: four (telomerase peptides) are better than one. Hum Vaccin Immunother. 2015;11:838–50.
53. Rebucci-Peixoto M, et al. A phase II study evaluating the interest to combine UCPVax, a telomerase CD4 T(H)1-Inducer Cancer Vaccine, and Atezolizumab for the treatment of HPV positive cancers: VolATIL Study. Front Oncol. 2022;12:957580.
54. Haakensen VD, et al. UV1 telomerase vaccine with ipilimumab and nivolumab as second line treatment for pleural mesothelioma – a phase II randomised trial. Eur J Cancer. 2024;202:113973.
55. Vahidi S, Zabeti Touchaei A. Telomerase-based vaccines: a promising frontier in cancer immunotherapy. Cancer Cell Int. 2024;24:421.
56. Ramachandran C, et al. Curcumin inhibits telomerase activity through human telomerase reverse transcritpase in MCF-7 breast cancer cell line. Cancer Lett. 2002;184(1):1–6.
57. Singh M, Singh N. Curcumin counteracts the proliferative effect of estradiol and induces apoptosis in cervical cancer cells. Mol Cell Biochem. 2011;347(1–2):1–11.

58. Chakraborty SMN, et al. Curcumin-induced apoptosis in human leukemia cell HL-60 is associated with inhibition of telomerase activity. Mol Cell Biochem. 2007;297(1–2):31–9.
59. Lou M, et al. Quercetin nanoparticles induced autophagy and apoptosis through AKT/ERK/Caspase-3 signaling pathway in human neuroglioma cells: *in vitro* and *in vivo*. Biomed Pharmacother. 2016;84:1–9.
60. Ren KW, et al. Quercetin nanoparticles display antitumor activity via proliferation inhibition and apoptosis induction in liver cancer cells. Int J Oncol. 2017;50(4):1299–311.
61. Ganesan K, Xu B. Telomerase inhibitors from natural products and their anticancer potential. Int J Mol Sci. 2017;19(1):13.
62. Chen RJ, et al. P53-dependent downregulation of hTERT protein expression and telomerase activity induces senescence in lung cancer cells as a result of pterostilbene treatment. Cell Death Dis. 2017;8(8):e2985.
63. Tippani R, et al. Pterostilbene as a potential novel telomerase inhibitor: molecular docking studies and its *in vitro* evaluation. Curr Pharm Biotechnol. 2014;14(12):1027–35.
64. Platella C, et al. Antitumour activity of resveratrol on human melanoma cells: a possible mechanism related to its interaction with malignant cell telomerase. Biochim Biophys Acta Gen Subj. 2017;1861(11 Pt A):2843–51.
65. Cosan DT, et al. Effects of various agents on DNA fragmentation and telomerase enzyme activities in adenocarcinoma cell lines. Mol Biol Rep. 2011;38(4):2463–9.
66. Chung SS, et al. Combination of resveratrol and 5-flurouracil enhanced anti-telomerase activity and apoptosis by inhibiting STAT3 and Akt signaling pathways in human colorectal cancer cells. Oncotarget. 2018;9(68):32943–57.
67. Zhai XX, et al. Effects of resveratrol on the proliferation, apoptosis and telomerase ability of human A431 epidermoid carcinoma cells. Oncol Lett. 2016;11(5):3015–8.
68. Colomer R, et al. Natural polyphenols and their synthetic analogs as emerging anticancer agents. Curr Drug Targets. 2017;18(2):147–59.
69. Zhang W, et al. Effects of epigallocatechin gallate on the proliferation and apoptosis of the nasopharyngeal carcinoma cell line CNE2. Exp Ther Med. 2014;8(6):1783–8.
70. Gurung RL, et al. MST-312 alters telomere dynamics, gene expression profiles and growth in human breast cancer cells. J Nutrigenet Nutrigenomics. 2014;7(4–6):283–98.
71. Liu L, et al. Epigallocatechin-3-gallate suppresses cell proliferation and promotes apoptosis in Ec9706 and Eca109 esophageal carcinoma cells. Oncol Lett. 2017;14(4):4391–5.
72. Naponelli V, et al. Apigenin: molecular mechanisms and therapeutic potential against cancer spreading. Int J Mol Sci. 2024;25(10):5569.
73. Jayasooriya RG, et al. Apigenin decreases cell viability and telomerase activity in human leukemia cell lines. Food Chem Toxicol. 2012;50(8):2605–11.
74. Huang L, et al. Luteolin inhibits cell cycle progression and induces apoptosis of breast cancer cells through downregulation of human telomerase reverse transcriptase. Oncol Lett. 2019;17(4):3842–50.
75. Wang Z, et al. Dietary luteolin: a narrative review focusing on its pharmacokinetic properties and effects on glycolipid metabolism. J Agric Food Chem. 2021;69(5):1441–54.
76. Yang R, et al. Regulation and clinical potential of telomerase reverse transcriptase (TERT/hTERT) in breast cancer. Cell Commun Signal. 2023;21(1):218.
77. Li Y, et al. Genistein depletes telomerase activity through cross-talk between genetic and epigenetic mechanisms. Int J Cancer. 2009;125(2):286–96.
78. Xie Q, et al. Genistein inhibits DNA methylation and increases expression of tumor suppressor genes in human breast cancer cells. Genes Chromosomes Cancer. 2014;53(5):422–31.
79. Prasad R, et al. Silibinin and non-melanoma skin cancers. J Tradit Complement Med. 2020;10(3):236–44.
80. Faezizadeh Z, et al. The effect of silymarin on telomerase activity in the human leukemia cell line K562. Planta Med. 2012;78(9):899–902.
81. Thelen P, et al. Inhibition of telomerase activity and secretion of prostate specific antigen by silibinin in prostate cancer cells. J Urol. 2004;171(5):1934–8.

82. Yurtcu., E., et al. Effects of silymarin and silymarin-doxorubicin applications on telomerase activity of human hepatocellular carcinoma cell line HepG2. J BUON. 2015;20(2):555–61.
83. Nasiri M, et al. Curcumin and silibinin inhibit telomerase expression in T47D human breast cancer cells. Asian Pac J Cancer Prev. 2013;14(6):3449–53.
84. Abbas A, et al. Sulforaphane modulates telomerase activity via epigenetic regulation in prostate cancer cell lines. Biochem Cell Biol. 2015;94(1):71–81.
85. Moon D, et al. Gossypol suppresses telomerase activity in human leukemia cells via regulating hTERT. FEBS Lett. 2008;582(23–24):3367–73.
86. Paunovic D, et al. The potential roles of gossypol as anticancer agent: advances and future directions. Chin Med. 2023;18(1):163.
87. Kim GY, et al. Pectenotoxin-2 from marine sponges: a potential anti-cancer agent-a review. Mar Drugs. 2011;9(11):1490–1.
88. Kim MO, et al. Pectenotoxin-2 represses telomerase activity in human leukemia cells through suppression of hTERT gene expression and Akt-dependent hTERT phosphorylation. FEBS Lett. 2008;582(23–24):3263–9.
89. Hassig, C. A., Zeng, F. Y., Kung, P., Kiankarimi, M., Kim, S., Diaz, P. W., Zhai, D., Welsh, K., Morshedian, S., Su, Y., O'Keefe, B., Newman, D. J., Rusman, Y., Kaur, H., Salomon, C. E., Brown, S. G., Baire, B., Michel, A. R., Hoye, T. R., Francis, S., … Reed, J. C. (2014). Ultra-High-Throughput Screening of Natural Product Extracts to Identify Proapoptotic Inhibitors of Bcl-2 Family Proteins. Journal of biomolecular screening, 19(8), 1201–1211. https://doi.org/10.1177/1087057114536227
90. Ueno T, et al. Inhibition of human telomerase by rubromycins: implication of spiroketal system of the compounds as an active moiety. Biochemistry. 2000;39(20):5995–6002.
91. Powell RG, et al. Alkaloids of Cephalotaxus harringtonia var. drupacea. 11-Hydroxycephalotaxine and drupacine. J Org Chem. 1974;39(5):676–80.
92. Xie WZ, et al. Homoharringtonine-induced apoptosis of human leukemia HL-60 cells is associated with down-regulation of telomerase. Am J Chin Med. 2006;34(2):233–44.
93. Kazemi Noureini S, Tanavar F. Boldine, a natural aporphine alkaloid, inhibits telomerase at non-toxic concentrations. Chem Biol Interact. 2015;231:27–34.
94. Gerhardt D, et al. Boldine: a potential new antiproliferative drug against glioma cell lines. Investig New Drugs. 2009;27(6):517–25.
95. Noureini SK, Wink M. Dose-dependent cytotoxic effects of boldine in HepG-2 cells-telomerase inhibition and apoptosis induction. Molecules. 2015;20(3):3730–43.
96. Imanshahidi M, Hosseinzadeh H. Pharmacological and therapeutic effects of Berberis vulgaris and its active constituent, berberine. Phytother Res. 2008;22(8):99–1012.
97. Franceschin M, et al. Natural and synthetic G-quadruplex interactive berberine derivatives. Bioorg Med Chem Lett. 2006;16(6):1707–11.
98. Naasani, et al. FJ5002: a potent telomerase inhibitor identified by exploiting the disease-oriented screening program with COMPARE analysis. Cancer Res. 1999;59(16):4004–11.
99. Wu HL, et al. Berberine-induced apoptosis of human leukemia HL-60 cells is associated with down-regulation of nucleophosmin/B23 and telomerase activity. Int J Cancer. 1999;81(6):923–9.
100. Ji ZN, et al. Inhibition of telomerase activity and bcl-2 expression in berbamine-induced apoptosis in HL-60 cells. Planta Med. 2002;68(7):596–600.
101. Guittat L, et al. Interactions of cryptolepine and neocryptolepine with unusual DNA structures. Biochimie. 2003;85(5):535–47.
102. Kitahara Y, et al. Synthesis of meridine, cystodamine, and related compounds including iminoquinolinequinone structure. Tetrahedron. 1998;54(29):8421–32.
103. Kobayash J, et al. Ascididemin, a novel pentacyclic aromatic alkaloid with potent antileukemic activity from the okinawan tunicate didemnum sp. Tetrahedron Lett. 1988;29(10):1177–80.
104. Guittat L, et al. Ascididemin and meridine stabilise G-quadruplexes and inhibit telomerase *in vitro*. Biochim Biophys Acta. 2005;1724(3):375–84.
105. Lai Shi Min S, et al. Plant terpenoids as the promising source of cholinesterase inhibitors for anti-AD therapy. Biology. 2022;11(2):307.

106. Kamran S, et al. Therapeutic potential of certain terpenoids as anticancer agents: a scoping review. Cancers. 2022;14(5):1100.
107. Noureini SK, Wink M. Antiproliferative effects of crocin in HepG2 cells by telomerase inhibition and hTERT down-regulation. Asian Pac J Cancer Prev. 2012;13(5):2305–9.
108. Naz I, et al. An overview of the anti-cancer actions of Tanshinones, derived from Salvia miltiorrhiza (Danshen). Explor Target Antitumor Ther. 2020;1(3):153–70.
109. Lee KH, et al. Antitumor agents. 3. Synthesis and cytotoxic activity of helenalin amine adducts and related derivatives. J Med Chem. 1972;15(6):609–11.
110. Schröder HL, et al. Helenalin and 11 alpha,13-dihydrohelenalin, two constituents from Arnica montana L., inhibit human platelet function via thiol-dependent pathways. Thromb Res. 1990;57(6):839–45.
111. Hall IH, et al. Antitumor agents XXX: evaluation of alpha-methylene-gamma-lactone-containing agents for inhibition of tumor growth, respiration, and nucleic acid synthesis. J Pharm Sci. 1978;67(9):1235–9.
112. Scotti MT, et al. Quantitative structure-activity relationship of sesquiterpene lactones with cytotoxic activity. Bioorg Med Chem. 2007;15(8):2927–34.
113. Huang PR, et al. Potent inhibition of human telomerase by helenalin. Cancer Lett. 2005;227(2):169–74.
114. Shin BK, et al. Chemical diversity of ginseng saponins from Panax ginseng. J Ginseng Res. 2015;39(4):287–98.
115. Kim YJ, et al. Anti-tumor activity of the ginsenoside Rk1 in human hepatocellular carcinoma cells through inhibition of telomerase activity and induction of apoptosis. Biol Pharm Bull. 2008;31(5):826–30.
116. Bardone MR, et al. Structure determination of purpuromycin, a new antibiotic. Tetrahedron. 1974;30(16):2747–54.
117. Warabi K, et al. Dictyodendrins A-E, the first telomerase-inhibitory marine natural products from the sponge Dictyodendrilla verongiformis. J Org Chem. 2003;68(7):2765–70.
118. Kumar R, et al. Novel combination of tanshinone I and lenalidomide induces chemo-sensitivity in myeloma cells by modulating telomerase activity and expression of shelterin complex and its associated molecules. Mol Biol Rep. 2018;45(6):2429–39.
119. Ruchawapol C, et al. Natural products and their derivatives against human herpesvirus infection. Molecules. 2021;26(20):6290.
120. Gurung RL, et al. Thymoquinone induces telomere shortening, DNA damage and apoptosis in human glioblastoma cells. PLoS One. 2010;5(8):e12124.
121. Deeb D, et al. Inhibition of hTERT/telomerase contributes to the antitumor activity of pristimerin in pancreatic ductal adenocarcinoma cells. Oncol Rep. 2015;34(1):518–24.
122. Deeb DG, et al. Inhibition of cell proliferation and induction of apoptosis by oleanane triterpenoid (CDDO-Me) in pancreatic cancer cells is associated with the suppression of hTERT gene expression and its telomerase activity. Biochem Biophys Res Commun. 2012;422(4):561–7.
123. Guo QL, et al. Inhibition of human telomerase reverse transcriptase gene expression by gambogic acid in human hepatoma SMMC-7721 cells. Life Sci. 2006;78(11):1238–45.
124. Tahara H, et al. G-Quadruplex stabilization by telomestatin induces TRF2 protein dissociation from telomeres and anaphase bridge formation accompanied by loss of the 3′ telomeric overhang in cancer cells. Oncogene. 2006;25(13):1955–66.
125. Gomez D, et al. The G-quadruplex ligand telomestatin inhibits POT1 binding to telomeric sequences *in vitro* and induces GFP-POT1 dissociation from telomeres in human cells. Cancer Res. 2006;66(14):6908–12.
126. Oda M, et al. Inhibition of telomerase by linear-chain fatty acids: a structural analysis. Biochem J. 2002;367(Pt 2):329–34.
127. Warabi K, et al. Axinelloside A, an unprecedented highly sulfated lipopolysaccharide inhibiting telomerase, from the marine sponge, *Axinella infundibula*. J Am Chem Soc. 2005;127(38):13262–70.
128. Marconett CN, et al. Indole-3-carbinol downregulation of telomerase gene expression requires the inhibition of estrogen receptor-alpha and Sp1 transcription factor interactions

within the hTERT promoter and mediates the G1 cell cycle arrest of human breast cancer cells. Carcinogenesis. 2011;32(9):1315–23.
129. Tabata Y, et al. Diazaphilonic acid, a new azaphilone with telomerase inhibitory activity. J Antibiot. 1999;52(4):412–4.
130. Ye Y, et al. Z-ajoene causes cell cycle arrest at G2/M and decrease of telomerase activity in HL-60 cells. Zhonghua Zhong Liu Za Zhi. 2005;27(9):516–20.
131. Moujir L, et al. Applications of sesquiterpene lactones: a review of some potential success cases. Appl Sci. 2020;10:3001.
132. Cevatemre BE, et al. A promising natural product, pristimerin, results in cytotoxicity against breast cancer stem cells *in vitro* and xenografts *in vivo* through apoptosis and an incomplete autophagy in breast cancer. Pharmacol Res. 2018;129:500–14.
133. Han QB, Xu HX. Caged Garcinia xanthones: development since 1937. Curr Med Chem. 2009;16(28):3775–96.
134. Shin-ya K, et al. Telomestatin, a novel telomerase inhibitor from Streptomyces anulatus. J Am Chem Soc. 2001;123(6):1262–3.
135. Rezler EM, et al. Telomestatin and diseleno sapphyrin bind selectively to two different forms of the human telomeric G-quadruplex structure. J Am Chem Soc. 2005;127(26):9439–47.
136. Baker ES, et al. Cyclo[n]pyrroles: size and site-specific binding to G-quadruplexes. J Am Chem Soc. 2006;128(8):2641–8.
137. Tera M, et al. Macrocyclic hexaoxazoles as sequence- and mode-selective G-quadruplex binders. Angew Chem Int Ed. 2008;47(30):5557–60.

Molecular Targets of Epigenetic Regulators

Farzaneh Amirmahani, Saurav Kumar, and Sree Deepthi Muthukrishnan

1 Introduction

Epigenetic dysregulation is increasingly linked with cancer initiation, progression, and resistance. Post-translational modifications (PTMs) of histone tails and DNA methylation are the most widely studied epigenetic alterations in cancer. These epigenetic alterations regulate chromatin structure dynamics and gene expression patterns contributing to the molecular heterogeneity, plasticity, and evolution of cancer [1]. Gain or loss of function, promoter hypermethylation, chromosomal translocations, and mutations in histone-modifying enzymes and chromatin remodeling complexes are frequently observed in many cancers [2]. Heterozygous missense mutations in key residues of the histone tails generate "oncohistones" that dysregulate the epigenome and transcriptome, promoting tumorigenesis. These oncohistone mutations include H3.3K27M, H3.3G34R/V/W/L, and H3.3K36M and are mainly found in pediatric midline gliomas [3]. Noncanonical oncohistones are less common and are found in a broader range of cancers affecting the core histones H3, H4, H2A, and H2B. They disrupt nucleosome stability and chromatin structure, alter gene expression and contribute to genomic instability [4]. Emerging evidence indicates that disruptions in the 3D chromatin structure are also linked to tumor development. For instance, in breast cancer, about 12% of genomic regions show compartment switching compared to normal cells influencing gene expression [5]. In prostate cancer and multiple myeloma, disruptions in Topologically Associating Domains (TADs), the large, self-interacting regions of the genome, lead to abnormal gene expression and enhancer hijacking [6]. Structural variations in chromatin,

F. Amirmahani · S. Kumar · S. D. Muthukrishnan (✉)
Department of Oncology Science, University of Oklahoma Health Sciences Center, Oklahoma City, OK, USA
e-mail: Farzaneh-Amirmahani@ouhsc.edu; Saurav-Kumar@ouhsc.edu; sreedeepthi-muthukrishnan@ouhsc.edu

A. S. Ramteke et al. (eds.), *Molecular Targets in Cancer Therapy*,
https://doi.org/10.1007/978-981-96-7336-0_4

including insertions, deletions, and translocations, also contribute to cancer. In gliomas, hypermethylation of CTCF (CCCTC-binding factor) binding sites at TAD boundaries results in the loss of CTCF binding and aberrant activation of the oncogene PDGFRA through persistent enhancer interactions [7]. In neuroblastoma and breast cancer, structural variations induce neo-TADs that include oncogenes like MYC and ERBB2, promoting tumorigenesis [8]. Thus, perturbations in the epigenetic landscape and its regulators play a fundamental role in driving cancer development and progression.

In this chapter, we will provide a comprehensive overview of the most common as well as rare histone modifications and their functional contribution to cancer initiation, resistance, and progression in various cancers. We will also discuss recent findings from targeting epigenetic modifiers with small molecule inhibitors as monotherapy and in combination with standard therapies and immunotherapies, as well as their promise as a novel anti-cancer therapeutic strategy to improve cancer outcomes.

2 Histone Modifications and Their Mediators

Histones are highly conserved basic proteins that wrap DNA tightly around nucleosomes and condense the chromatin into a compact shape. They are divided into two major groups: core and linker histones. Core histones form an octamer consisting of two copies of each of the histone variants H3, H4, H2A, and H2B, while histones of the H1 family form the linker histones [9]. The interaction between H1 linker, the octameric nucleosome, and additional DNA stretches results in the formation of the beads-on-a-string chromosome and contributes to the establishment of a higher-order chromatin structure. Histones possess tails rich in basic amino acids like lysine and arginine that serve as prime sites for a diverse range of covalent post-translational modifications (Fig. 1). These chemical alterations include but are not limited to acetylation, methylation, phosphorylation, lactylation, ubiquitination, sumoylation, neddylation, citrullination, ADP-ribosylation, crotonylation, etc. These modifications not only mediate histone-DNA interactions and reshape the chromatin architecture but also create binding sites for proteins that regulate chromatin activity. In the next section, we will discuss the most-well studied histone modifications, such as acetylation, methylation, phosphorylation, and ubiquitination, and their functional roles in mediating the development, resistance, and progression of various cancers.

2.1 Histone Acetylation and Deacetylation

The acetylation and deacetylation of histones refers to the incorporation or removal of acetyl groups from lysine residues on histone tails catalyzed by the enzymes histone or lysine (K) acetyltransferases (HATs/KATs) or histone deacetylases (HDACs), respectively. These modifications play an important role in mediating

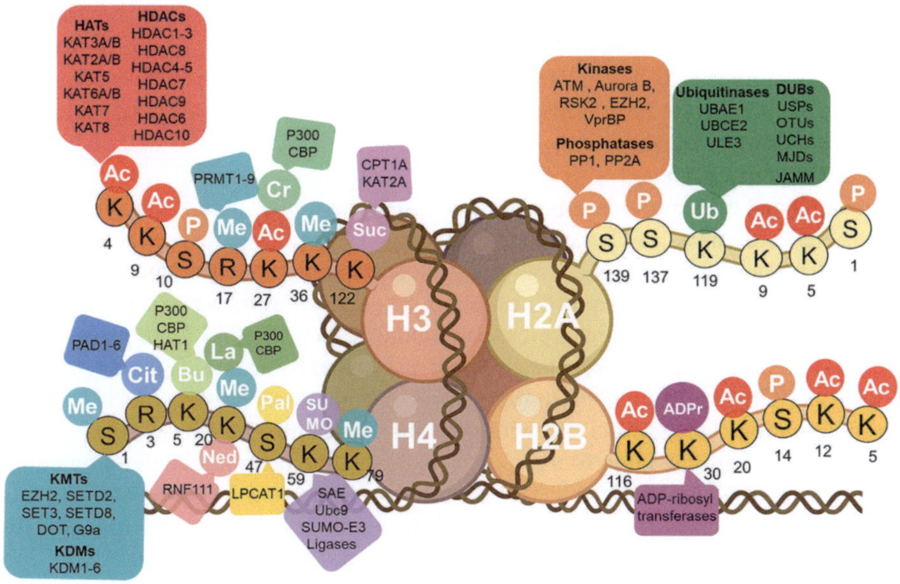

Fig. 1 Histone post-translational modifications and their mediators. Common PTMs such as acetylation (Ac), methylation (Me), phosphorylation (P) and ubiquitination (Ub) occur on the tails of the four core histones: H2A, H2B, H3, and H4 to regulate chromatin dynamics and gene transcription. Enzymes that write and erase these PTMs include histone acetyltransferases (HATs), Histone deacetylases (HDAC), lysine methyl transferases (KMT), lysine demethylases (KDM), ubiquitinases such as UBAE1: ubiquitin- activating enzyme (E1), UBCE2: ubiquitin conjugation enzyme (E2), ULE3: ubiquitin ligase enzyme (E3), and DUBs (Deubiquitinases). Less common modifications such as Citrullination (Cit), SUMOylation (SUMO), Butyrylation (Bu), Neddylation (Ned), Lactylation (La), ADP-ribosylation (ADP-r), palmitoylation (pal), Succinylation (Suc), and Cr: crotonylation (Cr) also occur on the core histones and regulate gene expression

DNA replication and repair, regulation of gene expression, and maintenance of the genome integrity [10]. Based on cellular localization, HATs are divided into two types: Type A HATs, which are localized in the nucleus, and Type B HATs, which reside in the cytoplasm. Type A HATs are responsible for acetylating both histone and non-histone proteins. They are further classified into several families based on their amino acid sequence homology: the general control non-repressible/GCN5-related N-acetyltransferases (GNAT) family, the MYST family, and the P300/CBP family [11]. The GNAT family includes enzymes like KAT2A (GCN5) and KAT2B (PCAF). The MYST family features enzymes such as KAT5 (TIP60), KAT6A (MOZ/MYST3), KAT6B (MORF/MYST4), KAT7 (HBO1/MYST2), and KAT8 (MOF/MYST1). The P300/CBP family is represented by KAT3A (CBP) and KAT3B (P300). Type B HATs include HAT1, HAT2, and HAT4, which acetylate free histones before they are transported into the nucleus and incorporated into newly synthesized DNA [12].

Numerous studies have highlighted a dual role for HATs in cancer as both tumor suppressors and oncogenes [13]. Mutations in *CBP* result in Rubenstein-Taybi

Syndrome, which leads to an increased risk of developing childhood tumors. Similarly, in a mouse model of myelodysplastic syndrome, deletion of *EP300* accelerates leukemogenesis by enhancing hematopoietic progenitor cell self-renewal [14]. Loss of *CBP or EP300* also results in the overexpression of oncogenes like NOTCH1, BMI1, MYC, CCNE, and SKP2, leading to lymphomas and histiocytic sarcomas [15]. KAT5 and KAT7 are also known to play tumor suppressor roles by regulating cell cycle progression and chromatin accessibility in head and neck squamous cell carcinomas and breast cancer. On the other hand, overexpression of P300/CBP in breast cancer correlates with advanced malignancy and tumor recurrence [16]. KAT5 overexpression promotes tumor progression by enhancing activation of MYC target genes in malignant pleural mesothelioma [17]. KAT7 enhances proliferation and cancer stem-like properties in breast cancer by activating oncogenes like E2F1 and RRM2 and silencing tumor suppressor genes such as IL-24 [18]. KAT7 overexpression is also associated with poor survival in bladder cancer [19].

HDACs are grouped into four main classes based on their structure and function. HDAC1, HDAC2, HDAC3, and HDAC8 are classified as class I HDACs. Class II is divided into subclass IIa (HDAC4, HDAC5, HDAC7, HDAC9) and subclass IIb (HDAC6, HDAC10). Class III (SIR2-like) encompasses SIRT1 through SIRT7 and requires NAD as a cofactor. Class IV consists of a single enzyme HDAC11, which exhibits features of both Class I and II HDACs. Class I HDACs are nuclear enzymes that deacetylate core histones as well as various non-histone proteins to regulate transcription. Class II HDACs are found in the nucleus and cytoplasm, with HDAC6 and HDAC10 featuring unique catalytic domains. Class III HDACs, commonly referred to as sirtuins, are NAD+ dependent, vary in their subcellular localizations ranging from nucleus to mitochondria, and possess diverse functions including deacetylation and ADP-ribosylation. Class IV HDACs share characteristics with both Class I and II enzymes, are less well-studied, and are involved in regulating protein stability and gene expression [20].

HDACs are highly over-expressed in several malignancies including breast, colon, gastric, renal, ovarian, pancreatic, prostate, medulloblastoma, neuroblastoma, lymphoma, and chronic lymphocytic leukemia. HDACs are known to perform different functions depending on the type of cancer. For instance, HDAC1, 2, and 3 are known to promote cell cycle progression in breast cancer, facilitate metastasis and invasion in prostate cancer, and enhance angiogenesis in non-small cell lung cancer (NSCLC) [21, 22]. In glioblastoma (GBM), HDAC1, 2, 3, and 8 are associated with cancer progression and resistance to the chemotherapeutic drug temozolomide (TMZ) [23]. HDAC8 is also overexpressed in esophageal squamous cell carcinoma, cervical cancer, and neuroblastoma, correlating with poor prognosis [24]. HDAC10 displays tumor suppressive function by inhibiting migration and metastasis in cervical cancer, while its low expression is associated with worse outcomes in lung cancer [25]. HDAC4 and HDAC5 have been noted for their altered expression in various tumors, but their specific roles in tumorigenesis are less well-defined [26, 27]. HDAC11 exhibits aberrant expression in numerous cancers and regulates cell growth and differentiation, immune evasion, and therapy resistance [28, 29]. For example, HDAC11 suppresses JAK2-driven proliferation and survival in myeloproliferative neoplasms, colon, prostate, and ovarian cancer, and inhibits lymph node metastasis in breast cancer [30–33].

2.2 Histone Methylation and Demethylation

The process of histone methylation is mediated by lysine methyltransferases (KMTs) or arginine methyltransferases (PRMTs), which transfer methyl groups from S-adenosyl methionine to lysine (K) and arginine (R) residues on H3 or H4 tails, respectively. Lysine demethylases (KDMs) reverse the methylation on histone lysine residues [34]. KMTs are categorized into several families based on their substrate specificity. The **SUV39H1 family** includes SUV39H1, which primarily methylates lysine 9 on histone H3 (H3K9me3) and is crucial for heterochromatin formation and gene silencing [35]. The **SET domain family** is associated with a range of histone and non-histone methyltransferases. This family features enzymes like Enhancer of zeste homolog 2 (EZH2), which trimethylates lysine 27 on histone H3 (H3K27me3) and is involved in transcriptional repression. SETD2 exclusively methylates H3K36, while SETD8 methylates H4K20, and SETD8 is responsible for monomethylation of H4K20 [36]. Other enzymes such as MLL5, SETD3, SETDB2, SMYD3, and SETMAR have been reported to methylate specific histone residues, although these studies need to be further validated. For instance, SETD3 is a histidine methyltransferase, and SETMAR, although it methylates free histones, does not act on nucleosomes [37]. SMYD2, known for its role in P53 regulation, shows promiscuous behavior towards substrates but lacks specificity for histones. The DOT1 family includes DOT1L, which specifically methylates lysine 79 on histone H3 (H3K79) and the G9a/GLP family encompasses G9a and GLP, which methylate lysine 9 on histone H3 (H3K9me2) [38].

KMTs play diverse roles in cancer development and progression. EZH2 is involved in transcriptional repression and has a significant role in regulating cell cycle, DNA repair, differentiation, and immune response. Mutations like Y641 in EZH2 enhance its catalytic activity, promoting oncogenesis in diffuse large B-cell lymphoma (DLBCL) and pancreatic cancer [39]. In GBM, EZH2 regulates tumor growth through PTEN and AKT signaling pathways, although prolonged inhibition can potentially promote tumorigenesis [40]. EZH2 has also been observed to act as a tumor suppressor in certain contexts. Its loss can synergize with other mutations (e.g., JAK2-V617F) to initiate cancer or promote KRAS-driven lung adenocarcinoma [41]. SETD2 also has a dual role in cancer, acting as a tumor suppressor and an oncogene. As a tumor suppressor, SETD2 mutations or loss of function are linked to poor outcomes in renal-cell carcinoma (RCC) and lung adenocarcinoma due to disruptions in DNA repair and transcription regulation. Conversely, mutations in SETD2 lead to reduced levels of H3K36me3 promote oncogenesis in high-risk gastrointestinal cancers and acute myeloid leukemia (AML) [42]. DOT1L drives oncogenesis in MLL fusion-mediated leukemia by promoting overexpression of genes like HOXA9. In hormone-responsive cancers, such as ovarian, breast, and prostate cancers, DOT1L enhances proliferation and contributes to tumorigenesis.

Conversely, DOT1L exhibits a protective effect in melanoma induced by ultraviolet radiation [43]. G9a mediates H3K9 methylation and represses tumor suppressor genes to promote proliferation, migration, and invasiveness in hepatocellular carcinoma (HCC) and melanoma [44, 45]. In lung cancer, G9a promotes invasion and metastasis, which is crucial for maintaining cellular adhesion and suppressing invasive behavior. In breast cancer, G9a is involved in hypoxia-mediated gene repression, which supports cancer cell survival and tumor growth under low-oxygen conditions [46]. Other H3K36 methyltransferases such as NSD2 (Nuclear Receptor Binding SET Domain Protein 2), along with NSD1, NSD3, and ASH1L are associated with oncogenesis in both solid tumors and hematological malignancies. In solid tumors, NSD2 overexpression is implicated in metastatic progression of prostate cancer [47]. Furthermore, a gain-of-function variant of NSD2 (E1099K) enhances its catalytic activity and increases H3K36me2 levels, contributing to the oncogenic reprogramming and progression of pediatric Acute Lymphoblastic Leukemia (ALL) [48].

PRMTs are categorized into three types based on their catalytic end products and whether they promote mono-methylation, asymmetric, or symmetric di-methylation. Type I PRMTs (PRMT1, 2, 3, 4, 6, and 8) produce monomethylarginine and asymmetric dimethylarginine. Type II PRMTs (PRMT5 and 9) generate monomethylarginine and symmetrical dimethylarginine. Type III PRMTs (PRMT7) exclusively create monomethylarginine. PRMTs are localized in both the nucleus and cytoplasm, although some, like PRMT3 and 9, are mainly cytoplasmic, and others, like PRMT2, shift localization between nuclei and cytoplasm. These enzymes play crucial roles in maintaining genome integrity, regulate gene expression, RNA splicing, and ribosomal biogenesis [49]. For example, PRMT1 functions in an oncogenic role in breast cancer by influencing the methylation of ERα and EZH2, which enhances IGF-1 signaling and promotes EMT. In pancreatic cancer, PRMT1 methylates HSP70 to stabilize BCL2 mRNA, supporting tumor cell survival [50]. PRMT5 enhances WNT/β-catenin signaling in lymphoma and regulates alternative splicing in AML [51]. Furthermore, PRMT6 promotes oncogenic signaling in gastric and lung cancers by modulating histone methylation and interacting with AKT/mTOR signaling [52].

KDMs are divided into two groups based on their catalytic mechanisms of demethylation: amine oxidases or Jumonji C domain containing KDMs. The amine-oxidase lysine specific demethylases 1 and 2 (LSD1 and 2), also known as KDM1A and KDM1B, belong to class I KDMs [53]. LSD1/KDM1A is the first characterized KDM that mediates H3K4 demethylation [54]. KDM1A (LSD1) and KDM1B (LSD2) both demethylate H3K4me1/me2, however, KDM1A also targets H3K9me1/me2 [55]. KDM1A has been shown to regulate the cell cycle in pancreatic cancer, inhibit apoptosis of breast cancer and tumor growth through modulation of P53, MYPT1, Hypoxia-inducible factors (HIFs)-1α, and immune checkpoint regulators in HCC [56, 57]. KDM1B is associated with poor prognosis in cancers of the breast, colorectal, and lung, primarily through its role in inhibiting apoptosis and promoting tumor growth [58]. KDM2A and KDM2B, members of the KDM2 family with JmjC and PHD domains, are implicated in cancer due to their demethylation activities. KDM2A specifically demethylates H3K36me2, while KDM2B targets H3K4me3 and H3K36me2. Both enzymes are overexpressed in colorectal, gastric, ovarian, and cervical cancers and contribute to tumor growth and progression. KDM2A promotes carcinogenesis in lung cancer by demethylating H3K36me2 at the HDAC3 and DUSP3 promoters, which increases tumor aggressiveness and enhances the ERK1/2 signaling pathway [59]. KDM2B participates in transforming growth factor beta (TGFβ)-induced EMT in lung and pancreatic cancers, facilitating invasion and metastasis [60]. The KDM3 family, including KDM3A, KDM3B, and KDM3C, exhibits diverse and context-dependent roles in cancer. KDM3A is overexpressed in breast, colorectal, lung, liver, pancreatic, and prostate cancers, acting as an oncogene by enhancing pathways like WNT/β-catenin and estrogen receptor signaling. KDM3B also acts as a tumor suppressor in colorectal cancer and leukemia. KDM3C is overexpressed in breast, colorectal, and esophageal cancers, where it supports tumor growth and metastasis [61].

The KDM4 family contains catalytic JmjN and JmjC domains and includes KDM4A, KDM4B, KDM4C, and KDM4D that demethylate histones at H3K9 and H3K36. These enzymes are overexpressed in various cancers. KDM4A promotes tumor growth and invasion in prostate, lung, and gastric cancers by activating androgen receptors, WNT/β-catenin signaling, and suppressing apoptosis. KDM4B drives cancer cell proliferation by influencing cell cycle regulators and glucose uptake in breast and colorectal cancers [62]. KDM4C contributes to prostate carcinogenesis by modulating androgen receptors, P53, and MYC. KDM4D supports tumor progression in liver cancer and other malignancies by activating HIF and Hedgehog/β-catenin signaling pathways and by directly antagonizing P53 [63]. The KDM5 family, which includes KDM5A, KDM5B, KDM5C, and KDM5D, play diverse roles in different cancers, from promoting tumorigenesis, drug resistance, genomic instability, to therapeutic resistance. KDM5A is overexpressed in multiple cancers such as breast, lung, and gastric cancer, where it enhances tumor growth by promoting cell cycle progression and inhibits tumor suppressors like P27 and NOTCH2. KDM5B is also oncogenic, contributing to poor outcomes in breast, lung, liver, and bladder cancers by upregulating genes that promote cell cycle progression and tumor growth [64]. Conversely, KDM5C exhibits dual roles. It acts as a tumor suppressor in clear-cell renal cell carcinoma and cholangiocarcinoma, where its loss correlates with poor outcomes. On the other hand, it promotes tumor growth in lung, liver, and colon cancers through H3K4 demethylation and miRNA regulation. KDM5D is predominantly tumor suppressive as genetoc knockdown enhances tumor proliferation and reduces apoptosis in prostate cancer. It also inhibits tumor invasion by demethylating H3K4 in prostate, gastric, and lung cancers [65]. The KDM6 family, including KDM6A, KDM6B, and KDM6C, plays significant and context-dependent roles in cancer by modulating gene expression through histone demethylation and scaffolding. KDM6A is often aberrantly expressed across a range of cancers. Its oncogenic role has been established in breast and cervical cancer, whereas it was found to function as a tumor suppressor in AML, NSCLC, and other malignancies [66, 67]. KDM6B shows both tumor-suppressive and oncogenic roles depending on the cancer type. KDM6B suppresses neuroblastoma, colorectal cancer, and AML but promotes cancer progression in multiple myeloma, prostate, and gastric cancer by upregulating genes involved in cell cycle and survival [67]. KDM6C, while generally a tumor suppressor in AML and squamous-like pancreatic cancer, has less well-defined roles in other cancers due to its lower demethylase activity [67].

2.3 Histone Phosphorylation

Histone phosphorylation involves the addition of phosphate groups to serine, threonine, and tyrosine residues by kinases and reversed by phosphatases. This modification alters chromatin accessibility to facilitate transcription factor binding and initiates downstream signaling cascades critical for cell division, DNA damage response, and cancer progression. Key modifications include phosphorylation of γH2AX at serine 139, crucial for DNA damage sensing. Phosphorylation of H3 and H2B residues activates gene transcription in various cancers, including colorectal cancer (CRC), under UV radiation. Dysregulated phosphorylation like H2Bser32p by RSK2 in skin cancer promotes oncogenic processes. Overexpression of Aurora B, a serine/threonine kinase, enhances histone phosphorylation in CRC and breast cancer, which is critical for chromosome segregation and therapeutic resistance [68]. EZH2 and anti-silencing factor 1 have been identified as key players in phosphorylating histone H2B and H4, respectively, and activate autophagy-related gene transcription and foster tumor growth. The kinase VprBP, overexpressed in CRC, orchestrates epigenetic gene silencing through histone H2A phosphorylation to enhance cancer cell proliferation [69].

2.4 Histone Ubiquitination and Deubiquitination

Histone ubiquitination plays a significant role in cancer by influencing protein stability. It primarily occurs at lysine 119 on histone H2A (H2AK119ub1) and lysine 120 on histone H2B (H2BK120ub1). Ubiquitination involves a cascade of reactions beginning with the ubiquitin-activating enzyme (E1), which transfers ubiquitin to the ubiquitin-conjugating enzyme (E2), and then to the substrate protein via the ubiquitin ligase (E3) [70]. These enzymes are grouped into four families: HECT, RING-finger, U-box, and PHD-finger. The RING-finger family is the largest, encompassing complexes like the anaphase-promoting complex (APC) and the SCF complex (Skp1-Cullin-F-box protein complex). The SCF complex, for instance, consists of RBX1, CUL1, SKP1, and a variable F-box protein that provides substrate specificity [71]. RING finger E3 ligases can function independently as monomers or homodimers, such as TRIM, TRAF6, cIAP, XIAP, and MDM2, or associate with other proteins to form heterodimers or multi-subunit complexes. Notable heterodimeric examples include TRAF6, MDM2/MDMX, BRCA1/BARD1, and RING1B/BMI-1, where the associated proteins, like MDMX, BARD1, and BMI-1, lack E3 ligase activity but enhance ubiquitination by partnering with E3 ligase-containing domains [72]. Dysregulation of histone ubiquitination is associated with altered expression of tumor suppressors and oncogenes. In pancreatic cancer, H2A ubiquitination at lysine 119 (H2AK119ub1) mediated by RING1/2 is associated with gene silencing and tumor development. In breast cancer, K63-linked ubiquitination pathways involving TRAF6, UBC13, and TRIM59 influence metastasis by modulating various proteins, including AEP and PDCD10. Additionally, in lung cancer and melanoma, ubiquitination pathways involving RNF8 and TRAF6 influence EMT and metastasis, highlighting the diverse roles of histone and protein ubiquitination in cancer [73]. UBR7 inhibits breast cancer metastasis by monoubiquitinating histone H2B [70]. Additionally, histone H2AX monoubiquitination regulated by Ubiquitin-Conjugating Enzyme E2T (UBE2T) enhances radiation-resistance in HCC by promoting the phosphorylation of CHK1 [74].

Deubiquitinases (DUBs) reverse ubiquitination, preventing protein degradation. DUBs are classified into five types: ubiquitin-specific proteases (USPs), ovarian tumor proteases (OTUs), ubiquitin C-terminal hydrolases (UCHs), Machado-Joseph disease protein domain proteases (MJDs) are cysteine-dependent proteases, and JAMM motif proteases are metal-dependent. These enzymes primarily function by reversing the polyubiquitination or monoubiquitination of target proteins. Dysregulation of DUBs can impair the ubiquitin system, leading to enhanced oncogene effects or reduced tumor suppressor gene activity [75]. USP49, a deubiquitinase targeting histone H2B, has been shown to influence CRC proliferation. In HCT116 cells, USP49 knockout leads to increased cell proliferation and migration by elevating MDM2 levels and reducing *TP53* target gene expression [76].

2.5 Other Histone Modifications

Some understudied modifications such as neddylation, lactylation, succinylation, palmitoylation, SUMOylation, citrullination, crotonylation, and butyrylation also contribute to tumorigenesis.

Neddylation involves the attachment of the ubiquitin-like protein NEDD8 to target proteins, predominantly Cullin-RING ligases (CRLs), enhancing their activity in protein degradation pathways. Unlike ubiquitin, NEDD8-conjugated proteins are mainly found in the nucleus and are regulated by the COP9 signalosome. NEDD8 modification of histone H4 and H2A is regulated by the E3 ligases RNF111 and RNF168, respectively. RNF111 mediates increased NEDD8 attachment to H4 in response to DNA damage, while RNF168 facilitates H2A ubiquitination by reducing its neddylation. This regulation is crucial for recruiting DNA damage repair factors like γ-H2AX and BRCA1. Additionally, neddylation affects apoptosis; for example, NEDP1, a de-neddylating enzyme, has been shown to reverse the inhibition of apoptosis executors like caspase 7 by removing NEDD8 [77]. MLN4924, an inhibitor of NEDD8-activating enzyme, has been shown to reduce tumor growth and angiogenesis in various cancers such as urothelial cell carcinoma, cervical cancer, renal cell carcinoma, and pharyngeal squamous cell carcinoma [78].

Histone lactylation (Kla) involves the addition of lactate molecules to lysine residues on histone proteins. Enzymes known as lactyltransferases facilitate this modification by transferring lactate from lactyl-CoA to histones. Histone delactylation, the removal of lactate modifications, is mediated by de-lactylases such as HDACs (HDAC1-3) and sirtuins (SIRT1-3) which help maintain the balance of lactylation levels. Histone lactylation is known to promote the malignant progression of endometrial carcinoma by enhancing USP39 expression and activation of the PI3K/AKT/HIF-1α signaling pathway [79]. In CRC, hypoxia-induced glycolysis stimulates β-catenin lactylation, boosting its stability and expression to drive malignant proliferation [54]. Conversely, reduced histone lactylation levels in CRC promotes M1 macrophage polarization and inhibit CRC progression and metastasis [80].

Histone succinylation occurs when a succinyl group derived from succinyl-CoA is attached to lysine residues on proteins by the enzymes carnitine palmitoyltransferase 1A (CPT1A) and KAT2A. SIRT5 and SIRT7 function as de-succinylases to remove succinyl groups from proteins. In pancreatic ductal adenocarcinoma (PDAC), histone succinylation mediated by KAT2A is associated with enhanced tumor cell proliferation and worsened patient survival. In GBM, KAT2A succinylation activity is linked to increased cell proliferation [81]. In addition, succinylation of non-histone proteins such as K99 succinylation of PGAM1 in liver cancer and K672 succinylation FBN1 in gastric cancer is associated with tumor promoting roles [82].

Histone palmitoylation involves the addition of palmitic acid to histone proteins and includes S-palmitoylation/S-acylation and O-palmitoylation. S-palmitoylation, the better-characterized form, attaches palmitic acid to cysteine residues via a labile thioester bond. O-palmitoylation, while less well-studied, involves palmitic acid attachment to serine residues via a more stable oxygen ester bond. Lipidation of various histone H3 variants has been identified and includes cysteine 110 as a primary site for S-acylation. O-palmitoylation of histone 4 at serine 47 is catalyzed by the enzyme lysophosphatidylcholine acyltransferase (LPCAT1) in a calcium-dependent manner and regulates gene transcription in lung epithelial cells [83]. Histone palmitoylation by ZDHHC5 activates EZH2, increasing H3K27me3 levels and promoting GBM malignancy [84].

SUMOylation is a process where small ubiquitin-related modifier (SUMO) proteins are attached to histones. There are four SUMO isoforms: SUMO1, SUMO2, SUMO3, and SUMO4. The sumoylation process involves the E1 enzyme (comprising SAE1 and SAE2/UBA2) and the E2 enzyme. SUMO1/2/3 proteins are found to be over-expressed in high-grade gliomas such as GBM, and increased SUMOylation is associated with cell cycle regulation, glycolysis, and DNA repair [85]. In breast cancer, SUMOylation also promotes the interaction of transcriptional repressors such as SnoN with epigenetic modifiers like HDAC1 and P300/CBP to regulate EMT [86].

Histone citrullination is the modification of the amino acid arginine to citrulline by a family of enzymes called peptidylarginine deiminases (PADIs). These enzymes include PAD1, PAD2, PAD3, PAD4, and PAD6 and are known to be involved in various cellular functions. For instance, PAD4 has been demonstrated to play a key role in inducing chromatin de-condensation and controlling gene expression. In lung cancer, PAD4-mediated citrullination of histone H3 at Arg3 impairs P53 transcriptional activity, while in NSCLC, it plays a role in DNA damage response [87]. In multiple myeloma, PAD2-mediated citrullination of histone H3 at Arg6 contributes to resistance to ubiquitin-proteasome inhibitor bortezomib. In prostate cancer, PAD2-mediated citrullination of histone H3 at Arg26 upregulates estrogen-suppressing genes [88].

Histone lysine crotonylation (Kcr) is a relatively newly discovered epigenetic modification akin to histone acetylation implicated in various cellular processes and diseases, including cancer. In the liver, prostate, and cervical cancer models, crotonate treatment increases pan-Kcr levels and promotes the proliferation and invasion of tumor cells [89]. In colon cancer, Liao et al. found that H3K27cr levels decrease during DNA damage, highlighting a complex interplay between crotonylation and the cellular response to genotoxic stress [90]. Additionally, recent studies indicate that P300/CBP-mediated histone Kcr may play a role in the hypoxic induction, functioning in a HIF-2α-dependent manner in colon cancer cells [91].

Histone lysine butyrylation (Kbu) involves the transfer of a butyryl group from Butyrl-coA to lysines residues on histones. Butyrylation is structurally like other histone acylations such as acetylation, crotonylation and propionylation that marks active transcriptional start sites and directly stimulates gene transcription [92]. Early studies identified P300 and CBP as mediators of histone butyrylation in cells. H4K5bu and H4K8bu were identified in spermatogenic cells whereas H3K9bu was found to be regulated by a high fat diet. Additionally, H3K27bu was identified as a regulator of gene expression in intestinal epithelial cells [93]. Recent research has also identified a related modification, histone isobutyrylation (KiBu) on H3K18 and H3K23 as well as H4 and was mediated by P300 and HAT1 in HEK293T cells [94]. Despite the identification of several Kbu sites that can regulate gene expression in different cell types, their function in cancer remains poorly understood. One study in bladder cancer identified that the level of H4K20 butyrylation (H4K20Bu) levels was altered by heat shock protein inhibitors and dependent on lysine coenzyme A. However, the precise mechanisms by which Kbu influences cancer progression remain unclear [95].

Histone biotinylation is a very rare modification with scarce evidence of its role in cancer. Initial studies indicated that enzymes biotinidase and holocarboxylase synthetase mediate the covalent attachment of vitamin biotin to histones. Biotinylation sites in histone proteins were identified using synthetic peptides and include K9, 13, 125, 127, and 129 in histone H2A, K9 and K18 in histone H3 and K8 and K12 in histone H4 [96]. Biotinylation increases significantly during cell proliferation, peaking in the G1 phase and remaining elevated throughout the cell cycle. Additionally, biotinylated histones are found in transcriptionally silent chromatin and may be involved in DNA damage response, with changes in biotinylation patterns observed after UV exposure and DNA breaks [97]. The exact role of histone biotinylation in DNA repair and in cancer is yet to be determined.

3 Functions of Histone Modifications in Solid Tumors and Hematological Malignancies

3.1 Breast Cancer

Several histone modifications and their mediators are implicated in tumor initiation and progression of breast cancer. P300/CBP family of HATs have been shown to acetylate estrogen receptor α (ERα) and enhance its activity promoting breast tumorigenesis. MYST family members KAT8 and KAT7 are associated with the regulation of replication initiation in breast cancer cells [98]. Dysregulation of KMTs, particularly overexpression of G9a, has been implicated in promoting breast cancer progression by suppressing tumor suppressor genes and enhancing the expression of oncogenic factors like T-Box 2 (TBX2) [99]. Overexpression of JMJD6, a Jumonji C domain-containing epigenetic modifier that has both arginine demethylase and lysine hydroxylase enzyme activity, promotes autophagy and is associated with breast cancer tumorigenesis [100]. Histone phosphorylation also significantly impacts breast cancer by affecting DNA repair, mitosis, and gene regulation [101]. BRCA1, a RING finger E3 ubiquitin ligase, maintains heterochromatin structure through H2A monoubiquitination, and its loss disrupts this process, leading to satellite DNA de-repression and tumorigenesis [102]. Similarly, RNF20, another E3 ligase responsible for H2B ubiquitination, acts as a tumor suppressor, and its depletion results in enhanced proto-oncogene expression and increased tumorigenesis. BMI-1, a key regulator of H2A ubiquitination, is associated with advanced breast cancer stages and drives EMT and stemness [103].

3.2 Lung Cancer

Lung cancer is a leading cause of cancer-related deaths in both men and women, with NSCLC comprising approximately 85% of all lung cancer cases. Histone modifiers such as KDM1A (LSD1) and EZH2 play pivotal roles in lung cancer through their regulation of histone methylation patterns. KDM1A is frequently overexpressed in NSCLC, which impacts gene expression dynamics that are critical for tumor progression. EZH2, which mediates H3K27 trimethylation, induces chromatin condensation and transcriptional repression of tumor suppressor genes, promoting tumorigenesis [104]. G9a-mediated H3K9me2 recruits transcription factors that inhibit tumor suppressor genes, contributing to WNT activation and lung cancer progression [105]. Conversely, reduced G9a or H3K9me2 levels can promote lung adenocarcinoma by allowing the expression of genes like MMP10 and KRAS [106]. Histone demethylase UTX, which regulates H3K27me3 through EZH2, acts as a tumor suppressor, with its loss leading to increased H3K27me3 and enhanced tumor proliferation [107]. HDAC1 and HDAC2 deacetylate histones and non-histone proteins, such as P53 and CDKN family proteins, thus impairing apoptosis and cell cycle regulation [108]. PAD4, which facilitates histone citrullination and regulates

DNA damage response, is associated with reduced tumor size in NSCLC samples [87].

3.3 Colorectal Cancer

Colorectal cancer (CRC) develops through a multifaceted process involving genetic and epigenetic alternations influenced by the microenvironment. Histone modifications have gained significant attention in CRC research due to their association with disease progression and aggressiveness [109]. Acetylation of histone H3 at specific lysine residues such as H3K12ac and H3K18ac is significantly altered in CRC. Increased levels of these acetylation marks have been observed in moderately to well-differentiated CRC, while poorly differentiated cancers exhibit decreased acetylation patterns. Notably, HDAC2 is upregulated in adenocarcinomas, suggesting its role in promoting CRC development by suppressing tumor suppressor genes [110]. Histone methylation alterations, including increased H3K9me3 and H3K27me3, are also implicated in CRC pathogenesis. The enzymes responsible for these modifications, such as SUV39H1 and EZH2, are overexpressed in CRC tissues and correlate with advanced tumor stages, lymph node metastasis, and poorer patient prognosis [111]. Other histone modifications like neddylation of HuR and RNF40-mediated H2B monoubiquitination regulate cell survival and anti-apoptotic gene expression [69].

3.4 Prostate Cancer

Both genomic and epigenetic alterations contribute significantly to the development and progression of prostate cancer (PC), a leading cause of cancer-related deaths in men. Particularly, KMTs have been shown to play diverse roles in PC malignancy. KDM1A and KDM4 family members (KDM4A, KDM4B, KDM4D) regulate histone methylation at H3K4me1, H3K4me2, H3K4me3, H3K9me2 and H3K9me3, which promotes proliferation, survival, EMT and DNA repair mechanisms [112]. KDM6B affects H3K27me2/me3 marks, enhancing cancer cell aggressiveness [113]. In addition, histone phosphorylation at threonine 6 (H3T6ph) by PKCβ prevents LSD1-mediated demethylation of H3K4 during androgen receptor (AR)-activated gene expression promoting tumor cell proliferation in PC [114]. HATs like P300/CBP and HAT1 regulate the acetylation of H3K18, H3K27, H4K5, and H4K12, impacting processes like EMT, tumor progression, and response to chemotherapy [115, 116]. P300/CBP function as transcriptional co-activators for androgen receptors (AR) and are frequently upregulated in AR+ and castration-resistant prostate cancer, and selective inhibition of these HATs reduced PC tumor growth in xenograft models [117].

3.5 Gastric Cancer

Gastric cancer (GC) is the fifth most frequently diagnosed cancer and ranks third in terms of cancer-related mortality. Various histone modifications play crucial roles in GC progression and prognosis. Overexpression of KMTs like G9a leads to activation of oncogenic pathways such as mTOR signaling and correlates with poor prognosis. EZH2 promotes GC development by silencing tumor suppressors like PTEN and activating oncogenes like FOXM1 [118]. Dysregulation of HATs such as P300/CBP and HAT1 enhances acetylation of H3K18 and H3K27 to influence EMT and chemotherapy responses [119]. Other modifications, such as phosphorylation and ubiquitination, significantly influence GC progression by altering chromatin dynamics and gene expression. For instance, H3 Ser10 phosphorylation by MSK1 promotes cell proliferation via the NFATC2 pathway and serves as a prognostic marker in GC [120]. Additionally, Aurora B-mediated H1.4 Ser27 phosphorylation, influenced by RAS-ERK1/2 signaling, contributes to GC progression [121]. Additionally, H2AK119 ubiquitination mediated by the Cullin4B-RING E3 ligase complex (CRL4B) is associated with increased tumorigenesis [122].

3.6 Liver Cancer

Hepatocellular carcinoma (HCC) is the predominant malignant neoplasm affecting the liver, characterized by a generally unfavorable prognosis. Abnormal histone modifications have been implicated in initiating and advancing liver cancer by driving dysregulated transcriptional programs. KMTs like G9a are upregulated in HCC and promote tumorigenesis by repressing tumor suppressor genes to facilitate uncontrolled cell proliferation and tumor growth. Histone demethylases such as KDM1A, KDM5B, and KDM3A also contribute significantly to HCC development by regulating gene expression patterns critical for cell survival, stemness, and proliferation [123]. HATs such as P300/CBP acetylate key metabolic enzymes, promoting glycolysis and thus HCC proliferation. The overexpression of P300/CBP correlates with poor survival [124]. HDACs are also implicated in tumor progression and drug resistance in HCC. For instance, elevated SIRT1 levels correlate with larger tumor size, increased P53 expression, and poorer overall survival in HCC patients. HDAC1 and HDAC2 overexpression is linked with decreased fructose-1,6-bisphosphatase (FBP1) levels, impairing glucose metabolism and contributing to a poor prognosis [125]. Ubiquitin-specific protease 15 (USP15) impacts tumorigenesis by stabilizing key proteins involved in cell growth and repair, such as TGF-β receptors, MDM2, and HECTD1. Histone phosphorylation at H3S10 by Aurora kinases correlates with HCC aggressiveness, affecting cell cycle regulation and response to therapies [126].

3.7 Glioblastoma

Glioblastoma (GBM) is the deadliest and most common primary adult brain tumor, with a median survival of 15 months [127]. Histone modifications play crucial roles in regulating GBM tumorigenesis, plasticity, and resistance [128]. HDACs, particularly class I HDACs, are extensively investigated in GBM tumorigenesis and are associated with therapeutic resistance. While HATs are less studied in GBM, they have been shown to influence PI3K/AKT signaling and promote tumorigenesis [129]. P300 has been implicated in promoting radiation resistance and recurrence of GBM by regulating GSC conversion to vascular-like cells and chemo-resistance via regulation of DNA repair and pro-survival genes [130, 131]. KMTs and HDMs are also implicated in GBM. Upregulated KMTs like SUV39H1 and SETDB1 regulate proliferation and migration in GBM [132]. G9a, a KMT, is highly expressed in GBM cells and its inhibition enhances sensitivity to chemo-radiation [133]. PRMT2, a protein arginine methyltransferase (PRMT), is elevated in GBM and crucial for the renewal of GSCs and tumor growth by activating oncogenic transcriptional programs [134]. PRMT3 and PRMT6 also contribute to GBM progression and radiation response by regulating key cellular processes [135]. Histone demethylases such as KDM5B, KDM2A, KDM5A, KDM6B, KDM4C, KDM7A, and LSD1 (KDM1A), are overexpressed in GBM and impact various aspects of tumor behavior, including proliferation, migration, resistance to TMZ chemotherapy, and stemness. Inhibition of specific demethylases can induce DNA damage, differentiation, or sensitize GSCs to therapy [136]. EZH2 has been extensively studied in GBM for its role in maintaining stem cell characteristics and promoting resistance to TMZ [137]. SUMOylation of histones facilitated by enzymes like SAE1 impacts signaling pathways like AKT and contributes to GBM growth [138]. Histone phosphorylation, particularly at sites like H3T6 and H3S10, correlates with poor survival outcomes in GBM patients and affects response to radiation and chemotherapy [139]. Aberrant histone ubiquitination, including modifications at H2AK119ub1 and H2BK120ub1, alters the regulation of tumor suppressors and oncogenes, further influencing GBM development and treatment resistance [140].

3.8 Hematopoietic Cancers

Hematopoietic malignancies, which include leukemias, lymphomas, and myelomas, are driven by diverse histone modifications that influence oncogenesis and disease progression. In AML, disruptions in histone acetylation are significant. For instance, the recruitment of HDACs to gene promoters is frequently observed to contribute to the pathology of AML. Additionally, acetylation patterns are disrupted by the oncogenic drivers MYC and BCL-6, which recruit HATs like CBP and P300 in malignant B cells [141]. Loss-of-function mutations in CBP, in conjunction with BCL2, reduce histone acetylation and affect germinal center development, B-cell signaling, and immune evasion in follicular lymphoma (FL). This aberrant acetylation promotes lymphomagenesis by accelerating cellular growth and downregulating MHCII, underscoring the role of these modifications as tumor suppressors [142]. Histone methylation also plays a pivotal role in hematopoietic malignancies. In AML, the removal of H3K4 methyl groups by KDM1A suppresses tumor suppressor gene expression. Similarly, mutations in EZH2 are associated with immune evasion in follicular lymphoma. In multiple myeloma (MM), alterations in H3K36 methylation mediated by NSD2 are central to disease progression. NSD2 interacts with various histone-modifying enzymes, influencing chromatin structure and gene expression in myeloma cells. Overexpression of EZH2 in MM enhances cell proliferation and regulates tumor suppressor miRNAs targeting MYC, BLIMP-1, and IRF4. KDM3A supports IRF4-driven cell growth in myeloma [143]. In myeloproliferative neoplasms, phosphorylation of histone H3 at tyrosine 41 by JAK2 kinase impacts chromatin structure and activates oncogenes such as LMO2, contributing to tumor progression [144].

In summary, dysregulation of the epigenetic modifiers and their activity influences numerous processes critical for cancer initiation, resistance, and progression, and selective targeting of these modifiers has emerged as a promising anti-cancer approach to improve the efficacy of standard therapies as well as in combination with other molecular targeted therapies.

4 Small Molecule Inhibitors Targeting Histone Modifications as Potential Mono- and Combinatorial Therapies

Over the last decade, several small molecule inhibitors targeting histone modifiers have been developed as novel anti-cancer therapeutics (Table 1). Some of these inhibitors have shown promise in pre-clinical and clinical settings in improving cancer outcomes. For instance, HDAC inhibitors such as Vorinostat (SAHA), Panobinostat (LBH-589), and Romidepsin (FK2280) have shown effectiveness in treating hematologic malignancies such as cutaneous T-cell lymphoma (CTCL) and multiple myeloma by inducing histone acetylation and apoptosis [145, 146]. Valproic Acid (VPA) has demonstrated promise in metastatic Non-Hodgkin Lymphoma (NHL), while benzamide derivatives like Entinostat (MS-275) and Tacedinaline (CI-994) are utilized for both lymphoid malignancies and solid tumors [147, 148]. The histone methyltransferase inhibitor Tazemetostat (EPZ-6438) has proven effective in diffuse large B cell lymphoma (DLBCL) and follicular lymphoma [149]. EZH2 inhibitors like CPI-1205 and Tazemetostat are also in clinical trials for B-cell lymphomas and various solid tumors [150]. Selective inhibitors of PRMT enzymes such as DB75 (PRMT1i) and SGC707 (PRMT3i) in pancreatic cancer, TP-064 and EZM2302 (PRMT4i) in melanoma, GSK3326595, JNJ-64619178, LLY-283 and PF-06939999 (PRMT5i) in breast and gastric cancer and hematological tumors have shown promise in clinical trials by targeting key mechanisms involved in tumor cell proliferation, metastasis and oncogenic processes mediated by arginine methylation [151, 152].

Table 1 An overview of the extensively validated small molecule inhibitors targeting epigenetic modifiers in various cancers

Category	Drugs	Targets	Cancer	References
HDAC inhibitors	Vorinostat (SAHA)	Class I, II, and IV HDACs	CTCL, DLBCL, GBM, NSCLC, prostate, and breast cancer	[155–157]
	Panobinostat (LBH-589)	Class I, II, and IV HDACs	Multiple myeloma, CTCL, Hodgkin's lymphoma, leukemia, prostate, thyroid, breast, and pancreatic cancer	[158–161]
	Romidepsin (FK2280)	Class I HDACs	CTCL, peripheral T cell lymphoma, bladder, and colon cancer	[162–165]
	Valproic acid	Class I and IIa HDACs	Melanoma, mesothelioma, renal cell carcinoma, and breast cancer	[166, 167]
	Entinostat (MS-275)	Class I HDACs	Hodgkin's lymphoma, lung, breast, and prostate cancer	[168–170]
	Tacedinaline (CI-994)	Class I HDACs	Medulloblastoma	[171]
HAT inhibitors	CPI-1612	P300/CBP	GBM and breast cancer	[131, 172]
	CCS1477	P300/CBP	GBM, prostate, pancreatic cancer, and AML	[173, 174]
	PF-07248144	KAT6A/B	Breast cancer	[175]
	C646	P300/CBP	AML, GBM, cervical, gastric, prostate, pancreatic, and lung cancer	[130, 176–178]
HMT inhibitors	Tazemetostat (EPZ-6438)	EZH2	DLBCL and follicular lymphoma	[179, 180]
	UNC1999	EZH2	Multiple myeloma, bladder, colorectal, and colon cancer	[181, 182]
	GSK126	EZH2	Multiple myeloma, lung, and gastric cancer	[183, 184]
	DB75	PRMT1	Pancreatic cancer	[50]
	SGC707	PRMT3	Acute leukemia and GBM	[185, 186]
	TP-064	PRMT4	Multiple myeloma, endometrial, and breast cancer	[187–189]
	EZM2302	PRMT4	AML, multiple myeloma, and lung cancer	[190, 191]
	GSK3326595	PRMT5	Melanoma, liver, pancreatic, and colorectal cancer	[192–195]
	LLY-283	PRMT5	GBM, pancreatic, gastric and breast cancer	[196, 197]
	BRD4770	G9a	Pancreatic cancer	[198]
	UNC0646	G9a	Melanoma, ALL, liver cancer	[199–201]
	Verticillin A	SUV39H1, SUV39H2, G9a, GLP, NSD2, and KMT2A	Ovarian, gastric, cervical, and pancreatic cancer	[202–204]
	Chaetocin	SUV39H1	GBM, esophageal squamous cell carcinoma, colorectal, lung, and liver cancer	[205–208]
HDM inhibitors	GSK2879552	LSD1	AML and SCLC	[209, 210]
	Pargyline	LSD1	Prostate and breast cancer	[211, 212]

While monotherapies are effective in suppressing the growth of tumors, they often fail in the long-term as tumors develop resistance. Therefore, combining epigenetic modifier-drugs with standard chemotherapy and immunotherapy has emerged as a more potent strategy to enhance anti-tumoral effects and address drug resistance challenges [153]. Combination therapies with HDACis like Vorinostat and Olaparib (PARP inhibitor) have demonstrated potent anti-cancer activity in leukemia, colon adenocarcinoma, and breast cancer. HDACi with other epigenetic-drugs like JQ1 or EZH2 inhibitors has also shown synergistic effects in prostate cancer. HDACi also enhances the efficacy of immune checkpoint inhibitors (ICIs) such as PD-1/PD-L1 and CTLA-4 inhibitors. HDACi improves antigen presentation and T cell responses by increasing the expression of immune checkpoint ligands and tumor-associated antigens (TAAs) on tumor cells [154]. Additionally, combining HDACi with immune checkpoint blockers has demonstrated improved therapeutic outcomes in various cancer models, such as melanoma and ovarian tumors, where these combinations have led to enhanced tumor regression and prolonged survival. Preclinical studies have shown that LBH589 improves the efficacy of radiotherapy by inducing apoptosis and increasing DNA damage [62]. The combination of Valproic acid with epirubicin has shown efficacy in treating anthracycline-resistant tumors, while combinations of HDACis with cisplatin or carboplatin have increased sensitivity and efficacy in bladder and NSCLC. Combining HDACi with hormonal therapies has improved outcomes in hormone-resistant breast and prostate cancers, with vorinostat and tamoxifen showing significant efficacy in patients with advanced breast cancer [57].

Combining HAT inhibitors (HATi) with chemoradiation therapy has also emerged as a promising strategy for some cancers. For instance, the P300/CBP HAT inhibitor, CPI-1612, enhances the sensitivity of GBM cells to TMZ [131]. C646, another P300/CBP inhibitor, was effective in blocking radiation-induced epigenetic alterations and conversion of glioma stem cells to vascular-like cells to prevent tumor growth and recurrence [130]. CCS1477, combined with immune checkpoint inhibitors like anti-PD-L1 or anti-CTLA4 antibodies, has shown increased efficacy in prostate and melanoma cancers. CCS1477 is also in Phase 1/2a trials with drugs such as pomalidomide, dexamethasone, azacitidine, and venetoclax for both hematologic and solid tumors [174]. KAT6A/B inhibitor PF-07248144 is being evaluated in Phase 1 trials for metastatic breast cancer, prostate cancer, and NSCLC, combined with treatments like fulvestrant, letrozole, CDK4 inhibitors, abiraterone acetate, AR antagonists, Olaparib, and atezolizumab. These studies underscore the potential of HAT inhibitors to enhance the efficacy effectiveness of standard chemoradiation therapy as well as immunotherapies [213].

Histone methyltransferase inhibitors (HMTis) targeting enzymes like EZH2 and KDM1A have shown potential in reversing resistance to conventional treatments. EZH2 inhibitors such as tazemetostat combined with chemotherapy agents like doxorubicin or immunotherapy agents like rituximab, vincristine, and cyclophosphamide are being evaluated for their synergistic effects in treating different cancers. For example, EZH2 inhibitors such as UNC1999 and GSK126 have been effective in increasing sensitivity to drugs like 5-fluorouracil in CRC and promoting apoptosis in PC cells by reducing H3K27 methylation [214]. KDM1A inhibitors such as pargyline and GSK2879552 resensitize sorafenib-resistant HCC cells by modulating the WNT/β-catenin signaling pathway [215]. G9a inhibitors, including BRD4770 and UNC0646, have shown efficacy in reducing H3K9 methylation and enhancing the effects of drugs like gemcitabine in pancreatic cancer [216]. Additionally, novel agents like verticillin A and chaetocin are being explored for their ability to target multiple HMTs, such as SUV39H1 and KMT2D, to improve therapeutic outcomes and mitigate resistance in pancreatic cancer [204]. Overall, combining epigenetic drugs with conventional therapies and immunotherapies offers a multifaceted approach to cancer treatment, addressing chemo-radiation resistance and immune evasion while providing synergistic therapeutic benefits to improve cancer outcomes.

5 Conclusions

This chapter has provided an overview of the histone modifications and their important role in mediating cancer development and progression. While we have gained a great deal of knowledge on the common histone modifications such as methylation and acetylation and their dynamic regulation in cancer, there is a dearth of studies on the functional role of the less common modifications such as lactylation, propionlylation, and neddylation in tumorigenesis. A better understanding of the histone code and its interactions with other chromatin regulators will be vital for their development as rational monotherapies or as combinatorial therapies with standard treatments and immunotherapy to improve patient outcomes.

References

1. Yu X, et al. Cancer epigenetics: from laboratory studies and clinical trials to precision medicine. Cell Death Discov. 2024;10(1):28.
2. Espinoza Pereira KN, et al. Histone mutations in cancer. Biochem Soc Trans. 2023;51(5):1749–63.
3. Huang TY, et al. Effects of H3.3G34V mutation on genomic H3K36 and H3K27 methylation patterns in isogenic pediatric glioma cells. Acta Neuropathol Commun. 2020;8(1):219.
4. Kang TZE, et al. The elevated transcription of ADAM19 by the oncohistone H2BE76K contributes to oncogenic properties in breast cancer. J Biol Chem. 2021;296:100374.
5. Kim T, et al. Comparative characterization of 3D chromatin organization in triple-negative breast cancers. Exp Mol Med. 2022;54(5):585–600.
6. Campbell MJ. Tales from topographic oceans: topologically associated domains and cancer. Endocr Relat Cancer. 2019;26(11):R611–r26.
7. Tarjan DR, et al. Epigenome editing strategies for the functional annotation of CTCF insulators. Nat Commun. 2019;10(1):4258.
8. Dixon JR, et al. Integrative detection and analysis of structural variation in cancer genomes. Nat Genet. 2018;50(10):1388–98.
9. Ma F, et al. Recent advances in histone modification and histone modifying enzyme assays. Expert Rev Mol Diagn. 2019;19(1):27–36.

10. Chan JC, Maze I. Nothing is yet set in (hi)stone: novel post-translational modifications regulating chromatin function. Trends Biochem Sci. 2020;45(10):829–44.
11. Wapenaar H, Dekker FJ. Histone acetyltransferases: challenges in targeting bi-substrate enzymes. Clin Epigenetics. 2016;8(1):59.
12. Sun XJ, et al. The role of histone acetyltransferases in normal and malignant hematopoiesis. Front Oncol. 2015;5:108.
13. Wu D, et al. Small molecules targeting HATs, HDACs, and BRDs in cancer therapy. Front Oncologia. 2020;10:560487.
14. Cheng G, et al. Loss of p300 accelerates MDS-associated leukemogenesis. Leukemia. 2017;31(6):1382–90.
15. Kang-Decker N, et al. Loss of CBP causes T cell lymphomagenesis in synergy with p27Kip1 insufficiency. Cancer Cell. 2004;5(2):177–89.
16. Xiao XS, et al. High expression of p300 in human breast cancer correlates with tumor recurrence and predicts adverse prognosis. Chin J Cancer Res. 2011;23(3):201–7.
17. Cregan S, et al. KAT5 (Tip60) is a potential therapeutic target in malignant pleural mesothelioma. Int J Oncol. 2016;48(3):1290–6.
18. Chen G, et al. Acetylation regulates ribonucleotide reductase activity and cancer cell growth. Nat Commun. 2019;10(1):3213.
19. Chen Z, et al. HBO1 promotes cell proliferation in bladder cancer via activation of Wnt/β-catenin signaling. Mol Carcinog. 2018;57(1):12–21.
20. Seto E, Yoshida M. Erasers of histone acetylation: the histone deacetylase enzymes. Cold Spring Harb Perspect Biol. 2014;6(4):a018713.
21. Wawruszak A, et al. Histone deacetylase inhibitors and phenotypical transformation of cancer cells. Cancers. 2019;11(2):148.
22. Shanmugam G, et al. HDAC inhibitors: targets for tumor therapy, immune modulation and lung diseases. Transl Oncol. 2022;16:101312.
23. Hanisch D, et al. Class I HDAC overexpression promotes temozolomide resistance in glioma cells by regulating RAD18 expression. Cell Death Dis. 2022;13(4):293.
24. Li Y, Seto E. HDACs and HDAC inhibitors in cancer development and therapy. Cold Spring Harb Perspect Med. 2016;6(10):a026831.
25. Gu YY, et al. HDAC10 inhibits cervical cancer progression through downregulating the HDAC10-microRNA-223-EPB41L3 Axis. J Oncol. 2022;2022:8092751.
26. Cuttini E, et al. HDAC4 in cancer: a multitasking platform to drive not only epigenetic modifications. Front Mol Biosci. 2023;10:1116660.
27. Yang J, et al. Insights into the function and clinical application of HDAC5 in cancer management. Front Oncol. 2021;11:661620.
28. Mostofa A, et al. Plasma cell dependence on histone/protein deacetylase 11 reveals a therapeutic target in multiple myeloma. JCI Insight. 2021;6(24):e151713.
29. Bora-Singhal N, et al. Novel HDAC11 inhibitors suppress lung adenocarcinoma stem cell self-renewal and overcome drug resistance by suppressing Sox2. Sci Rep. 2020;10(1):4722.
30. Liu S-S, et al. HDAC11: a rising star in epigenetics. Biomed Pharmacother. 2020;131:110607.
31. Zhao H, et al. HDAC11 is related to breast cancer prognosis and inhibits invasion and proliferation of breast cancer cells. Int J Clin Exp Pathol. 2023;16(7):172.
32. Wen Y, et al. Histone deacetylase (HDAC) 11 inhibits matrix metalloproteinase (MMP) 3 expression to suppress colorectal cancer metastasis. J Cancer. 2022;13(6):1923.
33. Núñez-Álvarez Y, Suelves M. HDAC11: a multifaceted histone deacetylase with proficient fatty deacylase activity and its roles in physiological processes. FEBS. 2022;289(10):2771–92.
34. Saleh R, et al. Role of epigenetic modifications in inhibitory immune checkpoints in cancer development and progression. Front Immunol. 2020;11:1469.
35. Levinsky AJ, et al. Targets of histone H3 lysine 9 methyltransferases. Front Cell Dev Biol. 2022;10:1026406.
36. Husmann D, Gozani O. Histone lysine methyltransferases in biology and disease. Nat Struct Mol Biol. 2019;26(10):880–9.

37. Wilkinson AW, et al. SETD3 is an actin histidine methyltransferase that prevents primary dystocia. Nature. 2019;565(7739):372–6.
38. Albert M, Helin K. Histone methyltransferases in cancer. Semin Cell Dev Biol. 2010;21(2):209–20.
39. Cardenas H, et al. EZH2 inhibition promotes epithelial-to-mesenchymal transition in ovarian cancer cells. Oncotarget. 2016;7(51):84453–67.
40. Yang R, et al. E2F7-EZH2 axis regulates PTEN/AKT/mTOR signalling and glioblastoma progression. Br J Cancer. 2020;123(9):1445–55.
41. Shimizu T, et al. Loss of Ezh2 synergizes with JAK2-V617F in initiating myeloproliferative neoplasms and promoting myelofibrosis. J Exp Med. 2016;213(8):1479–96.
42. Walter DM, et al. Systematic *in vivo* inactivation of chromatin-regulating enzymes identifies Setd2 as a potent tumor suppressor in lung adenocarcinoma. Cancer Res. 2017;77(7):1719–29.
43. Alexandrova E, et al. Histone Methyltransferase DOT1L as a promising epigenetic target for treatment of solid tumors. Front Genet. 2022;13:864612.
44. Kelly GM, et al. G9a inhibition enhances checkpoint inhibitor blockade response in melanoma. Clin Cancer Res. 2021;27(9):2624–35.
45. Nakatsuka T, et al. Inhibition of histone methyltransferase G9a attenuates liver cancer initiation by sensitizing DNA-damaged hepatocytes to p53-induced apoptosis. Cell Death Dis. 2021;12(1):99.
46. Casciello F, et al. G9a drives hypoxia-mediated gene repression for breast cancer cell survival and tumorigenesis. Proc Natl Acad Sci USA. 2017;114(27):7077–82.
47. Cheong CM, et al. Twist-1 is upregulated by NSD2 and contributes to tumour dissemination and an epithelial-mesenchymal transition-like gene expression signature in t(4;14)-positive multiple myeloma. Cancer Lett. 2020;475:99–108.
48. Swaroop A, et al. An activating mutation of the NSD2 histone methyltransferase drives oncogenic reprogramming in acute lymphocytic leukemia. Oncogene. 2019;38(5):671–86.
49. Ning J, et al. The protein arginine methyltransferase family (PRMTs) regulates metastases in various tumors: from experimental study to clinical application. Biomed Pharmacother. 2023;167:115456.
50. Wang L, et al. Methylation of HSP70 orchestrates its binding to and stabilization of BCL2 mRNA and renders pancreatic cancer cells resistant to therapeutics. Cancer Res. 2020;80(20):4500–13.
51. Chung J, et al. Protein arginine methyltransferase 5 (PRMT5) promotes survival of lymphoma cells via activation of WNT/β-catenin and AKT/GSK3β proliferative signaling. J Biol Chem. 2019;294(19):7692–710.
52. Hwang JW, et al. Protein arginine methyltransferases: promising targets for cancer therapy. Exp Mol Med. 2021;53(5):788–808.
53. Punnia-Moorthy G, et al. Lysine demethylases: promising drug targets in melanoma and other cancers. Front Genet. 2021;12:680633.
54. AbuHammad S, et al. Regulation of PRMT5-MDM4 axis is critical in the response to CDK4/6 inhibitors in melanoma. Proc Natl Acad Sci USA. 2019;116(36):17990–8000.
55. Martinez-Gamero C, Malla S, Aguilo F. LSD1: expanding functions in stem cells and differentiation. Cells. 2021;10(11):3252.
56. Hou X, et al. KDM1A and KDM3A promote tumor growth by upregulating cell cycle-associated genes in pancreatic cancer. Exp Biol Med (Maywood). 2021;246(17):1869–83.
57. Wang Y, Cao K. KDM1A promotes immunosuppression in hepatocellular carcinoma by regulating PD-L1 through Demethylating MEF2D. J Immunol Res. 2021;2021:9965099.
58. Diao W, et al. Targeting histone demethylases as a potential cancer therapy (review). Int J Oncol. 2022;61(3):103.
59. Dhar SS, et al. Transcriptional repression of histone deacetylase 3 by the histone demethylase KDM2A is coupled to tumorigenicity of lung cancer cells. J Biol Chem. 2014;289(11):7483–96.

60. Wanna-Udom S, et al. KDM2B is involved in the epigenetic regulation of TGF-β-induced epithelial-mesenchymal transition in lung and pancreatic cancer cell lines. J Biol Chem. 2021;296:100213.
61. Yoo J, et al. Epigenetic roles of KDM3B and KDM3C in tumorigenesis and their therapeutic implications. Cell Death Dis. 2024;15(6):451.
62. Li H, et al. KDM4B facilitates colorectal cancer growth and glucose metabolism by stimulating TRAF6-mediated AKT activation. J Exp Clin Cancer Res. 2020;39(1):12.
63. Lee DH, et al. Histone demethylase KDM4C controls tumorigenesis of glioblastoma by epigenetically regulating p53 and c-Myc. Cell Death Dis. 2021;12(1):89.
64. Jose A, et al. Histone demethylase KDM5B as a therapeutic target for cancer therapy. Cancers (Basel). 2020;12(8):2121.
65. Yang GJ, et al. The emerging role of KDM5A in human cancer. J Hematol Oncol. 2021;14(1):30.
66. Revia S, et al. Histone H3K27 demethylase KDM6A is an epigenetic gatekeeper of mTORC1 signalling in cancer. Gut. 2022;71(8):1613–28.
67. Hua C, et al. KDM6 demethylases and their roles in human cancers. Front Oncol. 2021;11:779918.
68. Shanmugam MK, et al. Role of novel histone modifications in cancer. Oncotarget. 2018;9(13):11414–26.
69. An X, et al. Histone modification: biomarkers and potential therapies in colorectal cancer. Ann Hum Genet. 2023;87(6):274–84.
70. Scholz N, et al. Targeting the ubiquitin system in glioblastoma. Front Oncol. 2020;10:574011.
71. Duan S, Pagano M. Ubiquitin ligases in cancer: functions and clinical potentials. Cell Chem Biol. 2021;28(7):918–33.
72. Sampson C, et al. The roles of E3 ubiquitin ligases in cancer progression and targeted therapy. Clin Transl Med. 2023;13(3):e1204.
73. Xu Y, et al. RNF8-mediated regulation of Akt promotes lung cancer cell survival and resistance to DNA damage. Cell Rep. 2021;37(3):109854.
74. Sun J, et al. UBE2T-regulated H2AX monoubiquitination induces hepatocellular carcinoma radioresistance by facilitating CHK1 activation. J Exp Clin Cancer Res. 2020;39(1):222.
75. Young M-J, et al. The role of ubiquitin-specific peptidases in cancer progression. J Biomed Sci. 2019;26(1):42.
76. Shi L, et al. USP49-mediated histone H2B Deubiquitination regulates HCT116 cell proliferation through MDM2-p53 Axis. Mol Cell Biol. 2022;42(3):e0043421.
77. Zheng Y-C, et al. Targeting neddylation E2s: a novel therapeutic strategy in cancer. J Hematol Oncol. 2021;14(1):57.
78. Shi CS, et al. Neddylation inhibitor, MLN4924 suppresses angiogenesis in huvecs and solid cancers: *in vitro* and *in vivo* study. Am J Cancer Res. 2020;10(3):953–64.
79. Wei S, et al. Histone lactylation promotes malignant progression by facilitating USP39 expression to target PI3K/AKT/HIF-1α signal pathway in endometrial carcinoma. Cell Death Discov. 2024;10(1):121.
80. Li X-M, et al. Histone lactylation inhibits RARγ expression in macrophages to promote colorectal tumorigenesis through activation of TRAF6-IL-6-STAT3 signaling. Cell Rep. 2024;43(2):113688.
81. Lu K, Han D. A review of the mechanism of succinylation in cancer. Medicine. 2022;101(45):e31493.
82. Dai X, et al. Succinylation and redox status in cancer cells. Front Oncol. 2022;12:1081712.
83. Zou C, et al. Acyl-CoA: lysophosphatidylcholine acyltransferase I (Lpcat1) catalyzes histone protein O-palmitoylation to regulate mRNA synthesis. J Biol Chem. 2011;286(32):28019–25.
84. Gong M, et al. Loss of p53 concurrent with RAS and TERT activation induces glioma formation. Mol Neurobiol. 2023;60(6):3452–63.
85. Fox BM, et al. SUMOylation in glioblastoma: a novel therapeutic target. Int J Mol Sci. 2019;20(8):1853.

86. Chanda A, et al. Sumoylated SnoN interacts with HDAC1 and p300/CBP to regulate EMT-associated phenotypes in mammary organoids. Cell Death Dis. 2023;14(7):405.
87. Tanikawa C, et al. Regulation of histone modification and chromatin structure by the p53-PADI4 pathway. Nat Commun. 2012;3:676.
88. Wang L, et al. PADI2-mediated Citrullination promotes prostate cancer progression. Cancer Res. 2017;77(21):5755–68.
89. Zhang X-Y, et al. SEPT2 crotonylation promotes metastasis and recurrence in hepatocellular carcinoma and is associated with poor survival. Cell Biosci. 2023;13(1):63.
90. Liao M, et al. Reduction of H3K27cr modification during DNA damage in colon cancer. Front Oncol. 2022;12:924061.
91. Qu M, et al. Hypoxia increases ATX expression by histone Crotonylation in a HIF-2α-dependent manner. Int J Mol Sci. 2023;24(8):7031.
92. Chen Y, et al. Lysine propionylation and butyrylation are novel post-translational modifications in histones. Mol Cell Proteomics. 2007;6(5):812–9.
93. Gates LA, et al. Histone butyrylation in the mouse intestine is mediated by the microbiota and associated with regulation of gene expression. Nat Metab. 2024;6(4):697–707.
94. Zhu Z, et al. Identification of lysine isobutyrylation as a new histone modification mark. Nucleic Acids Res. 2021;49(1):177–89.
95. Xue Q, et al. Functions and mechanisms of protein lysine butyrylation (Kbu): therapeutic implications in human diseases. Genes Dis. 2023;10(6):2479–90.
96. Kuroishi T, et al. Biotinylation is a natural, albeit rare, modification of human histones. Mol Genet Metab. 2011;104(4):537–45.
97. Kothapalli N, et al. Biological functions of biotinylated histones. J Nutr Biochem. 2005;16(7):446–8.
98. Feng J, Meng X. Histone modification and histone modification-targeted anti-cancer drugs in breast cancer: fundamentals and beyond. Front Pharmacol. 2022;13:946811.
99. Crawford NT, et al. TBX2 interacts with heterochromatin protein 1 to recruit a novel repression complex to EGR1-targeted promoters to drive the proliferation of breast cancer cells. Oncogene. 2019;38(31):5971–86.
100. Janaki Ramaiah M, Vaishnave S. BMI1 and PTEN are key determinants of breast cancer therapy: a plausible therapeutic target in breast cancer. Gene. 2018;678:302–11.
101. Julien M, et al. Intrinsic disorder and phosphorylation in BRCA2 facilitate tight regulation of multiple conserved binding events. Biomol Ther. 2021;11(7):1060.
102. Witus SR, Stewart MD, Klevit RE. The BRCA1/BARD1 ubiquitin ligase and its substrates. Biochem J. 2021;478(18):3467–83.
103. Tarcic O, et al. RNF20 and histone H2B ubiquitylation exert opposing effects in basal-like versus luminal breast cancer. Cell Death Differ. 2017;24(4):694–704.
104. Bajbouj K, et al. Histone modification in NSCLC: molecular mechanisms and therapeutic targets. Int J Mol Sci. 2021;22(21):11701.
105. Zhang K, et al. Targeting histone methyltransferase G9a inhibits growth and Wnt signaling pathway by epigenetically regulating HP1α and APC2 gene expression in non-small cell lung cancer. Mol Cancer. 2018;17(1):153.
106. Rowbotham SP, et al. H3K9 methyltransferases and demethylases control lung tumor-propagating cells and lung cancer progression. Nat Commun. 2018;9(1):4559.
107. Wu Q, et al. *In vivo* CRISPR screening unveils histone demethylase UTX as an important epigenetic regulator in lung tumorigenesis. Proc Natl Acad Sci USA. 2018;115(17):E3978–e86.
108. Contreras-Sanzón E, et al. Histone deacetylases modulate resistance to the therapy in lung cancer. Front Genet. 2022;13:960263.
109. Ashktorab H, et al. Global histone H4 acetylation and HDAC2 expression in colon adenoma and carcinoma. Dig Dis Sci. 2009;54(10):2109–17.
110. Kim JK, et al. Targeted inactivation of HDAC2 restores p16INK4a activity and exerts antitumor effects on human gastric cancer. Mol Cancer Res. 2013;11(1):62–73.
111. Li B, et al. The oncogenic potential of SUV39H2: a comprehensive and perspective view. J Cancer. 2019;10(3):721–9.

112. Willmann D, et al. Impairment of prostate cancer cell growth by a selective and reversible lysine-specific demethylase 1 inhibitor. Int J Cancer. 2012;131(11):2704–9.
113. Liu XY, et al. Histone methylation in pancreatic cancer and its clinical implications. World J Gastroenterol. 2021;27(36):6004–24.
114. Metzger E, et al. Phosphorylation of histone H3T6 by PKCbeta(I) controls demethylation at histone H3K4. Nature. 2010;464(7289):792–6.
115. Welti J, et al. Targeting the p300/CBP Axis in lethal prostate cancer. Cancer Discov. 2021;11(5):1118–37.
116. Hong Z, et al. Histone acetyltransferase 1 upregulates androgen receptor expression to modulate CRPC cell resistance to enzalutamide. Clin Transl Med. 2021;11(7):e495.
117. Lasko LM, et al. Discovery of a selective catalytic p300/CBP inhibitor that targets lineage-specific tumours. Nature. 2017;550(7674):128–32.
118. Gan L, et al. The polycomb group protein EZH2 induces epithelial-mesenchymal transition and pluripotent phenotype of gastric cancer cells by binding to PTEN promoter. J Hematol Oncol. 2018;11(1):9.
119. Park YS, et al. The global histone modification pattern correlates with cancer recurrence and overall survival in gastric adenocarcinoma. Ann Surg Oncol. 2008;15(7):1968–76.
120. Qi H, et al. STAT3 activates MSK1-mediated histone H3 phosphorylation to promote NFAT signaling in gastric carcinogenesis. Oncogenesis. 2020;9(2):15.
121. Xu J, et al. Ras-ERK1/2 signaling participates in the progression of gastric cancer through repressing Aurora B-mediated H1. 4 phosphorylation at Ser27. J Cell Physiol. 2020;237(4):1–9.
122. Tang SY, et al. Gastric cancer: An epigenetic view. World J Gastrointest Oncol. 2022;14(1):90–109.
123. Charidemou E, et al. Histone methylation in pre-cancerous liver diseases and hepatocellular carcinoma: recent overview. Clin Transl Oncol. 2023;25(6):1594–605.
124. Xia J-K, et al. Roles and regulation of histone acetylation in hepatocellular carcinoma. Front Genet. 2022;13:982222.
125. Oura K, et al. The roles of epigenetic regulation and the tumor microenvironment in the mechanism of resistance to systemic therapy in hepatocellular carcinoma. Int J Mol Sci. 2023;24(3):2805.
126. Braghini MR, et al. Epigenetic remodelling in human hepatocellular carcinoma. J Exp Clin Cancer Res. 2022;41(1):107.
127. Wu W, et al. Glioblastoma multiforme (GBM): An overview of current therapies and mechanisms of resistance. Pharmacol Res. 2021;171:105780.
128. McCornack C, et al. The function of histone methylation and acetylation regulators in GBM pathophysiology. Front Oncol. 2023;13:1144184.
129. Lv D, et al. Histone acetyltransferase KAT6A upregulates PI3K/AKT signaling through TRIM24 binding. Cancer Res. 2017;77(22):6190–201.
130. Muthukrishnan SD, et al. P300 promotes tumor recurrence by regulating radiation-induced conversion of glioma stem cells to vascular-like cells. Nat Commun. 2022;13(1):6202.
131. Mladek AC, et al. RBBP4-p300 axis modulates expression of genes essential for cell survival and is a potential target for therapy in glioblastoma. Neuro-Oncology. 2022;24(8):1261–72.
132. Spyropoulou A, et al. Role of histone lysine methyltransferases SUV39H1 and SETDB1 in gliomagenesis: modulation of cell proliferation, migration, and colony formation. NeuroMolecular Med. 2014;16(1):70–82.
133. Ciechomska IA, et al. BIX01294, an inhibitor of histone methyltransferase, induces autophagy-dependent differentiation of glioma stem-like cells. Sci Rep. 2016;6:38723.
134. Dong F, et al. PRMT2 links histone H3R8 asymmetric dimethylation to oncogenic activation and tumorigenesis of glioblastoma. Nat Commun. 2018;9(1):4552.
135. Huang T, et al. PRMT6 methylation of RCC1 regulates mitosis, tumorigenicity, and radiation response of glioblastoma stem cells. Mol Cell. 2021;81(6):1276–91.e9.

136. Alejo S, et al. Lysine-specific histone demethylase 1A (KDM1A/LSD1) inhibition attenuates DNA double-strand break repair and augments the efficacy of temozolomide in glioblastoma. Neuro-Oncology. 2023;25(7):1249–61.
137. Yu T, et al. EZH2 interacts with HP1BP3 to epigenetically activate WNT7B that promotes temozolomide resistance in glioblastoma. Oncogene. 2023;42(6):461–70.
138. Kunadis E, et al. Targeting post-translational histone modifying enzymes in glioblastoma. Pharmacol Ther. 2021;220:107721.
139. Pacaud R, et al. Histone H3 phosphorylation in GBM: a new rational to guide the use of kinase inhibitors in anti-GBM therapy. Theranostics. 2015;5(1):12–22.
140. Azab MA. The potential role of histone modifications in glioblastoma therapy: review article. J Mol Pathol. 2023;4(4):196–212.
141. Pasqualucci L, et al. Inactivating mutations of acetyltransferase genes in B-cell lymphoma. Nature. 2011;471(7337):189–95.
142. Issah MA, et al. Epigenetic modifications in acute myeloid leukemia: the emerging role of circular RNAs (review). Int J Oncol. 2021;59(6):107.
143. Harrop S, et al. Epigenetic modifications in lymphoma and their role in the classification of lymphomas. Hematology. 2022;3(1):174–87.
144. Dawson MA, et al. Three distinct patterns of histone H3Y41 phosphorylation mark active genes. Cell Rep. 2012;2(3):470–7.
145. Lu G, et al. Update on histone deacetylase inhibitors in peripheral T-cell lymphoma (PTCL). Clin Epigenetics. 2023;15(1):124.
146. Imai Y, et al. HDAC inhibitors exert anti-myeloma effects through multiple modes of action. Cancers (Basel). 2019;11(4):475.
147. Ridwansyah H, et al. Rationale behind using valproic acid for non-Hodgkin lymphoma: a biomolecular perspective. Eur Rev Med Pharmacol Sci. 2021;25(23):7486–500.
148. Lu Y, et al. Epigenetic regulation in human cancer: the potential role of epi-drug in cancer therapy. Mol Cancer. 2020;19(1):79.
149. von Keudell G, Salles G. The role of tazemetostat in relapsed/refractory follicular lymphoma. Ther Adv Hematol. 2021;12:20406207211015882.
150. Liu Y, Yang Q. The roles of EZH2 in cancer and its inhibitors. Med Oncol. 2023;40(6):167.
151. Xiao W, et al. Small-molecule inhibitors overcome epigenetic reprogramming for cancer therapy. Front Pharmacol. 2021;12:702360.
152. Feustel K, Falchook GS. Protein arginine Methyltransferase 5 (PRMT5) inhibitors in oncology clinical trials: a review. J Immunother Precis Oncol. 2022;5(3):58–67.
153. Morel D, et al. Combining epigenetic drugs with other therapies for solid tumours – past lessons and future promise. Nat Rev Clin Oncol. 2020;17(2):91–107.
154. Huang KC, et al. Decitabine augments chemotherapy-induced PD-L1 upregulation for PD-L1 blockade in colorectal cancer. Cancers (Basel). 2020;12(2):462.
155. Kumagai T, et al. Histone deacetylase inhibitor, suberoylanilide hydroxamic acid (Vorinostat, SAHA) profoundly inhibits the growth of human pancreatic cancer cells. Int J Cancer. 2007;121(3):656–65.
156. Galanis E, et al. Phase II trial of vorinostat in recurrent glioblastoma multiforme: a north central cancer treatment group study. J Clin Oncol. 2009;27(12):2052–8.
157. Richon V. Cancer biology: mechanism of antitumour action of vorinostat (suberoylanilide hydroxamic acid), a novel histone deacetylase inhibitor. Br J Cancer. 2006;95(1):S2–6.
158. Kauh J, et al. C-FLIP degradation mediates sensitization of pancreatic cancer cells to TRAIL-induced apoptosis by the histone deacetylase inhibitor LBH589. PLoS One. 2010;5(4):e10376.
159. Wang X, Yin X. Panobinostat inhibits breast cancer progression via Vps34-mediated exosomal pathway. Hum Cell. 2023;36(1):366–76.
160. Eleutherakis-Papaiakovou E, et al. Efficacy of panobinostat for the treatment of multiple myeloma. J Oncol. 2020;2020(1):7131802.
161. Lopez AT, Bates S, Geskin L. Current status of HDAC inhibitors in cutaneous T-cell lymphoma. Am J Clin Dermatol. 2018;19:805–19.

162. Grant C, et al. Romidepsin: a new therapy for cutaneous T-cell lymphoma and a potential therapy for solid tumors. Expert Rev Anticancer Ther. 2010;10(7):997–1008.
163. Yamasaki S, et al. Phase II trial of romidepsin as consolidation therapy after gemcitabine, dexamethasone, and cisplatin in elderly transplant-ineligible patients with relapsed/refractory peripheral T-cell lymphoma. Hematol Rep. 2024;16(2):336–46.
164. Okubo K, et al. Simvastatin-romidepsin combination kills bladder cancer cells synergistically. Transl Oncol. 2021;14(9):101154.
165. Shi Y, et al. Romidepsin (FK228) regulates the expression of the immune checkpoint ligand PD-L1 and suppresses cellular immune functions in colon cancer. Cancer Immunol Immunother. 2021;70:61–73.
166. Shyu Y-M, et al. Synergistic effect of simultaneous versus sequential combined treatment of histone deacetylase inhibitor valproic acid with etoposide on melanoma cells. Int J Mol Sci. 2021;22(18):10029.
167. Vandermeers F, et al. Valproate, in combination with pemetrexed and cisplatin, provides additional efficacy to the treatment of malignant mesothelioma. Clin Cancer Res. 2009;15(8):2818–28.
168. Jóna Á, et al. The histone deacetylase inhibitor entinostat (SNDX-275) induces apoptosis in Hodgkin lymphoma cells and synergizes with Bcl-2 family inhibitors. Exp Hematol. 2011;39(10):1007–17. e1.
169. Ruiz R, et al. Entinostat (SNDX-275) for the treatment of non-small cell lung cancer. Expert Opin Investig Drugs. 2015;24(8):1101–9.
170. Lin J, et al. Phase I study of entinostat in combination with enzalutamide for treatment of patients with castration-resistant prostate cancer. Oncologist. 2021 Dec;26(12):e2136–42.
171. Marquardt V, et al. Tacedinaline (CI-994), a class I HDAC inhibitor, targets intrinsic tumor growth and leptomeningeal dissemination in MYC-driven medulloblastoma while making them susceptible to anti-CD47-induced macrophage phagocytosis via NF-kB-TGM2 driven tumor inflammation. J Immunother Cancer. 2023;11(1):e005871.
172. Bommi-Reddy A, et al. CREBBP/EP300 acetyltransferase inhibition disrupts FOXA1-bound enhancers to inhibit the proliferation of ER+ breast cancer cells. PLoS One. 2022;17(3):e0262378.
173. Bugbee T, et al. Inhibition of p300 increases cytotoxicity of cisplatin in pancreatic cancer cells. Gene. 2023;888:147762.
174. Nicosia L, et al. Therapeutic targeting of EP300/CBP by bromodomain inhibition in hematologic malignancies. Cancer Cell. 2023;41(12):2136–53. e13.
175. Mukohara T, et al. Inhibition of lysine acetyltransferase KAT6 in ER+ HER2− metastatic breast cancer: a phase 1 trial. Nat Med. 2024;30:1–9.
176. Gao X-N, et al. A histone acetyltransferase p300 inhibitor C646 induces cell cycle arrest and apoptosis selectively in AML1-ETO-positive AML cells. PLoS One. 2013;8(2):e55481.
177. He H, et al. Selective p300 inhibitor C646 inhibited HPV E6-E7 genes, altered glucose metabolism and induced apoptosis in cervical cancer cells. Eur J Pharmacol. 2017;812:206–15.
178. Wang YM, et al. Histone acetyltransferase p300/CBP inhibitor C646 blocks the survival and invasion pathways of gastric cancer cell lines. Int J Oncol. 2017;51(6):1860–8.
179. Sarkozy C, et al. A LYSA phase Ib study of tazemetostat (EPZ-6438) plus R-CHOP in patients with newly diagnosed diffuse large B-cell lymphoma (DLBCL) with poor prognosis features. Clin Cancer Res. 2020;26(13):3145–53.
180. Ribrag V, et al. Phase 1 study of tazemetostat (EPZ-6438), an inhibitor of enhancer of zeste-homolog 2 (EZH2): preliminary safety and activity in relapsed or refractory non-Hodgkin lymphoma (NHL) patients. Blood. 2015;126(23):473.
181. Chen Z, et al. EZH2 inhibition suppresses bladder cancer cell growth and metastasis via the JAK2/STAT3 signaling pathway. Oncol Lett. 2019;18(1):907–15.
182. Hsieh Y-Y, et al. EZH2 inhibitors transcriptionally upregulate cytotoxic autophagy and cytoprotective unfolded protein response in human colorectal cancer cells. Am J Cancer Res. 2016;6(8):1661.

183. Chen Y-T, et al. The novel EZH2 inhibitor, GSK126, suppresses cell migration and angiogenesis via down-regulating VEGF-A. Cancer Chemother Pharmacol. 2016;77:757–65.
184. Guo S, et al. EZH2 inhibition induces senescence via ERK1/2 signaling pathway in multiple myeloma. Acta Biochim Biophys Sin. 2024;56:1055–64.
185. Zou W, et al. Discovery of PRMT3 degrader for the treatment of acute leukemia. Adv Sci. 2024;11:2405963.
186. Liao Y, et al. PRMT3 drives glioblastoma progression by enhancing HIF1A and glycolytic metabolism. Cell Death Dis. 2022;13(11):943.
187. Nakayama K, et al. TP-064, a potent and selective small molecule inhibitor of PRMT4 for multiple myeloma. Oncotarget. 2018;9(26):18480.
188. Inoue F, et al. Histone arginine methyltransferase CARM1 selective inhibitor TP-064 induces apoptosis in endometrial cancer. Biochem Biophys Res Commun. 2022;601:123–8.
189. Peng B-L, et al. A CARM1 inhibitor potently suppresses breast cancer both *in vitro* and *in vivo*. J Med Chem. 2024;67(10):7921–34.
190. Mookhtiar AK, et al. CARM1 inhibition: evaluation of response and efficacy in acute myeloid leukemia. Blood. 2018;132:2719.
191. Zhang L-X, et al. The circular RNA circHMGB2 drives immunosuppression and anti-PD-1 resistance in lung adenocarcinomas and squamous cell carcinomas via the miR-181a-5p/CARM1 axis. Mol Cancer. 2022;21(1):110.
192. Chan LH, et al. PRMT5 and CDK4/6 inhibition result in distinctive patterns of alternative splicing in melanoma. PLoS One. 2023;18(11):e0292278.
193. Luo Y, et al. Myelocytomatosis-protein arginine N-Methyltransferase 5 Axis defines the tumorigenesis and immune response in hepatocellular carcinoma. Hepatology. 2021;74(4):1932–51.
194. Bedard GT, et al. Combined inhibition of MTAP and MAT2a mimics synthetic lethality in tumor models via PRMT5 inhibition. J Biol Chem. 2024;300(1):105492.
195. Sun Y, et al. MST2 methylation by PRMT5 inhibits hippo signaling and promotes pancreatic cancer progression. EMBO. 2023;42(23):e114558.
196. Degorre C, et al. Targeting PRMT5 enhances the radiosensitivity of tumor cells grown *in vitro* and *in vivo*. Sci Rep. 2024;14(1):17316.
197. Bonday ZQ, et al. LLY-283, a potent and selective inhibitor of arginine Methyltransferase 5, PRMT5, with antitumor activity. ACS Med Chem Lett. 2018;9(7):612–7.
198. Urrutia G, et al. Combined targeting of G9a and checkpoint kinase 1 synergistically inhibits pancreatic cancer cell growth by replication fork collapse. Mol Cancer Res. 2020;18(3):448–62.
199. Filiú-Braga LDdC, et al. Molecular and functional anticancer effects of GLP/G9a inhibition by UNC0646 in MeWo melanoma cells. Heliyon. 2024;10(5):e27085.
200. Silva-Carvalho A, et al. Epigenetic priming by EHMT1/EHMT2 in acute lymphoblastic leukemia induces TP53 and TP73 overexpression and promotes cell death. Toxicol In Vitro. 2020;69:104992.
201. Thng DKH, et al. Histone-lysine N-methyltransferase EHMT2 (G9a) inhibition mitigates tumorigenicity in Myc-driven liver cancer. Mol Oncol. 2023;17(11):2275–94.
202. Salvi A, et al. Verticillin a causes apoptosis and reduces tumor burden in high-grade serous ovarian cancer by inducing DNA damage. Mol Cancer Ther. 2020;19(1):89–100.
203. Lu J, et al. Verticillin a suppresses HGF-induced migration and invasion via repression of the c-met/FAK/Src pathway in human gastric and cervical cancer cells. Onco Targets Ther. 2019;12:5823–33.
204. Lu C, et al. Contrasting roles of H3K4me3 and H3K9me3 in regulation of apoptosis and gemcitabine resistance in human pancreatic cancer cells. BMC Cancer. 2018;18(1):149.
205. Xin DE, et al. Chaetocin-mediated SUV39H1 inhibition targets stemness and oncogenic networks of diffuse midline gliomas and synergizes with ONC201. Neuro-Oncology. 2024;26(4):735–48.

206. Zhang Q, et al. Natural compound chaetocin induced DNA damage and apoptosis through reactive oxygen species-dependent pathways in A549 lung cancer cells and *in vitro* evaluations. IET Nanobiotechnol. 2023;17(5):465–75.
207. Li L, et al. Chaetocin exhibits anticancer effects in esophageal squamous cell carcinoma via activation of hippo pathway. Aging (Albany NY). 2023;15(12):5426–44.
208. Lin TC. Perturbation analysis of a prognostic DDX3X-mediated gene expression signature identifies the Antimetastatic potential of Chaetocin in hepatocellular carcinoma. Cells. 2023;12(12):1628.
209. Roboz GJ, et al. Phase I trials of the lysine-specific demethylase 1 inhibitor, GSK2879552, as mono- and combination-therapy in relapsed/refractory acute myeloid leukemia or high-risk myelodysplastic syndromes. Leuk Lymphoma. 2022;63(2):463–7.
210. Bauer TM, et al. Phase I, open-label, dose-escalation study of the safety, pharmacokinetics, pharmacodynamics, and efficacy of GSK2879552 in relapsed/refractory SCLC. J Thorac Oncol. 2019;14(10):1828–38.
211. Wang M, et al. Inhibition of LSD1 by Pargyline inhibited process of EMT and delayed progression of prostate cancer *in vivo*. Biochem Biophys Res Commun. 2015;467(2):310–5.
212. Lee HT, et al. Effects of pargyline on cellular proliferation in human breast cancer cells. Mol Cell Toxicol. 2012;8(4):393–9.
213. White J, et al. Histone lysine acetyltransferase inhibitors: an emerging class of drugs for cancer therapy. Trends Pharmacol Sci. 2024;45(3):243–54.
214. Lima-Fernandes E, et al. Targeting bivalency de-represses Indian hedgehog and inhibits self-renewal of colorectal cancer-initiating cells. Nat Commun. 2019;10(1):1436.
215. Huang M, et al. Targeting KDM1A attenuates Wnt/β-catenin signaling pathway to eliminate sorafenib-resistant stem-like cells in hepatocellular carcinoma. Cancer Lett. 2017;398:12–21.
216. Paschall AV, et al. H3K9 Trimethylation silences Fas expression to confer colon carcinoma immune escape and 5-fluorouracil Chemoresistance. J Immunol. 2015;195(4):1868–82.

Signaling Pathways in Cancer Drug Resistance: Potential Targets for Therapeutic Intervention

Anjana Sajeev, Mukesh Kumar Manickasamy, Ravichandran Vishwa, and Ajaikumar B. Kunnumakkara

1 Introduction

Cancer continues to be one of the leading causes of mortality worldwide, despite the significant advancements in therapeutic strategies, including chemotherapy, radiotherapy, immunotherapy, and targeted treatments [1–3]. A major obstacle in achieving successful outcomes in cancer treatment is drug resistance, a complex phenomenon that undermines the efficacy of anti-cancer therapies and contributes significantly to poor patient prognosis [4]. Drug resistance is characterized by the ability of cancer cells to adapt and evade the cytotoxic effects of therapeutic agents, rendering treatments ineffective and leading to disease progression or recurrence [4]. Drug resistance can be broadly categorized into two types: intrinsic and acquired resistance [5, 19]. Intrinsic resistance occurs due to inherent characteristics of the tumor cells, such as genetic mutations or molecular abnormalities, present even before treatment initiation. These pre-existing factors render the therapy ineffective from the outset. On the other hand, acquired resistance develops as a result of adaptive mechanisms triggered by cancer cells during the course of treatment, often in response to prolonged exposure to chemotherapeutic drugs. This adaptability enables cancer cells to withstand therapies that were initially effective [5]. Multidrug resistance (MDR), a particularly challenging form of resistance, arises when cancer cells become resistant to multiple classes of drugs, further complicating treatment efforts [6, 18]. An in-depth understanding of the mechanisms driving drug resistance is essential to devise effective therapeutic strategies for combating this challenge. Over the years, research has uncovered a wide range of mechanisms underlying drug resistance. These include drug inactivation, alterations in drug

A. Sajeev · M. K. Manickasamy · R. Vishwa · A. B. Kunnumakkara (✉)
Cancer Biology Laboratory, Department of Biosciences and Bioengineering, Indian Institute of Technology Guwahati (IITG), Guwahati, Assam, India
e-mail: kunnumakkara@iitg.ac.in

A. S. Ramteke et al. (eds.), *Molecular Targets in Cancer Therapy*,
https://doi.org/10.1007/978-981-96-7336-0_5

targets, increased drug efflux, inhibition of apoptosis, enhanced DNA damage repair, epithelial-mesenchymal transition (EMT), cancer cell heterogeneity, genetic mutations, and epigenetic modifications [7] (Fig. 1). Among these, one of the most extensively studied contributors to chemotherapy-induced MDR is the overexpression of ATP-binding cassette (ABC) transporters. These transporters utilize energy from ATP hydrolysis to actively expel chemotherapeutic agents from cancer cells, thereby reducing intracellular drug accumulation and diminishing therapeutic efficacy [6]. Another critical factor in drug resistance is the dysregulation or abnormal activation of cellular signaling pathways. Cancer cells exploit these pathways to promote survival, proliferation, and evasion of apoptosis in the face of therapeutic interventions. Dysregulated signaling pathways, such as NF-κB, PI3K/Akt, Wnt/β-catenin, and MAPK play pivotal roles in enhancing cellular defense mechanisms, including drug efflux, DNA repair, and anti-apoptotic signaling. These alterations enable cancer cells to thrive despite exposure to chemotherapeutic drugs, further emphasizing the need for targeted interventions [8]. Importantly, NF-κB pathway is frequently overactivated in various cancers and is associated with promoting cell survival and resistance to apoptosis [9]. Similarly, the PI3K/Akt pathway, known for its role in cell growth and survival, contributes to resistance by inhibiting pro-apoptotic signals [10]. The Wnt pathway, often linked to cancer stem cell (CSC) maintenance, drives EMT and tumor progression, both of which are associated with enhanced resistance to therapies [11]. The MAPK pathway, a key regulator of cell proliferation and differentiation, also significantly promotes resistance by modulating stress response mechanisms [12]. To overcome drug resistance and improve therapeutic outcomes, novel strategies that target these signaling pathways should be explored. Combination therapies involving conventional drugs and natural compounds, such as phytochemicals with anti-cancer properties, have shown promise in sensitizing resistant cancer cells to treatment [13–15]. Additionally, silencing specific proteins within these signaling pathways using RNA interference (RNAi) or small molecule inhibitors is emerging as a powerful approach to mitigate resistance [16, 17]. Such strategies enhance the efficacy of existing therapies and minimize the likelihood of resistance development.

This chapter explores the role of major cellular signaling pathways in cancer drug resistance and discusses recent innovative therapeutic strategies for overcoming chemoresistance. Additionally, it provides an overview of the primary mechanisms underlying drug resistance, emphasizing the importance of a comprehensive understanding of these processes in the development of effective cancer treatments. By addressing the multifaceted nature of drug resistance, this chapter aims to contribute to the ongoing efforts to improve therapeutic outcomes and enhance the quality of life for cancer patients.

2 Major Mechanism of Drug Resistance-Drug Efflux Pathways

Chemoresistance is a significant barrier in cancer treatment, often leading to the failure of conventional therapies. Among the various mechanisms contributing to resistance, drug efflux mediated by ATP-binding cassette (ABC) transporters plays a crucial role [18]. Overexpression of ABC transporters leads to reduced drug concentrations inside the cell thereby limiting chemotherapy efficiency [18]. The major role of drug efflux proteins is to remove xenobiotics and toxic endogenous substances from cells and organs and maintain interstitial homeostasis. However, cancer cells exploit the efflux mechanisms of these transporters to expel therapeutic drugs and evade their cytotoxic effects [19] (Fig. 2). ABC transporter mechanisms are driven by ATP, with drug-resistant cancer cells maintaining higher ATP levels than their parental counterparts. Cancer cells uptake extracellular ATP via micropinocytosis and enhance intracellular ATP concentration, rendering them resistant to multiple drugs [20]. Among the forty eight ABC transporter proteins identified in humans, major ABC transporters that are involved in drug resistance include P-glycoprotein (P-gp), breast cancer resistance protein (BCRP), and multidrug resistance-associated protein (MRP) [21, 22] (Fig. 1).

P-gp, encoded by ABCB1 gene was the first drug resistance-related protein to be identified and it was discovered in 1976 [23]. Overexpression of P-gp is frequently observed in drug-resistant cancer cells and is associated with MDR, as it can expel drugs from multiple classes. This overexpression not only reduces drug efficacy but also contributes to the survival and proliferation of resistant cancer cells. The ability of P-gp to confer resistance to diverse drugs signifies its critical role as a therapeutic target. For instance, a recent study reported that FRAX486, a p21-activated kinase inhibitor, reversed P-gp-mediated drug resistance in breast cancer cells. Further, FRAX486 also enhanced drug accumulation inside cells and reduced drug efflux [24]. In addition, another study identified a pyridoquinoxaline-based derivative as co-adjutant to chemotherapeutics drugs like vincristine and etopside, when co-administered helped in reversing MDR in human cancer cells [25]. Manoharan et al. revealed that annonacin, a phytochemical obtained from *Annona muricata* substantially attenuated the expression of P-gp and induced apoptosis of oxaliplatin-resistant colon cancer cells. This study indicated that annonacin could overcome P-gp-mediated drug resistance in colon cancer [26]. Another study discovered a novel pyrazolo [1,5-a] pyrimidine derivative as sensitizer of breast cancer cells towards paclitaxel by increasing intracellular accumulation of paclitaxel and thereby restoring its cytotoxic effects [27].

MRPs, particularly MRP1, encoded by the ABCC1 gene, are also key players in cancer drug resistance by actively transporting a variety of chemotherapeutic agents out of cells. MRP1 is capable of exporting a wide range of substrates, including vinca alkaloids, anthracyclines, and other drugs [19]. BCRP, encoded by ABCG2 gene is another important drug efflux protein that expels a variety of chemo drugs including camptothecins, epidophyllotoxin, anthracyclines, tyrosine kinase inhibitors, etc. from cancer cells [19]. BCRP/ABCG2 plays a critical role in resistance to

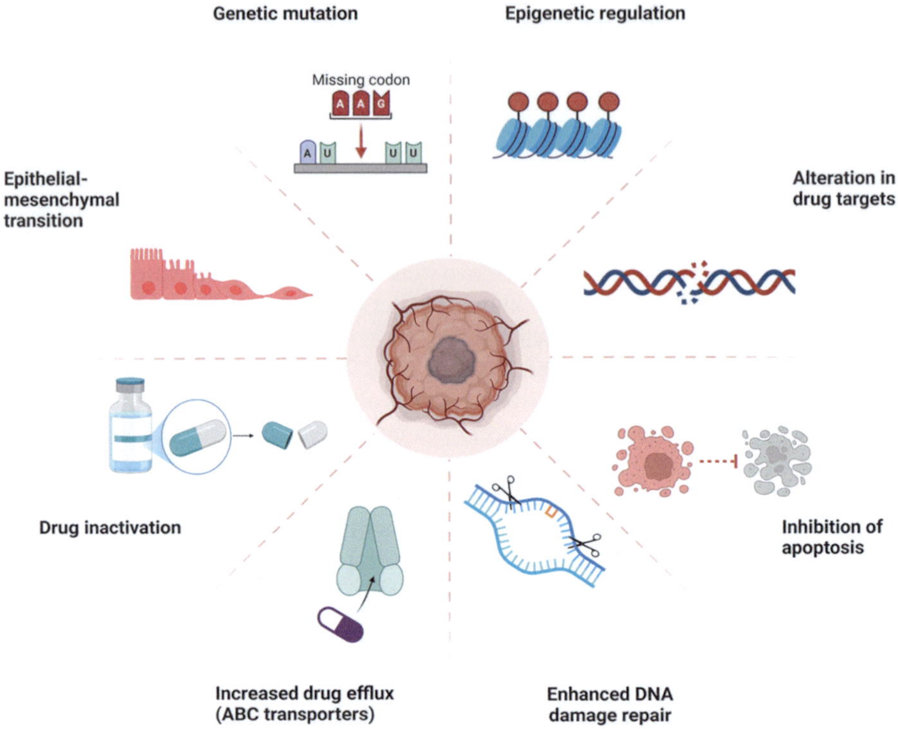

Fig. 1 This figure illustrates the various mechanisms of drug resistance in cancer, highlighting the multifaceted challenges in effective therapy. These mechanisms include genetic mutations, epigenetic regulation, alterations in drug targets, inhibition of apoptosis, enhanced DNA damage repair, increased drug efflux through ABC transporters, drug inactivation, and EMT. Each of these pathways contributes to cancer cells' ability to evade the cytotoxic effects of therapeutic agents, leading to treatment failure

chemotherapy, particularly in breast cancer, but its role in other cancers has also been delineated [19]. Overexpression of MRP1 and BCRP is frequently observed in drug-resistant cancers, correlating with poor therapeutic outcomes and disease progression. Targeted inhibition of these drug efflux-related proteins also represents a promising avenue to overcome MDR and enhance the efficacy of chemotherapeutic agents in resistant tumors. For instance, da Silva et al. demonstrated the reversal of chemoresistance towards SN-38, a neoplastic drug, by a magnolol derivative by specifically inhibiting BCRP [28]. Besides, Yahya et al. reported that the ectopic expression of the micro-RNA, miR-34a suppressed ABCC1 expression and increased the influx of doxorubicin (DOX) in DOX-resistant breast cancer cells [29]. In addition, another study utilized the repurposing of anti-depressant drugs such as citalopram, fluoxetine, and sertraline for overcoming drug resistance by inhibiting MRP1 and MRP7 transporters, thereby overcoming MDR in cancer therapy [30]. Another study revealed that P-gp, BCRP, and MRP4 mediated the efflux of ganciclovir at the blood-brain barrier, and inhibition of these transporters could be a plausible strategy for facilitating the entry of ganciclovir and treatment of brain malignancies [31]. These studies collectively indicate that the overexpression of drug efflux transporters is a major driver of drug resistance in cancer, highlighting their potential as critical therapeutic targets for overcoming chemoresistance and improving treatment outcomes in patients.

3 Cell Signaling Pathways as Potential Targets in Cancer Drug Resistance

MDR is a significant hurdle in successful chemotherapy and patient outcomes. A wide array of factors are responsible for the development of resistance to anticancer drugs, including genetic mutations, epigenetic modifications, increased drug efflux, and alterations in multiple molecular and cellular mechanisms, notably the activation of specific signaling pathways [32–35]. Understanding these signaling pathways and their associated proteins that regulate cancer cell proliferation, survival, invasion, angiogenesis, migration, etc. offers potential targets for developing therapies aimed at overcoming MDR (Fig. 2).

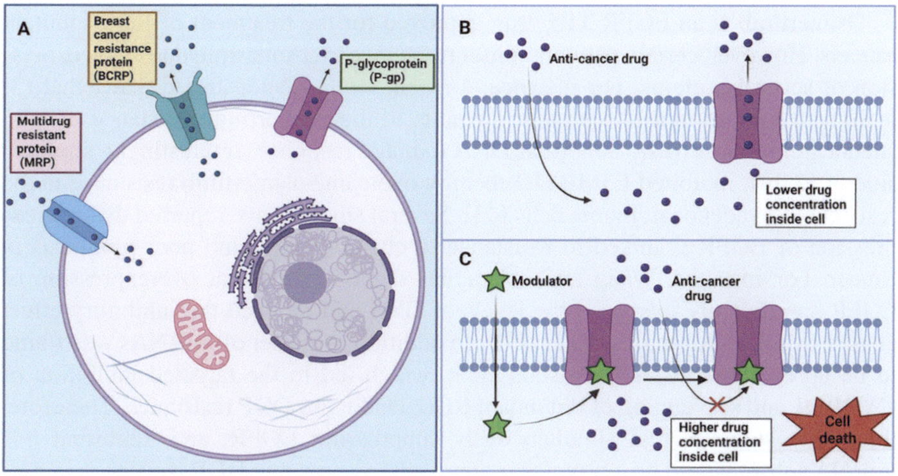

Fig. 2 This figure illustrates the mechanism of ABC transporters in therapy-resistant cancer cells (**a**) ABC transporters, including P-glycoprotein (P-gp), breast cancer resistance protein (BCRP), and multidrug resistance protein (MRP), actively expel anticancer drugs from cancer cells, leading to reduced intracellular drug concentrations. (**b**) In the absence of an inhibitor, anticancer drugs are efficiently removed by ABC transporters, resulting in lower drug accumulation inside the cell and reduced therapeutic efficacy. (**c**) When a modulator or inhibitor of ABC transporters is supplied, it blocks the drug efflux activity of these proteins, allowing anticancer drugs to accumulate inside the cell at higher concentrations. This enhances the cytotoxic effects of the drugs, leading to effective cancer cell death

3.1 EGFR Signaling Pathway

Stanley Cohen, the Nobel Prize Laureate in Physiology or Medicine, made a groundbreaking discovery 25 years ago when he identified epidermal growth factor (EGF) and elucidated its pivotal role in regulating cell growth and proliferation. Epidermal growth factor receptors (EGFRs) are a huge family of receptor tyrosine kinases (RTKs) which has been demonstrated to play a crucial role in tumor cell development and growth, influencing various cellular processes such as proliferation and apoptosis [36, 37].

It is well established that mutations in EGFR are an important reason for lung adenocarcinoma progression. Hence, Geng et al. revealed that mutated EGFR upregulated hypoxanthine phosphoribosyltransferase 1 (HPRT1) via the activation of HIF-1α which in turn increases purine metabolism and tumor proliferation. As a result, suppression of HPRT1 can be used as a strategy for increasing sensitivity of EGFR-mutant gastric cancer cells towards gefitinib [38]. In addition, several studies have reported that mutations in EGFR are also responsible for the development of drug resistance in non-small cell lung cancer (NSCLC) [39–41]. Tang et al. reported the inhibitory effects of a small molecule compound D6, isolated from Codonopsis pilosula, on the proliferation and migration of EGFR-tyrosine kinase inhibitor (TKI)-resistant NSCLC cells harboring EGFR-T790M mutations. Further, D6 in combination with erlotinib or osimertinib effectively sensitized NSCLC cells towards these drugs [42].

Osimertinib is an EGFR TKI drug approved for the treatment of EGFR mutant cancers. However, certain patients confer resistance to osimertinib due to the expression of several proteins. For instance, a recent study demonstrated that lethal (3) malignant brain tumor-like protein 1 results in the osimertinib resistance of lung adenocarcinoma cells by abnormal DNA damage response. Interestingly, suppression of EGFR, inhibited L3MBTL1 thereby overcame osimertinib resistance in the resistant lung adenocarcinoma cells [43]. Several studies have reported that overexpression of EGFR is linked to resistance to chemotherapy and poor prognosis in cancer. For instance, Wang and colleagues demonstrated that overexpression of EGFR significantly increased the levels of GSH and reduced the inhibitory effect caused by DDP in gastric cancer cells. In addition, a subset of miRNAs was found to be upregulated in DDP-resistant cells, which led to the targeted inhibition of CYBRD1 and scavenging of redundant ROS, leading to DDP resistance. Therefore, this study suggests that simultaneously suppressing EGFR, and inhibiting this miRNA subset would be a promising approach to overcome DDP resistance in gastric cancer [44]. Another study demonstrated that elevated interleukin-1 receptor antagonist (IL-1RA) expression in OSCC tumor tissues was linked to larger tumor size, advanced cancer stage, and decreased patient survival. IL-1RA overexpression enhanced OSCC cell growth, migration, invasion, and radio resistance via mitochondrial metabolism-driven EGFR/JNK activation and increased SOX2 expression. Conversely, IL-1RA knockdown led to significantly opposite effects [45].

Zhang et al. investigated the effect of cytochalasin H (CyH) in three types of NSCLC cell lines, such as A549 (wild type EGFR), HCC827 (EGFR mutation), and

NCI-H1975 (dual EGFR mutations and acquired gefitinib resistance). This study revealed that CyH suppressed the expression of PD-L1 and EGFR. Further, CyH enhanced the inhibitory effect of gefitinib on EGFR and PD-L1 in HCC827 and NCI-H1975 cells. Furthermore, CyH showed similar inhibitory effects in NSCLC xenografts by suppressing Ki-67, PD-L1, and EGFR. Hence, this study reported that CyH is a promising candidate for sensitizing resistant NSCLC cells towards gefitinib and thereby overcoming resistance [46]. These studies emphasize the critical role of EGFR signaling in chemoresistance and indicate that targeting this pathway could overcome drug resistance in cancer cells.

3.2 Hippo Signaling

The Hippo signaling pathway (HSP) was first identified and studied in *Drosophila melanogaster* and has been extensively studied over the past decades [47]. HSP plays a crucial role in modulating various physiological processes in mammals, like tissue regeneration, self-renewal, pluripotency, organ size control, and cell differentiation. Dysregulation of this pathway has been associated with a growing number of chronic diseases, including cancer [48]. This pathway comprises several critical components, such as the transcriptional co-activators Yes-associated protein 1 (YAP) and WW domain-containing transcription regulator 1 (TAZ), the TEA domain (TEAD) family of TFs, mammalian STE20-like kinases 1/2 (MST1/2), the adaptor protein Salvador homolog 1 (SAV1), MOB kinase activator-like 1A/B (MOB1A/B), and the large tumor suppressor kinases 1/2 (LATS1/2) [49].

Multiple studies have elucidated that any aberration or imbalance of the HSP in human cancer cells plays a significant role in MDR. Modulating key molecules of this pathway, either through activation or inhibition, has been shown to influence the development of MDR [50]. For instance, Ji et al. reported that LATS1 loss confers a protective effect against paclitaxel in the case of NSCLC cells [51]. Another study revealed that miR-363 is highly expressed in paclitaxel-resistant cells, and overexpressing miR-363 in paclitaxel-sensitive cells resulted in increased resistance to paclitaxel by inhibiting the expression of LATS2 and facilitating YAP nuclear accumulation in ovarian cancer cells [52]. Similarly, another study reported that AJUBA, a protein that plays a role in the HSP, is upregulated in cervical cancer, and increasing the levels of AJUBA led to cisplatin resistance by activating its downstream effectors, YAP and TAZ [53].

Further, modulating HSP holds potential clinical benefits for cancer patients by helping overcome the MDR challenge. For example, it was observed that the loss of FAT atypical cadherin 1 (FAT1), an upstream regulator of HSP, contributed to increased resistance to CDK4/6 inhibitors in estrogen receptor-positive (ER+) breast cancer. Therefore, modulating FAT1 could provide insights into the role of Hippo signaling in ER+ breast cancer and potentially enhance sensitivity to CDK4/6 inhibitors [54]. Further, an allosteric pan-TEAD inhibitor, GNE-7883, effectively blocked oncogenic YAP/TAZ signaling and inhibited all human TEAD paralogs by binding to the TEAD lipid pocket, thereby overcoming resistance to Kirsten rat sarcoma viral oncogene homolog (KRAS) G12C inhibitors [55]. Furthermore, it was revealed that an increased level of YAP expression is a significant alteration that is observed in 5-fluorouracil (5-FU)-resistant colorectal cancer (CRC) cells when compared to non-resistant cells. However, the knockdown of YAP was shown to enhance 5-FU toxicity in preclinical settings [56]. Few studies demonstrated that YAP/TAZ overexpression resulted in enhanced DOX resistance in hepatocellular carcinoma (HCC). However, the downregulation of YAP/TAZ via RNA interference restored sensitivity to DOX [57, 58]. These studies highlight how modulating HSP could effectively address and overcome the challenge of MDR.

3.3 JAK/STAT Pathway

The Janus kinase (JAK) and signal transducer and activator of transcription (STAT) pathway represent an evolutionarily conserved mechanism for transmembrane signal transduction, facilitating cellular communication with the extracellular environment [59]. It was initially identified through investigations into how interferons (IFNs) activate transcription [60]. The interferon-stimulated gene factor 3 (ISGF3), a TF responsive to IFN-α, was discovered in the year 1990 and was found to consist of multiple polypeptide chains with molecular weights of 48, 84, 91, and 113 kDa. In 1992, Fu and colleagues characterized ISGF3's 113, 91, and 84 kDa subunits (ISGF3α), revealing the presence of conserved Src homology 2 and Src homology 3 domains. They further demonstrated that IFN-α-induced cytoplasmic tyrosine kinase could phosphorylate and activate ISGF3α, supporting a model in which IFN-α signaling occurs via direct effector-mediated signal transduction, thereby revealing the functional cascade of the JAK/STAT pathway [60–62].

The JAK family consists of four members: TYK2, JAK1, 2, and 3, while the STAT family comprises seven proteins: STAT1, 2, 3, 4, 5a, 5b, and 6 [63]. JAK/STAT pathway is of two types, canonical and non-canonical. In the canonical pathway, the STAT tyrosine phosphorylation is done by the activated JAKs [64]. In the non-canonical pathway, unphosphorylated STAT localizes to heterochromatin in the nucleus, where it associates with heterochromatin protein 1 [65]. The JAK/STAT pathway is a key signaling mechanism involved in diverse physiological processes, like immune modulation, cellular differentiation, metabolism, and hematopoiesis, and it is essential for various immune functions such as infection defense, immune tolerance, and cancer prevention, as well as regulating cell division, apoptosis, and oncogenesis [59, 66].

The JAK/STAT signaling pathway is critically involved in the development of drug resistance in cancer, significantly undermining the efficacy of chemotherapy. It has been reported that dysregulation of this pathway resulted in elevation of chemoresistance [66]. Studies have revealed that aberrant activation of the JAK/STAT pathway in prostate cancer is associated with resistance to androgen receptor-targeted therapies, regulation of lineage plasticity, and unfavorable clinical outcomes [67, 68]. In addition, Ma et al. revealed that angiopoietin-1 activates the JAK/STAT5 pathway in chronic myeloid leukemia, resulting in resistance to tyrosine kinase inhibitors [69]. Another study demonstrated that TRIM14 serves as a positive regulator for the SPHK1/STAT3 signaling pathway, facilitating CRC cell migration and invasion [70].

Additionally, studies have also shown that modulating this pathway can lead to the sensitization of cancer cells, overcoming drug resistance. For example, aberrant JAK/STAT signaling has been shown to remodel chromatin, thereby conferring resistance to chidamide. In response, targeting TNFRSF8 (CD30), a downstream effector of the JAK/STAT pathway, effectively attenuated JAK-STAT activity. This inhibition prompted a reprogramming of chromatin from a resistant to a sensitive configuration, culminating in a synergistic anti-tumor response [71]. Further, Haselager et al., reported that interleukin-4 (IL-4) and interleukin-21 (IL-21) have divergent impacts on venetoclax sensitivity, which are mediated through the modulation of Bcl-XL levels by STAT3 and STAT6 signaling pathways. Both STAT3 and STAT6 directly regulate the transcription of Bcl-XL without interfering with the DNA-binding activity of NF-κB. Furthermore, it was observed that STAT6 enhanced Bcl-XL expression by facilitating non-canonical NF-κB signaling and targeting the NF-κB-inducing kinase in this signaling cascade can sensitize resistant CLL cells to venetoclax, offering a potential therapeutic strategy [72]. Another study demonstrated that cytokine expression is elevated in hemophagocytic lymphohistiocytosis (HLH), in turn activating the JAK/STAT signaling pathway. This cytokine-mediated activation of the JAK/STAT pathway might contribute to dexamethasone (DEX) resistance observed in HLH. However, ruxolitinib, a JAK1/2 inhibitor, was found to restore apoptotic capacity, thereby enhancing the sensitivity of CD8+ T cells to DEX-induced apoptosis in vitro. Additionally, it substantially ameliorated tissue immunopathology and clinical manifestations of HLH in vivo, suggesting that ruxolitinib could improve patient outcomes in HLH and related cytokine storm syndromes [73]. These findings suggest how altering the JAK/STAT signaling pathway could effectively help overcome the MDR challenge.

3.4 MAPK Pathway

The MAPK signaling pathways, which consist of sequential protein kinase cascades, play an essential role in regulating cell proliferation, differentiation, survival, and stress responses [74]. MAPK signaling has three main well-characterized pathways which include: classical MAPK or Extracellular signal-regulated kinase (ERK), c-Jun N-terminal kinase (JNK), and p38 MAPK pathway [74, 75]. Each subfamily is activated by specific extracellular stimuli and has distinct cellular functions. Each MAPK cascade comprises at least three enzymes activated sequentially. The activation begins when a signaling molecule binds to cell surface receptor, leading to the activation of a MAPK kinase kinase (MAPKKK), which in turn activates a MAPK kinase (MAPKK), which then activates a MAP kinase (MAPK). Once activated, MAPKs translocate to the nucleus, where they regulate the expression of target genes by activating various transcription factors [76]. These cascades together control a variety of cellular signals thereby regulating cellular responses. As a result, the aberrant expression of these cascades is frequently linked to diseases, including cancer [76, 77] (Fig. 3).

Fang et al. demonstrated that inhibition of VEGFA suppressed the proliferation, motility, and EMT of gastric cancer cells. Besides, this inhibition of VEGFA also reduced the expression of genes associated with MDR, such as P-gp, ATP binding cassette subfamily C member 5 (ABCC5), enhancer of zeste homolog 2 (EZH2), and syntaxin 1A (STX1A) via MAPK/Akt/PTEN/TGF-β signaling pathways [78]. Another study demonstrated that the fringe family member, radical fringe (RFNG) enhanced the resistance of CRC cells to oxaliplatin treatment. Mechanistically, chemotherapy-activated MAPK signaling prompts ERK to phosphorylate the Ser255 residue on RFNG, which then interacts with importin proteins and translocates into the nucleus. Once inside the nucleus, RFNG inhibits phosphorylation of p53, decreases the expression of CDKN1A, and increases the expression of SLC7A11, thereby inhibiting apoptosis and ferroptosis. Further, the inhibition of RFNG Ser255 elevated the sensitivity of CRC cells and patient-derived xenograft models toward oxaliplatin [79]. A recent study showed that inhibition of

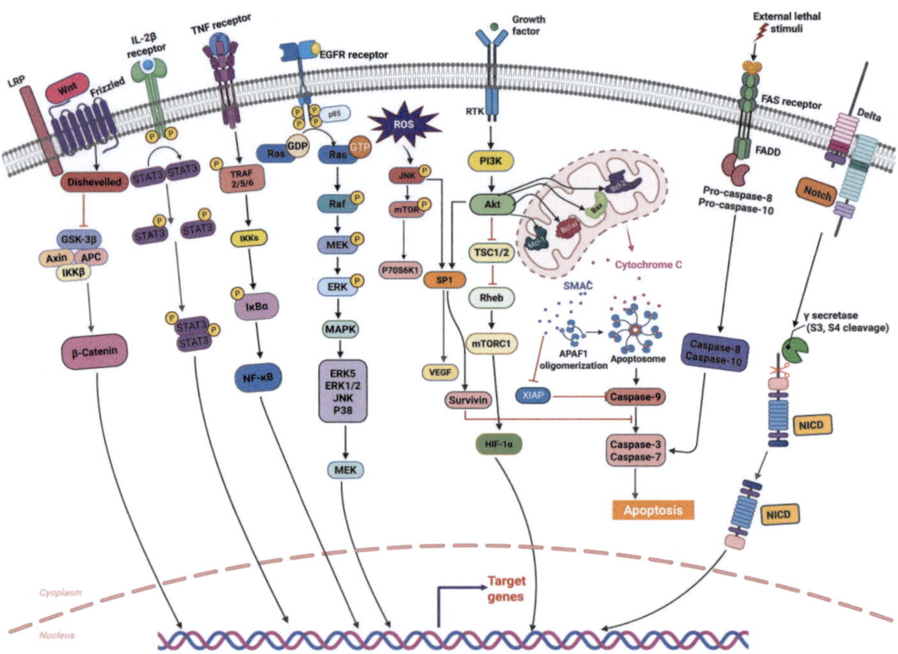

Fig. 3 The figure illustrates various signaling pathways involved in cancer cell drug resistance and highlights potential key molecules that can be targeted to overcome resistance. Key pathways include the Wnt/β-catenin, JAK/STAT, NF-κB, MAPK/ERK, PI3K/Akt/mTOR, Notch signaling, and apoptotic pathways, all of which contribute to cancer progression, survival, and resistance to therapeutic agents. Targeting key molecules, such as β-catenin, STAT3, NF-κB, mTOR, and apoptotic regulators (e.g., XIAP, Bcl-2), offers a promising strategy to disrupt these pathways, sensitize cancer cells to therapy, and restore drug efficacy

Rho-associated coiled-coil containing kinase 2 (ROCK2) resulted in increased sensitivity of acute myeloid leukemia (AML) cells towards DOX through downregulation of Akt/MAPK pathway. Interestingly, an E3 ubiquitin ligase, TRIM26 ubiquitylate ROCK2, thereby reducing its expression [80]. Additionally, Zhang et al. reported that tissue inhibitor metalloproteinase 2 (TIMP-2) is highly expressed in 5-FU-resistant CRC cells which in turn results in the constitutive activation of ERK/MAPK pathway. Hence, the suppression of TIMP-2 or inhibition of ERK/MAPK could be used as a novel strategy to overcome 5-FU resistance in CRC [81]. Another study investigated the effect of target of rapamycin complex 1 (TORC1) in BRAF-mutant pediatric brain tumor patient-derived xenograft (PDX) models, which revealed that resistance to trametinib, a MEK inhibitor, is caused due to the activation of TORC1 through the ERK/MAPK cascade. Further, the combinatorial treatment of trametinib and rapamycin, a TORC1 inhibitor, re-sensitized these resistant PDX models towards trametinib [82]. Further, another study reported that δ-tocotrienol, a vitamin E derivative synergistically enhanced the cytotoxicity and apoptotic effects of cisplatin in ovarian cancer cells and sensitized these cells to cisplatin treatment through the activation of p38 and JNK and the simultaneous production of mitochondrial ROS [83]. Furthermore, another study demonstrated that tanshinone I, a natural compound sensitized triple negative breast cancer (TNBC) cells towards paclitaxel by suppressing paclitaxel-induced autophagy. Besides, findings from KEGG analysis and molecular docking revealed that tanshinone I inhibited late autophagy by targeting p38/MAPK pathway. Hence, this study suggested that combination of tanshinone I and paclitaxel could be a novel approach in the treatment of TNBC by targeting the Akt/p38/MAPK/autophagy signaling axis [84]. Therefore, modulating the expression of key proteins in this pathway could be used as a potential strategy to overcome cancer drug resistance.

3.5 NF-κB Signaling Pathway

NF-κB plays a crucial role in many biological processes. The core structure of NF-κB signaling involves a sequence of both positive and negative regulatory components, which primarily include ligands, receptors, receptor-proximal signal adapter proteins, IκB kinase (IKK) complexes, IκB proteins, and NF-κB dimers [85]. NF-κB activation occurs through two main pathways: the canonical and non-canonical pathways. The canonical pathway is triggered by signals like TNF, IL-1, and LPS, leading to the activation of p50, RelA, and c-Rel. This pathway involves the phosphorylation and degradation of IκB proteins, allowing NF-κB dimers to move to the nucleus and activate gene transcription linked to cell survival, proliferation, and other key processes. The non-canonical pathway selectively activates p52 and RelB through IKKα-mediated phosphorylation of p100, which then forms p52/RelB dimers that also promote gene transcription [86, 87]. This pathway is a critical regulator of the expression of genes involved in cell survival, inflammation, and immune responses [87].

Importantly, NF-κB signaling pathway plays a significant role in the development of MDR in cancer, contributing to the failure of chemotherapy treatments. Dysregulated expression of NF-κB pathway leads to the upregulation of anti-apoptotic proteins, drug efflux transporters, and other survival pathways that help cancer cells resist the cytotoxic effects of chemotherapeutic agents. For instance, Yan et al. indicated that IL-33-mediated activation of NF-κB resulted in inhibition of apoptosis in AML cells, which also caused a reduction of drug sensitivity in these cells, which could be reversed by treatment with NF-κB inhibitor, Bay-117085. Further, activation of NF-κB also led to IL-6-mediated autocrine activation of p-Akt, paving the way for IL-33/NF-κB/p-Akt as potential targets for cancer therapeutics [88]. Besides, another study showed that homeobox A1 (HOXA1) gene is frequently overexpressed in oral squamous cell carcinoma (OSCC) and is associated with a low overall survival of OSCC patients. As a result, inhibition of HOXA1 reduced cisplatin resistance in OSCC cells through IκB/NF-κB signaling pathway [89]. In addition, Abdin et al. revealed that DOX-resistant breast cancer cells showed higher NF-κB activity thereby upregulating the expression of drug efflux proteins, ABCB1 and ABCC1. Interestingly, combinatorial treatment of NF-κB inhibitors, pentoxifylline, and bortezomib, re-sensitized resistant breast cancer cells to DOX by downregulating NF-κB and drug transporters [90]. Another study depicted that NSC35446. HCl, a compound that mimics the ability of BRCA1 by inhibiting ERα activity-induced apoptosis in antiestrogen-resistant breast cancer cells by inhibiting the mRNA and protein level expression of IKKB, IKKA, and IKKG/NEMO. Moreover, this compound also suppressed the expression of NF-κB and its target genes [91]. Further, Bharathiraja et al. demonstrated that solasodine, a steroidal alkaloid from Solanaceae family increased intracellular accumulation of DOX in drug-resistant HCC cells. The combination of this compound and DOX also induced apoptosis and G2/M phase cell cycle arrest in these cells. Mechanistically, solasodine sensitized HCC cells to DOX by targeting NF-κB pathway and by downregulating P-gp and its transport activity [92]. Another study showed the chemosensitizing effect of parthenolide, a sesquiterpene lactone from the plant *Tanacetum parthenium*, on pancreatic cancer cells towards gemcitabine by suppressing p65 activity through inhibition of its nuclear translocation and reducing the expression of multidrug-resistant protein 1 (MRP1) [93]. Further, Wang et al. demonstrated that abnormal activation of the NF-κB signaling pathway plays a pivotal role in mediating resistance to osimertinib in EGFR-mutant NSCLC. NF-κB activation promoted the recruitment of myeloid-derived suppressor cells (MDSCs) by inducing immunosuppressive cytokines such as IL-4 and IL-34, thereby contributing to an immune-excluded tumor microenvironment. Pharmacological inhibition of NF-κB re-sensitized resistant tumors and partially reversed immune evasion, highlighting NF-κB as a critical target for overcoming osimertinib resistance [94]. These studies highlight the role of NF-κB signaling in MDR and the importance of targeting this pathway as a strategy to overcome chemoresistance.

3.6 Notch Pathway

The Notch signaling pathway is a highly conserved cell communication system that plays a crucial role in the regulation of various cellular processes such as differentiation, proliferation, and apoptosis [95]. Central to this pathway are four Notch receptors (Notch 1-4), which get activated by interacting with its family of ligands (Delta-like and jagged ligands) on neighboring cell, which thereby leads to a cascade of events that determine cellular fate [96]. After ligand binding, the Notch receptor undergoes a series of proteolytic cleavages [97, 98]. The second cleavage by γ-secretase releases notch intracellular domain (NICD) into the cytoplasm which then translocates into the nucleus. Inside the nucleus, NICD forms a complex which then recruits additional transcription activators and results in the transcription of Notch target genes including Hes (Hairy enhance of split) family, NF-κB, VEGF, mammalian target of rapamycin (mTOR), cyclin D1, c-myc, p21, p27, Akt, etc. [97, 99–102].

Dysregulation of Notch signaling has been implicated in numerous diseases, including cancer, where this unique signaling has both tumor suppressor and oncogenic role in different cancers [98]. In addition, the role of Notch pathway in the formation of CSCs and the acquisition of an EMT phenotype has been studied widely, which delineated its influence on cancer drug resistance [98, 103]. In addition, Zhang ZP reported that Notch-1 expression is negatively correlated to cisplatin sensitivity in head and neck squamous cell carcinoma. Besides, treatment with a γ-secretase inhibitor, DAPT, increased the sensitivity of head and neck squamous cell carcinoma cells towards cisplatin by blocking Notch-1 pathway [104]. A recent study revealed that shRNA-mediated suppression of Notch-3 expression in colon cancer cells showed decreased levels of the DNA repair proteins PARP1 and APE1 and increased sensitivity to the chemotherapy drugs ara-c and cisplatin. This study indicated that Notch-3 may control DNA repair in colon cancer cells and targeting this protein could render these cells sensitive to chemotherapeutic drugs [105]. Further, another study demonstrated that inhibition of Notch-3 increased the sensitivity of pancreatic cancer cells towards gemcitabine and resulted in caspase-mediated apoptosis of these cells [106]. Furthermore, Patrawala et al. observed that glioma stem cells exhibit increased resistance to chemotherapy due to elevated expression of drug transporter proteins like ABCG2. Additionally, elevated ABCG2 levels correlate with cell proliferation and are associated with the expression of various "stemness" genes, including Notch-1 [107]. Besides, another study reported that upregulation of SOX8, Notch-4, and Hes-5 resulted in poor survival in castration-resistant prostate cancer (CRPC). Moreover, silencing SOX8 decreased the activation of Notch signaling in enzalutamide-resistant prostate cancer cells, thereby sensitizing these cells towards enzalutamide. This study indicated that targeting SOX8/Notch pathway could be a possible strategy for developing novel therapeutics for CRPC [108].

In another study, Sun et al. showed that oxyberberine (OBB), a major metabolite of berberine, sensitized liver cells to sorafenib by inhibiting cell growth and inducing apoptosis. Additionally, this compound attenuated the expression of ubiquitin-specific peptidase 7 (USP7), an oncogenic gene by inhibiting Notch signaling [109]. Further, Czerwonka et al. demonstrated that FLI-06, a small molecule inhibitor of Notch signaling downregulated key Notch target genes, thereby elevating chemosensitivity of head and neck cancer cells cells towards taxane-based chemotherapy [110]. Collectively, these studies shed light on the critical role of Notch signaling in cancer drug resistance.

3.7 PI3K/Akt/mTOR Signaling Pathway

PI3K/Akt/mTOR cascade is a crucial regulator of transcription, translation, cell proliferation, growth, and survival [111–113]. Even though it is indispensable for the cells' normal physiology, its dysregulation leads to the pathology of various diseases including cancer [112–116]. This cascade can be activated by extracellular stimuli from interleukins, cytokines, insulin, and other growth factors such as EGF, fibroblast growth factor (FGF), and insulin-like growth factor (IGF-1) [113, 114, 117]. However, in cancer, abnormal activation of this signaling is observed often, mainly due to the alterations in the components of this cascade and activation through RTKs [118].

For instance, Knuefermann et al. reported that increased expression of HER2 leads to PI3K-dependent activation of Akt and displayed resistance to several chemotherapeutics namely, DOX, 5-FU, etoposide, paclitaxel, and camptothecin in breast cancer cells. In addition, the modified MCF-7 cells having a high level of constitutively active Akt is resistant to all five mentioned chemotherapeutic agents which substantiated the direct relationship of Akt with MDR in breast cancer [119]. Besides, Chen et al. reported that heat shock protein family A member 12B overexpression is associated with increased cisplatin resistance in NSCLC in vitro and in vivo models. It was also demonstrated that this chemoresistance is mediated by the activation of PI3K/Akt/NF-κB signaling pathway and thereby helping the cancer cells to evade cisplatin-induced apoptosis [120]. Moreover, in human AML blasts, the expression of MRP1 is reported to be under the control of PI3K/Akt axis. This observation was validated by wortmannin, a PI3K inhibitor treatment which resulted in the reduction of p-Akt and MRP1 levels as well as the drug efflux activity in AML cells [121].

Interestingly PI3K/Akt pathway and its key proteins have been targeted to overcome MDR in cancer and have displayed a promising chemosensitizing efficiency. For instance, treating cisplatin-resistant small cell lung cancer (SCLC) cells with capivasertib inhibited proliferation and induced apoptosis and sensitized these cells to cisplatin via the suppression of Akt and MYC signaling [122]. Another study showed that PIM kinases are involved in PI3K/Akt/mTOR pathway and a multikinase inhibitor AUM302, which simultaneously targets PIM, PI3K and mTOR, overcomes gemcitabine resistance in pancreatic ductal adenocarcinoma cells [123]. Further, Zuo-Jin-Wan (ZJW) a Chinese medicinal formula was shown to overcome P-gp-mediated drug-resistance in CRC cells through inhibiting PI3K/Akt/NF-κB pathway [124]. The sensitivity to oxaliplatin treatment was enhanced by ZJW in both dose and time-dependent manner through suppressing the activation of Akt and NF-κB in resistant CRC cells [124]. Similarly, another Chinese medicine Babao dan, which has anti-gastric cancer activity was found to reverse MDR in gastric cancer. The Babao dan treatment was found to diminish the expression of ABCB1, ABCC1, and ABCG2 and induced autophagy and apoptosis through the inhibition of PI3K/Akt/mTOR pathway in MDR gastric cancer cells [125]. In addition, shikonin, a natural naphthoquinone pigment from *Lithospermum erythrorhizon* was shown to have significant apoptosis-inducing effects in MDR NSCLC cells. It was observed that shikonin downregulates p-Akt and p70s6k levels. Further, Shikonin in low doses displayed a synergistic effect with BEZ235, a PI3K-mTOR inhibitor, in hindering spheroid formation of chemoresistant lung cancer cells [126]. In conclusion, targeting the PI3K/Akt/mTOR pathway is a promising strategy to kill the MDR cancer cells. Therefore, combining inhibitors of the PI3K/Akt/mTOR pathway with conventional drugs will offer novel therapeutic strategies in combating MDR in cancer.

3.8 Wnt/β-Catenin Signaling Pathway

The Wnt (wingless-related integration site) gene was discovered in the 1980s when researchers observed that mouse mammary tumor virus (MMTV) caused tumors by integrating viral DNA into a specific site in the host genome, initially termed as int1 [127, 128]. It was later renamed Wnt due to the dissimilarities with other MMTV interaction-related genes [127, 129]. The Wnt pathway consists of two main classifications: the canonical (Wnt/β-catenin) pathway, which involves β-catenin-dependent transcriptional activation through TCF/LEF transcription factors, and non-canonical pathways, which are β-catenin-independent, including Wnt/Ca^{2+}, Wnt/planar cell polarity (PCP) and Wnt/receptor tyrosine kinase-like orphan receptors (ROR) signaling [11, 130, 131]. The non-canonical pathways regulate several cellular events including, cell motility, morphogenetic movements during developmental processes, cell polarity, etc. [130–134]. However, the canonical pathway primarily regulates embryonic stem cell maintenance, lung tissue repair, liver development, regeneration, etc. [135–137]. Apart from this, the Wnt signaling pathways are also involved in the pathogenesis of diabetes, coronary artery disease, schizophrenia, bipolar disorder, and cancer [136, 138–140].

Wnt signaling pathways regulate various hallmarks of cancer either by directly activating the transcription of various effector genes or through its interplay with various other signaling pathways [11, 141–146]. The Wnt signaling pathway and the expression of related proteins and factors were observed to be enhanced in various drug-resistant cancer cells [147–149]. In addition, this pathway has also been involved in the maintenance of CSCs and the development of therapeutic resistance in cancers. The CSCs are a subset of cells within the tumor, which is potentially responsible for the tumor initiation, and relapse and also have an inherent chemoresistance mechanism [142]. Further, in TNBC cells the expression of β-catenin is found to be crucial for stemness and sensitivity to the anticancer agents namely cisplatin and DOX [150]. Similarly, the activity of β-catenin was also reported to be an indispensable character of stemness in colon cancer cells [151, 152]. For instance, Lgr5, one of the targets of Wnt signaling was reported to promote stemness along with the development of chemoresistance through positively regulating the expression of ABCB1 in CRC cells [153]. This relationship between Wnt signaling, CSCs, and drug resistance was also substantiated in liver cancer cells [154]. Moreover, through the activation of HIF-1α mediated metabolic reprogramming, Wnt and Akt signaling pathways impart 5-FU resistance in CRC cells [155]. Further, Wnt/β-catenin pathway leads to the direct overexpression of glutathione peroxidase 4 (GPX4), the master regulator of a form of regulated cell death called ferroptosis, thereby leading to ferroptosis resistance in gastric cancer cells [156]. Therefore, it is clear that Wnt signaling plays an important role in the maintenance of the CSCs and the development of resistance to multiple chemotherapeutic drugs.

Several studies have shown that attenuating the abnormal activation of Wnt signaling aids in sensitizing cancer cells to the drugs to which they were resistant. For instance, Wang et al. observed that Wnt signaling is responsible for the ferroptosis resistance of gastric cancer cells, and the inhibition of this pathway with its specific inhibitor LF3, enhanced the ferroptosis sensitivity of gastric cancer cells. In addition, the selective inhibition of TCF4 has also been shown to enhance the sensitivity of gastric cancer cells to cisplatin-induced ferroptosis in both in vitro and in vivo models [156]. Similarly, in HCC cells, the knockdown of β-catenin was shown to increase the antitumor activity of sorafenib potentially through modulating ferroptosis [157]. In addition, the anti-tumor activity of sorafenib, an FDA approved multi-kinase inhibitor for the treatment of HCC, was improved by the combined administration with Wnt/β-catenin pathway inhibitors such as ICG-001, 2-[[[2-(4-ethylphenyl)-5-methyl-4-oxazolyl]methyl]thio]-N-(2-phenylethyl)acetamide and N-(2-Methyl-4-nitro)-2,4-dichlorosulfonamide in hepatoma and HCC cells [158, 159]. Similarly, another study showed that circFBXW7 through m6A, inhibits β-catenin by its ubiquitination and thereby reverses the tyrosine kinase inhibitors resistance of lung adenocarcinoma cells along with inhibiting the stemness [160]. Moreover, in another study it was shown that the treatment with dinaciclib, a multi-CDK inhibitor, reverted the resistance to bromo- and extra-terminal domain inhibitors by the supression of Wnt signaling pathway in multiple levels in AML [161]. This inhibition includes a reduction in the protein level of LRP6 and β-catenin following the treatment of dinaciclib [161]. Therefore, targeting Wnt signaling pathway for the reversal of drug resistance and the combinatorial treatment with this pathway inhibitors to enhance the treatment efficiency hold promise in the therapeutic regime of cancer.

3.9 Apoptotic Pathway

The apoptotic pathway is a fundamental cellular mechanism responsible for programmed cell death, which is an essential process for the maintenance of cellular homeostasis and normal development in multicellular organisms [162]. It is a tightly regulated energy-dependent, and evolutionarily conserved process that enables cells to undergo controlled death without damaging the neighboring cells [162]. This process is highly significant in various physiological contexts, including immune system function, and the elimination of damaged or potentially cancerous cells. Hence, aberrations in apoptosis are strongly associated with abnormal cell proliferation, thereby leading to diseases like cancer [162].

The regulation of apoptosis is highly complex and involves numerous pro-apoptotic and anti-apoptotic proteins orchestrated into two interconnected signaling routes, namely the intrinsic or mitochondrial and extrinsic or death receptor-mediated pathways [162, 163]. Caspases are the essential regulatory proteins involved in both pathways and are categorized into initiator caspases and effector caspases [162]. The intrinsic pathway of apoptosis is triggered by internal signals such as DNA damage and oxidative stress. caused by radiation, cytotoxic drugs, hypoxia, and so on [164, 165]. It leads to mitochondrial outer membrane permeabilization (MOMP) and the release of cytochrome c, which then activates downstream caspase proteins. This pathway is regulated by the B-cell lymphoma 2 (Bcl-2) family of proteins which initiates the release of pro-apoptotic proteins from the intra-membrane space of mitochondria [162, 166]. The extrinsic pathway of apoptosis is triggered by external signals such as binding of death-inducing ligands to the cell surface death receptors/ Membrane death receptors are part of the tumor necrosis factor (TNF) receptor superfamily and include receptors such as tumor necrosis factor receptor-1 (TNF-R1/DR1), Fas (Apo-1/CD95/DR2), death receptor-4 (DR4), and death receptor-5 (DR5), which are activated by the binding of specific ligands, TNF-α, FasL, and TNF-related apoptosis-inducing ligand (TRAIL) [162, 167]. Ligand binding to the receptors leads to the activation of initiator caspases and death-induced signaling complex formation [168]. Both pathways converge on the execution phase, where effector caspases, such as caspase-3 and caspase-7, degrade cellular components, ultimately leading to characteristic morphological changes such as cell shrinkage, chromatin condensation, membrane blebbing, and the formation of apoptotic bodies [167].

Several studies have shown the relationship between apoptosis and the associated proteins with resistance to chemotherapy. For instance, Devi et al. demonstrated that X-linked inhibitor of apoptosis protein (XIAP), the most potent inhibitor of apoptosis expression enhances chemoresistance and tumor aggressiveness in invasive breast cancer patients, and targeting this protein could open avenues for breast cancer chemotherapies [169]. Another study reported that Cullin7 and the anti-apoptotic protein survivin upregulation were associated with docetaxel resistance in lung adenocarcinoma cells. Silencing of these proteins in docetaxel-resistant lung adenocarcinoma cells enhanced the sensitivity of these cells towards docetaxel. This study revealed that Cullin7/survivin axis plays a significant role in conferring resistance to docetaxel through the inhibition of intrinsic apoptotic signaling [170]. Guo et al. demonstrated that loss of prostate apoptosis response-4 (PAR-4), a tumor suppressor protein, results in tumor recurrence and chemotherapy resistance in TNBC cell lines by preventing activation of caspase-8 and cellular inhibitor of apoptosis protein 1 (cIAP1) depletion. Hence, this study proposed that targeting cIAP1 along with using a genotoxic drug would activate caspase-8 mediated apoptosis and could be a promising approach in overcoming drug resistance in TNBC cells [171]. Further, DEBIO 1143, an inhibitor of IAP, was evaluated in ovarian cancer cells and mouse models where this compound was able to overcome carboplatin resistance and induced cell death by apoptosis or necroptosis [172]. An interesting study revealed that suppression of miR-34c-5p and overexpression of Bax in multiple myeloma cells resistant to bortezomib, a protease inhibitor, reduced Bcl-2 expression and re-sensitized these cells to bortezomib. This study indicated that this miRNA regulates apoptosis in multiple myeloma and could be a potential target to overcome resistance towards proteasome inhibitor drugs [173]. Another study demonstrated that Bcl-2 was significantly upregulated in osimertinib-resistant NSCLC cells. Further, a combinatorial treatment with osimertinib and Bcl-2 inhibitor, ABT263, remarkably reduced the expression of survivin and SQSTM1 and enhanced the expression of LC3-II, an autophagosome marker. This study suggested that osimertinib and ABT263 synergistically overcame the resistance of NSCLC cells towards osimertinib through various signaling pathways [174]. Hence, the development of drugs that target anti-apoptotic proteins is a promising approach for overcoming chemoresistance.

4 Perspectives

As mentioned earlier, a complex interplay of aberrantly activated signaling pathways including PIK/Akt/mTOR, Wnt/β-catenin, NF-κB, Notch, MAPK, etc., confers resistance to multiple cancer drugs, making single-target therapies insufficient to combat MDR effectively. Consequently, there is an urgent need to explore therapeutic strategies that target multiple signaling pathways simultaneously. For example, Wang et al. demonstrated that aberrant activation of NF-κB and mitogen-activated protein kinase (MAPK) signaling pathways leads to osimertinib resistance in tumors. Subsequently, this study explained that the inhibition of these pathways could be a potential solution for this [94]. Another study demonstrated that while MEK inhibitor treatment significantly reduced melanoma cell proliferation, this effect was reversed upon inhibitor removal. However, when the cells were treated with a combination of MEK and PI3K inhibitors, the growth and invasion of melanoma spheroids were effectively blocked [175]. Another study revealed that combinatorial treatment of three FDA-approved drugs, sunitinib, dasatinib, and everolimus synergistically increased cancer-killing efficacy in ovarian cancer cells by inhibiting STAT3, MAPK, and, PI3K/Akt/mTOR pathways. Further, this combination also showed enhanced effectiveness of paclitaxel treatment in these cells [176]. Besides, another study demonstrated that the overexpression of Bcl-2 and IKKβ contributed to the development of enzalutamide resistance in castration-resistant prostate cancer (CRPC) cells. Treatment with ABT-199, a Bcl-2 inhibitor, and IMD-0354, an IKKβ inhibitor, effectively reduced enzalutamide resistance in both cellular and animal models [177]. These studies highlight the potential of adopting a multi-pathway targeting approach in overcoming drug resistance in cancer.

5 Conclusion

The development of cancer drug resistance remains a major challenge in the field of oncology. While significant progress has been made in understanding the underlying mechanisms of drug resistance, there is still a pressing need to identify novel and effective strategies to overcome this obstacle. Targeting signaling pathways involved in drug resistance offers a promising approach to combat resistance, as these pathways are often central to the survival and proliferation of cancer cells. By inhibiting key components of pathways such as PI3K/Akt/mTOR, NF-κB, MAPK/ERK, Wnt/β-catenin, JAK/STAT, and so on, the sensitivity of cancer cells to therapy can be enhanced and the likelihood of resistance development can be reduced to a great extent. Furthermore, combination therapies targeting multiple pathways simultaneously are being explored to address the redundancy and compensatory mechanisms that cancer cells employ to evade chemotherapy.

This chapter envisages the need for a multifaceted approach that combines targeted therapies, immunotherapies, and novel drug delivery systems which is necessary to overcome the challenge of cancer drug resistance. Personalized medicine, which considers the unique genetic and molecular profiles of individual tumors, will also be crucial in developing more effective and tailored treatment strategies. Furthermore, a better understanding of the role of CSCs in drug resistance is needed. These cells have been shown to possess enhanced survival and self-renewal capabilities, making them particularly resistant to conventional therapies. Targeting the specific signaling pathways and epigenetic mechanisms that regulate the CSC phenotype may be a promising strategy for overcoming drug resistance. In conclusion, the continued investigation of the complex signaling networks that drive cancer drug resistance, as well as the development of novel therapeutic approaches that can effectively target these pathways, will be crucial in improving clinical outcomes for cancer patients.

Acknowledgements The figures were created using https://BioRender.com. Anjana Sajeev acknowledges the Council of Scientific and Industrial Research (CSIR), the Ministry of Education (MOE), Government of India for providing her with the fellowship. Ravichandran Vishwa acknowledges the Prime Minister's Research Fellowship (PMRF) program (PMRF ID: 1903324), Ministry of Education (MoE), Govt. of India for providing him the prestigious fellowship.The authors acknowledge the use of AI tools for improving the language and readability.

Declaration *Conflicts of Interest*: The authors declare that they have no known competing financial interests or personal relationships that could have appeared to influence the work reported in this paper.

References

1. Sajeev A, et al. Modulation of diverse oncogenic signaling pathways by oroxylin A: an important strategy for both cancer prevention and treatment. Phytomedicine. 2022;105:154369.
2. Girisa S, et al. Xanthohumol from Hop: Hope for cancer prevention and treatment. IUBMB Life. 2021;73(8):1016–44.

3. Bray F, et al. Global cancer statistics 2022: GLOBOCAN estimates of incidence and mortality worldwide for 36 cancers in 185 countries. CA Cancer J Clin. 2024;74(3):229–63.
4. Kunnumakkara AB, et al. Cancer cell Chemoresistance and Chemosensitization. World Scientific, Singapore; 2018.
5. Zhou HM, et al. Targeting cancer stem cells for reversing therapy resistance: mechanism, signaling, and prospective agents. Signal Transduct Target Ther. 2021;6(1):62.
6. Zahreddine H, Borden KL. Mechanisms and insights into drug resistance in cancer. Front Pharmacol. 2013;4:28.
7. Housman G, et al. Drug resistance in cancer: an overview. Cancers (Basel). 2014;6(3):1769–92.
8. Bou Antoun N, Chioni AM. Dysregulated signalling pathways driving anticancer drug resistance. Int J Mol Sci. 2023;24(15):12222.
9. Xia Y, et al. NF-kappaB, an active player in human cancers. Cancer Immunol Res. 2014;2(9):823–30.
10. Glaviano A, et al. PI3K/AKT/mTOR signaling transduction pathway and targeted therapies in cancer. Mol Cancer. 2023;22(1):138.
11. Menck K, et al. The WNT/ROR pathway in cancer: from signaling to therapeutic intervention. Cells. 2021;10(1):142.
12. Guo YJ, et al. ERK/MAPK signalling pathway and tumorigenesis. Exp Ther Med. 2020;19(3):1997–2007.
13. Sajeev A, et al. Unlocking the potential of Berberine: advancing cancer therapy through Chemosensitization and combination treatments. Cancer Lett. 2024;597:217019.
14. Manickasamy MK, et al. Synergistic enhancement: exploring the potential of piperine in cancer therapeutics through chemosensitization and combination therapies. Life Sci. 2024;354:122943.
15. Monisha J, et al. NF-κB blockers gifted by mother nature: Prospectives in cancer cell chemosensitization. Curr Pharm Des. 2016;22(27):4173–200.
16. Tian Z, et al. Insight into the prospects for RNAi therapy of cancer. Front Pharmacol. 2021;12:644718.
17. Liu GH, et al. Small molecule inhibitors targeting the cancers. Med Comm. 2020;3(4):e181.
18. Wilkens S. Structure and mechanism of ABC transporters. F1000Prime Rep. 2015;7:14.
19. Vaidya FU, et al. Molecular and cellular paradigms of multidrug resistance in cancer. Cancer Rep (Hoboken). 2022;5(12):e1291.
20. Wang X, et al. Extracellular ATP, as an energy and phosphorylating molecule, induces different types of drug resistances in cancer cells through ATP internalization and intracellular ATP level increase. Oncotarget. 2017;8(50):87860–77.
21. Dean M, et al. The human ATP-binding cassette (ABC) transporter superfamily. J Lipid Res. 2001;42(7):1007–17.
22. Choi YH, Yu AM. ABC transporters in multidrug resistance and pharmacokinetics, and strategies for drug development. Curr Pharm Des. 2014;20(5):793–807.
23. Juliano RL, Ling V. A surface glycoprotein modulating drug permeability in Chinese hamster ovary cell mutants. Biochim Biophys Acta. 1976;455(1):152–62.
24. Zhang M, et al. FRAX486, a PAK inhibitor, overcomes ABCB1-mediated multidrug resistance in breast cancer cells. Braz J Med Biol Res. 2024;57:e13357.
25. Ibba R, et al. Discovery of pyridoquinoxaline-based new P-gp inhibitors as coadjutant against Multi Drug Resistance in cancer. Eur J Med Chem. 2024;276:116647.
26. Manoharan JP, et al. Overcoming multi drug resistance mediated by ABC transporters by a novel acetogenin- annonacin from Annona muricata L. J Ethnopharmacol. 2024;322:117598.
27. Fu XJ, et al. Discovery of novel pyrazolo[1,5-a]pyrimidine derivatives as potent reversal agents against ABCB1-mediated multidrug resistance. Eur J Med Chem. 2024;277:116761.
28. da Silva Zanzarini I, et al. Magnolol derivatives as specific and noncytotoxic inhibitors of breast cancer resistance protein (BCRP/ABCG2). Bioorg Chem. 2024;146:107283.
29. Yahya SMM, et al. Restoring microRNA-34a overcomes acquired drug resistance and disease progression in human breast cancer cell lines via suppressing the ABCC1 gene. Breast Cancer Res Treat. 2024;204(1):133–49.

30. Bin Kanner Y, et al. Cytotoxicity and reversal effect of sertraline, fluoxetine, and citalopram on MRP1- and MRP7-mediated MDR. Front Pharmacol. 2023;14:1290255.
31. Shan Y, et al. The ATP-binding cassette transporter-mediated efflux transport of Ganciclovir at the blood-brain barrier. Eur J Drug Metab Pharmacokinet. 2024;49(5):609–17.
32. Asano T. Drug resistance in cancer therapy and the role of epigenetics. J Nippon Med Sch. 2020;87(5):244–51.
33. Cheng X, et al. A novel secondary ALK gene mutation which resistant to second-generation TKIs: a case report and literature review. Front Oncol. 2024;14:1430350.
34. Wang JQ, et al. ATP-binding cassette (ABC) transporters in cancer: a review of recent updates. J Evid Based Med. 2021;14(3):232–56.
35. Emran TB, et al. Multidrug resistance in cancer: understanding molecular mechanisms, immunoprevention and therapeutic approaches. Front Oncol. 2022;12:891652.
36. Wells A. EGF receptor. Int J Biochem Cell Biol. 1999;31(6):637–43.
37. Seshacharyulu P, et al. Targeting the EGFR signaling pathway in cancer therapy. Expert Opin Ther Targets. 2012;16(1):15–31.
38. Geng P, et al. HIF-1alpha-HPRT1 axis promotes tumorigenesis and gefitinib resistance by enhancing purine metabolism in EGFR-mutant lung adenocarcinoma. J Exp Clin Cancer Res. 2024;43(1):269.
39. Mirtavoos-Mahyari H, et al. Primary Erlotinib resistance in a patient with non-small cell lung cancer carrying simultaneous compound EGFR L718A, Q849H, and L858R mutations. Biomol Concepts. 2021;12(1):164–74.
40. Paez JG, et al. EGFR mutations in lung cancer: correlation with clinical response to gefitinib therapy. Science. 2004;304(5676):1497–500.
41. Costa DB, et al. Effects of erlotinib in EGFR mutated non-small cell lung cancers with resistance to gefitinib. Clin Cancer Res. 2008;14(21):7060–7.
42. Tang X, et al. A small-molecule compound D6 overcomes EGFR-T790M-mediated resistance in non-small cell lung cancer. Commun Biol. 2021;4(1):1391.
43. Zhang Z, et al. L3MBTL1, a polycomb protein, promotes Osimertinib acquired resistance through epigenetic regulation of DNA damage response in lung adenocarcinoma. Cell Death Dis. 2024;15(9):649.
44. Wang X, et al. EGFR upregulates miRNA subset to inhibit CYBRD1 and cause DDP resistance in gastric cancer. Gene. 2025;933:149005.
45. Yuan SF, et al. IL-1RA promotes oral squamous cell carcinoma malignancy through mitochondrial metabolism-mediated EGFR/JNK/SOX2 pathway. J Transl Med. 2023;21(1):473.
46. Zhang G, et al. Cytochalasin H enhances sensitivity to gefitinib in non-small-cell lung cancer cells through inhibiting EGFR activation and PD-L1 expression. Sci Rep. 2024;14(1):25276.
47. Kumar A, et al. Natural compounds targeting YAP/TAZ axis in cancer: current state of art and challenges. Pharmacol Res. 2024;203:107167.
48. Zheng Y, et al. The Hippo signaling pathway in development and disease. Dev Cell. 2019;50(3):264–82.
49. Cheng J, et al. The role and regulatory mechanism of Hippo signaling components in the neuronal system. Front Immunol. 2020;11:281.
50. Zeng R, Dong J. The Hippo signaling pathway in drug resistance in cancer. Cancers. 2021;13(2):318.
51. Ji D, et al. A screen of shRNAs targeting tumor suppressor genes to identify factors involved in A549 paclitaxel sensitivity. Oncol Rep. 2007;18(6):1499–505.
52. Mohamed Z, et al. miR-363 confers taxane resistance in ovarian cancer by targeting the Hippo pathway member, LATS2. Oncotarget. 2018;9(53):30053.
53. Bi L, et al. AJUBA increases the cisplatin resistance through hippo pathway in cervical cancer. Gene. 2018;644:148–54.
54. Li Z, et al. Loss of the FAT1 tumor suppressor promotes resistance to CDK4/6 inhibitors via the hippo pathway. Cancer Cell. 2018;34(6):893–905. e898
55. Hagenbeek TJ, et al. An allosteric pan-TEAD inhibitor blocks oncogenic YAP/TAZ signaling and overcomes KRAS G12C inhibitor resistance. Nat Cancer. 2023;4(6):812–28.

56. Song R, et al. Functional significance of Hippo/YAP signaling for drug resistance in colorectal cancer. Mol Carcinog. 2018;57(11):1608–15.
57. Zhang H, et al. TAZ-regulated expression of IL-8 is involved in chemoresistance of hepatocellular carcinoma cells. Arch Biochem Biophys. 2020;693:108571.
58. Huo X, et al. Overexpression of Yes-associated protein confers doxorubicin resistance in hepatocellullar carcinoma. Oncol Rep. 2013;29(2):840–6.
59. Xue C, et al. Evolving cognition of the JAK-STAT signaling pathway: autoimmune disorders and cancer. Signal Transduct Target Ther. 2023;8(1):204.
60. Fu X-Y, et al. ISGF3, the transcriptional activator induced by interferon alpha, consists of multiple interacting polypeptide chains. Proc Natl Acad Sci. 1990;87(21):8555–9.
61. Fu X-Y. A transcription factor with SH2 and SH3 domains is directly activated by an interferon α-induced cytoplasmic protein tyrosine kinase (s). Cell. 1992;70(2):323–35.
62. Fu XY. A direct signaling pathway through tyrosine kinase activation of SH2 domain-containing transcription factors. J Leukoc Biol. 1995;57(4):529–35.
63. Hu X, et al. The JAK/STAT signaling pathway: from bench to clinic. Signal Transduct Target Ther. 2021;6(1):402.
64. Majoros A, et al. Canonical and non-canonical aspects of JAK-STAT signaling: lessons from interferons for cytokine responses. Front Immunol. 2017;8:29.
65. Li WX. Canonical and non-canonical JAK–STAT signaling. Trends Cell Biol. 2008;18(11):545–51.
66. Hu Q, et al. JAK/STAT pathway: extracellular signals, diseases, immunity, and therapeutic regimens. Front Bioeng Biotechnol. 2023;11:1110765.
67. Mora LB, et al. Constitutive activation of Stat3 in human prostate tumors and cell lines: direct inhibition of Stat3 signaling induces apoptosis of prostate cancer cells. Cancer Res. 2002;62(22):6659–66.
68. Gao L, et al. Down-regulation of signal transducer and activator of transcription 3 expression using vector-based small interfering RNAs suppresses growth of human prostate tumor *in vivo*. Clin Cancer Res. 2005;11(17):6333–41.
69. Ma D, et al. Intracellular angiopoietin-1 promotes TKI-resistance via activation of JAK/STAT5 pathway in chronic myeloid leukemia. Oncogene. 2023;42(2):124–37.
70. Jin Z, et al. TRIM14 promotes colorectal cancer cell migration and invasion through the SPHK1/STAT3 pathway. Cancer Cell Int. 2018;18:202.
71. Chen J, et al. Aberrant JAK-STAT signaling-mediated chromatin remodeling impairs the sensitivity of NK/T-cell lymphoma to chidamide. Clin Epigenetics. 2023;15(1):19.
72. Haselager MV, et al. JAK–STAT signalling shapes the NF-κB response in CLL towards venetoclax sensitivity or resistance via Bcl-XL. Mol Oncol. 2023;17(6):1112–28.
73. Meyer LK, et al. JAK/STAT pathway inhibition sensitizes CD8 T cells to dexamethasone-induced apoptosis in hyperinflammation. Blood J Am Soc Hematol. 2020;136(6):657–68.
74. Zhang W, et al. MAPK signal pathways in the regulation of cell proliferation in mammalian cells. Cell Res. 2002;12(1):9–18.
75. Widmann C, et al. Mitogen-activated protein kinase: conservation of a three-kinase module from yeast to human. Physiol Rev. 1999;79(1):143–80.
76. Plotnikov A, et al. The MAPK cascades: signaling components, nuclear roles and mechanisms of nuclear translocation. Biochim Biophys Acta. 2011;1813(9):1619–33.
77. Dhillon AS, et al. MAP kinase signalling pathways in cancer. Oncogene. 2007;26(22):3279–90.
78. Fang H, et al. The VEGFA-induced MAPK-AKT/PTEN/TGFbeta signal pathway enhances progression and MDR in gastric cancer. Genes (Basel). 2024;15(10):1266.
79. Di Y, et al. MAPK signaling-mediated RFNG phosphorylation and nuclear translocation restrain Oxaliplatin-induced apoptosis and ferroptosis. Adv Sci (Weinh). 2024;11(38):e2402795.
80. Yan M, et al. ROCK2 increases drug resistance in acute myeloid leukemia via metabolic reprogramming and MAPK/PI3K/AKT signaling. Int Immunopharmacol. 2024;140:112897.
81. Zhang G, et al. TIMP-2 regulates 5-Fu resistance via the ERK/MAPK signaling pathway in colorectal cancer. Aging (Albany NY). 2022;14(1):297–315.

82. Li F, et al. Regulation of TORC1 by MAPK signaling determines sensitivity and acquired resistance to Trametinib in pediatric BRAFV600E brain tumor models. Clin Cancer Res. 2022;28(17):3836–49.
83. Fontana F, et al. delta-Tocotrienol sensitizes and re-sensitizes ovarian cancer cells to cisplatin via induction of G1 phase cell cycle arrest and ROS/MAPK-mediated apoptosis. Cell Prolif. 2021;54(11):e13111.
84. Tang X, et al. Tanshinone I improves TNBC chemosensitivity by suppressing late-phase autophagy through AKT/p38 MAPK signaling pathway. Biomed Pharmacother. 2024;177:117037.
85. Li Y, et al. Inhibition of NF-kappaB signaling unveils novel strategies to overcome drug resistance in cancers. Drug Resist Updat. 2024;73:101042.
86. Pai P, et al. HOX genes and the NF-κB pathway: a convergence of developmental biology, inflammation and cancer biology. Biochim Biophys Acta (BBA)-Rev Cancer. 2020;1874(2):188450.
87. Taniguchi K, Karin M. NF-κB, inflammation, immunity and cancer: coming of age. Nat Rev Immunol. 2018;18(5):309–24.
88. Yan M, et al. IL-33-dependent NF-kappaB activation inhibits apoptosis and drives chemoresistance in acute myeloid leukemia. Cytokine. 2024;180:156672.
89. Zhu R, et al. HOXA1 silencing inhibits cisplatin resistance of oral squamous cell carcinoma cells via IkappaB/NF-kappaB signaling pathway. Anti-Cancer Drugs. 2024;35(6):492–500.
90. Abdin SM, et al. Nuclear factor-kappaB signaling inhibitors revert multidrug-resistance in breast cancer cells. Chem Biol Interact. 2021;340:109450.
91. Nathan S, et al. BRCA1-mimetic compound NSC35446.HCl inhibits IKKB expression by reducing estrogen receptor-alpha occupancy in the IKKB promoter and inhibits NF-kappaB activity in antiestrogen-resistant human breast cancer cells. Breast Cancer Res Treat. 2017;166(3):681–93.
92. Bharathiraja P, et al. Solasodine targets NF-kappaB signaling to overcome P-glycoprotein mediated multidrug resistance in cancer. Exp Cell Res. 2024;441(1):114153.
93. Denda Y, et al. The natural product Parthenolide inhibits both angiogenesis and invasiveness and improves gemcitabine resistance by suppressing nuclear factor kappaB activation in pancreatic cancer cell lines. Nutrients. 2024;16(5):705.
94. Wang C, et al. Abnormal activation of NF-kappaB and MAPK signaling pathways affect osimertinib resistance and influence the recruitment of myeloid-derived suppressor cells to shape the immunosuppressive tumor immune microenvironment. Thorac Cancer. 2023;14(19):1843–56.
95. Miele L. Notch signaling. Clin Cancer Res. 2006;12(4):1074–9.
96. Miele L, Osborne B. Arbiter of differentiation and death: Notch signaling meets apoptosis. J Cell Physiol. 1999;181(3):393–409.
97. Fortini ME. Notch signaling: the core pathway and its posttranslational regulation. Dev Cell. 2009;16(5):633–47.
98. Wang Z, et al. Targeting Notch signaling pathway to overcome drug resistance for cancer therapy. Biochim Biophys Acta. 2010;1806(2):258–67.
99. Rizzo P, et al. Rational targeting of Notch signaling in cancer. Oncogene. 2008;27(38):5124–31.
100. Wang Z, et al. Down-regulation of Notch-1 inhibits invasion by inactivation of nuclear factor-κB, vascular endothelial growth factor, and matrix metalloproteinase-9 in pancreatic cancer cells. Cancer Res. 2006;66(5):2778–84.
101. Wang Z, et al. Down-regulation of Notch-1 contributes to cell growth inhibition and apoptosis in pancreatic cancer cells. Mol Cancer Ther. 2006;5(3):483–93.
102. Wang Z, et al. Exploitation of the Notch signaling pathway as a novel target for cancer therapy. Anticancer Res. 2008;28(6A):3621–30.
103. Wang Z, et al. Emerging role of Notch in stem cells and cancer. Cancer Lett. 2009;279(1):8–12.
104. Zhang ZP, et al. Correlation of Notch1 expression and activation to cisplatin-sensitivity of head and neck squamous cell carcinoma. Ai Zheng. 2009;28(2):100–3.

105. George DC, et al. Notch-3 affects chemoresistance in colorectal cancer via DNA base excision repair enzymes. Adv Biol Regul. 2024;91:101013.
106. Yao J, Qian C. Inhibition of Notch3 enhances sensitivity to gemcitabine in pancreatic cancer through an inactivation of PI3K/Akt-dependent pathway. Med Oncol. 2010;27(3):1017–22.
107. Patrawala L, et al. Side population is enriched in tumorigenic, stem-like cancer cells, whereas ABCG2+ and ABCG2− cancer cells are similarly tumorigenic. Cancer Res. 2005;65(14):6207–19.
108. Du Z, et al. SOX8 knockdown overcomes Enzalutamide resistance in castration-resistant prostate cancer by inhibiting the Notch signaling pathway. Biomed Res Int. 2022;2022:9235837.
109. Sun L, et al. Oxyberberine sensitizes liver cancer cells to sorafenib via inhibiting NOTCH1-USP7-c-Myc pathway. Hepatol Commun. 2024;8(4):e0405.
110. Czerwonka A, et al. The Notch inhibitor, FLI-06, increases the chemosensitivity of head and neck Squamous cell carcinoma cells to taxanes-based treatment. Biomed Pharmacother. 2024;177:116822.
111. Datta SR, et al. Cellular survival: a play in three Akts. Genes Dev. 1999;13(22):2905–27.
112. Porta C, et al. Targeting PI3K/Akt/mTOR signaling in cancer. Front Oncol. 2014;4:64.
113. Tewari D, et al. Natural products targeting the PI3K-Akt-mTOR signaling pathway in cancer: a novel therapeutic strategy. Semin Cancer Biol. 2022;80:1–17.
114. Huang X, et al. The PI3K/AKT pathway in obesity and type 2 diabetes. Int J Biol Sci. 2018;14(11):1483–96.
115. Marafie SK, et al. mTOR: its critical role in metabolic diseases, cancer, and the aging process. Int J Mol Sci. 2024;25(11):6141.
116. Roy T, et al. The PI3K-Akt-mTOR and associated signaling pathways as molecular drivers of immune-mediated inflammatory skin diseases: update on therapeutic strategy using natural and synthetic compounds. Cells. 2023;12(12):1671.
117. Wegiel B, et al. Interleukin-6 activates PI3K/Akt pathway and regulates cyclin A1 to promote prostate cancer cell survival. Int J Cancer. 2008;122(7):1521–9.
118. Khan KH, et al. Targeting the PI3K-AKT-mTOR signaling network in cancer. Chin J Cancer. 2013;32(5):253–65.
119. Knuefermann C, et al. HER2/PI-3K/Akt activation leads to a multidrug resistance in human breast adenocarcinoma cells. Oncogene. 2003;22(21):3205–12.
120. Chen W, et al. HSPA12B overexpression induces cisplatin resistance in non-small-cell lung cancer by regulating the PI3K/Akt/NF-kappaB signaling pathway. Oncol Lett. 2018;15(3):3883–9.
121. Tazzari PL, et al. Multidrug resistance-associated protein 1 expression is under the control of the phosphoinositide 3 kinase/Akt signal transduction network in human acute myelogenous leukemia blasts. Leukemia. 2007;21(3):427–38.
122. Long C, et al. Capivasertib augments chemotherapy via Akt inhibition in preclinical small cell lung cancer models. Fundam Clin Pharmacol. 2024;39:e13042.
123. Ingle K, et al. AUM302, a novel triple kinase PIM/PI3K/mTOR inhibitor, is a potent *in vitro* pancreatic cancer growth inhibitor. PLoS One. 2023;18(11):e0294065.
124. Sui H, et al. Wan reverses P-gp-mediated drug-resistance by inhibiting activation of the PI3K/Akt/NF-kappaB pathway. BMC Complement Altern Med. 2014;14:279.
125. Zhao J, et al. Babao Dan reverses multiple-drug resistance in gastric cancer cells via triggering apoptosis and autophagy and inhibiting PI3K/AKT/mTOR signaling. Evid Based Complement Alternat Med. 2021;2021:5631942.
126. Huang YH, et al. Attenuation of PI3K-Akt-mTOR pathway to reduce cancer Stemness on Chemoresistant lung cancer cells by Shikonin and synergy with BEZ235 inhibitor. Int J Mol Sci. 2024;25(1):616.
127. Nusse R, Varmus H. Three decades of Wnts: a personal perspective on how a scientific field developed. EMBO J. 2012;31(12):2670–84.
128. Nusse R, Varmus HE. Many tumors induced by the mouse mammary tumor virus contain a provirus integrated in the same region of the host genome. Cell. 1982;31(1):99–109.

129. Nusse R, et al. A new nomenclature for int-1 and related genes: the Wnt gene family. Cell. 1991;64(2):231.
130. Liu J, et al. Wnt/beta-catenin signalling: function, biological mechanisms, and therapeutic opportunities. Signal Transduct Target Ther. 2022;7(1):3.
131. Niehrs C. The complex world of WNT receptor signalling. Nat Rev Mol Cell Biol. 2012;13(12):767–79.
132. De A. Wnt/Ca 2+ signaling pathway: a brief overview. Acta Biochim Biophys Sin. 2011;43(10):745–56.
133. Simons M, Mlodzik M. Planar cell polarity signaling: from fly development to human disease. Annu Rev Genet. 2008;42:517–40.
134. Wang J, Sinha T, Wynshaw-Boris A. Wnt signaling in mammalian development: lessons from mouse genetics. Cold Spring Harb Perspect Biol. 2012;4(5):a007963.
135. Wray J, Hartmann C. WNTing embryonic stem cells. Trends Cell Biol. 2012;22(3):159–68.
136. Perugorria MJ, et al. Wnt-beta-catenin signalling in liver development, health and disease. Nat Rev Gastroenterol Hepatol. 2019;16(2):121–36.
137. Skronska-Wasek W, et al. Reduced Frizzled Receptor 4 expression prevents WNT/beta-Catenin-driven Alveolar lung repair in chronic obstructive pulmonary disease. Am J Respir Crit Care Med. 2017;196(2):172–85.
138. Nie X, et al. The complex role of Wnt ligands in type 2 diabetes mellitus and related complications. J Cell Mol Med. 2021;25(14):6479–95.
139. Weerackoon N, et al. Wnt signaling cascades and their role in coronary artery health and disease. J Cell Sig. 2021;2(1):52–62.
140. Zhao H, et al. Wnt signaling in colorectal cancer: pathogenic role and therapeutic target. Mol Cancer. 2022;21(1):144.
141. Chen Y, et al. The involvement of noncanonical Wnt signaling in cancers. Biomed Pharmacother. 2021;133:110946.
142. Pohl SG, et al. Wnt signaling in triple-negative breast cancer. Oncogenesis. 2017;6(4):e310.
143. Sajeev A, et al. Crosstalk between non-coding RNAs and Wnt/beta-Catenin signaling in head and neck cancer: identification of novel biomarkers and therapeutic agents. Noncod RNA. 2023;9(5):63.
144. Yu F, et al. Wnt/beta-catenin signaling in cancers and targeted therapies. Signal Transduct Target Ther. 2021;6(1):307.
145. Zhan T, et al. Wnt signaling in cancer. Oncogene. 2017;36(11):1461–73.
146. Zhu Y, Li X. Advances of Wnt signalling pathway in colorectal cancer. Cells. 2023;12(3):447.
147. Clark-Corrigall J, et al. Elevated expression of LGR5 and WNT signaling factors in neuroblastoma cells with acquired drug resistance. Cancer Investig. 2023;41(2):173–82.
148. Kim Y, et al. Wnt/beta-catenin pathway is a key signaling pathway to trastuzumab resistance in gastric cancer cells. BMC Cancer. 2023;23(1):922.
149. Zhang Z, et al. Wnt/beta-catenin signaling in the development and therapeutic resistance of non-small cell lung cancer. J Transl Med. 2024;22(1):565.
150. Xu J, et al. beta-Catenin is required for the tumorigenic behavior of triple-negative breast cancer cells. PLoS One. 2015;10(2):e0117097.
151. Fodde R, Brabletz T. Wnt/beta-catenin signaling in cancer stemness and malignant behavior. Curr Opin Cell Biol. 2007;19(2):150–8.
152. Medema JP. Cancer stem cells: the challenges ahead. Nat Cell Biol. 2013;15(4):338–44.
153. Liu YS, et al. Lgr5 promotes cancer stemness and confers chemoresistance through ABCB1 in colorectal cancer. Biomed Pharmacother. 2013;67(8):791–9.
154. Wang J, et al. N6-Methyladenosine-mediated up-regulation of FZD10 regulates liver cancer stem cells' properties and Lenvatinib resistance through WNT/beta-Catenin and Hippo signaling pathways. Gastroenterology. 2023;164(6):990–1005.
155. Dong S, et al. ROS/PI3K/Akt and Wnt/beta-catenin signalings activate HIF-1alpha-induced metabolic reprogramming to impart 5-fluorouracil resistance in colorectal cancer. J Exp Clin Cancer Res. 2022;41(1):15.

156. Wang Y, et al. Wnt/beta-catenin signaling confers ferroptosis resistance by targeting GPX4 in gastric cancer. Cell Death Differ. 2022;29(11):2190–202.
157. Li Z, et al. RNA helicase DDX5 modulates sorafenib sensitivity in hepatocellular carcinoma via the Wnt/beta-catenin-ferroptosis axis. Cell Death Dis. 2023;14(11):786.
158. Muche S, et al. Synergistic effects of beta-catenin inhibitors and sorafenib in hepatoma cells. Anticancer Res. 2014;34(9):4677–83.
159. Lin HH, et al. Inhibition of the Wnt/beta-catenin signaling pathway improves the anti-tumor effects of sorafenib against hepatocellular carcinoma. Cancer Lett. 2016;381(1):58–66.
160. Li K, et al. Enhancement of TKI sensitivity in lung adenocarcinoma through m6A-dependent translational repression of Wnt signaling by circ-FBXW7. Mol Cancer. 2023;22(1):103.
161. Marr AR, et al. The multi-CDK inhibitor dinaciclib reverses bromo- and extra-terminal domain (BET) inhibitor resistance in acute myeloid leukemia via inhibition of Wnt/beta-catenin signaling. Exp Hematol Oncol. 2024;13(1):27.
162. Jin Z, El-Deiry WS. Overview of cell death signaling pathways. Cancer Biol Ther. 2005;4(2):139–63.
163. Dogan E, et al. Targeting apoptosis to overcome chemotherapy resistance. In: Sergi CM, editor. Metastasis. Exon Publications, Brisbane; 2022.
164. Cao X, et al. Cadmium induced BEAS-2B cells apoptosis and mitochondria damage via MAPK signaling pathway. Chemosphere. 2021;263:128346.
165. Cao X, et al. Radiation induces apoptosis primarily through the intrinsic pathway in mammalian cells. Cell Signal. 2019;62:109337.
166. Warren CFA, et al. BCL-2 family isoforms in apoptosis and cancer. Cell Death Dis. 2019;10(3):177.
167. Elmore S. Apoptosis: a review of programmed cell death. Toxicol Pathol. 2007;35(4):495–516.
168. Humphreys LM, et al. A revised model of TRAIL-R2 DISC assembly explains how FLIP(L) can inhibit or promote apoptosis. EMBO Rep. 2020;21(3):e49254.
169. Devi GR, et al. Expression of X-Linked Inhibitor of Apoptosis Protein (XIAP) in breast cancer is associated with shorter survival and resistance to chemotherapy. Cancers (Basel). 2021;13(11):2807.
170. Li Y, et al. Cullin7 induces docetaxel resistance by regulating the protein level of the anti-apoptotic protein Survivin in lung adenocarcinoma cells. J Thorac Dis. 2023;15(9):5006–19.
171. Guo H, et al. PAR-4 overcomes chemo-resistance in breast cancer cells by antagonizing cIAP1. Sci Rep. 2019;9(1):8755.
172. Thibault B, et al. DEBIO 1143, an IAP inhibitor, reverses carboplatin resistance in ovarian cancer cells and triggers apoptotic or necroptotic cell death. Sci Rep. 2018;8(1):17862.
173. Matour E, et al. MiR-34c-5p inhibition affects Bax/Bcl2 expression and reverses Bortezomib resistance in multiple myeloma cells. Indian J Hematol Blood Transfus. 2024;40(4):596–603.
174. Liu Z, Gao W. Synergistic effects of Bcl-2 inhibitors with AZD9291 on overcoming the acquired resistance of AZD9291 in H1975 cells. Arch Toxicol. 2020;94(9):3125–36.
175. Smalley KS, et al. Multiple signaling pathways must be targeted to overcome drug resistance in cell lines derived from melanoma metastases. Mol Cancer Ther. 2006;5(5):1136–44.
176. Wen W, et al. Synergistic anti-tumor activity by targeting multiple signaling pathways in ovarian cancer. Cancers (Basel). 2020;12(9):2586.
177. Liang Y, et al. Emergence of enzalutamide resistance in prostate cancer is associated with BCL-2 and IKKB dependencies. Clin Cancer Res. 2021;27(8):2340–51.

Targeting Angiogenesis for Cancer Prevention and Therapy

Aleena Sinha, Praveen K. Kujur, and Rana P. Singh

1 Introduction

The development and sustenance of cellular tissues in the body necessitate an extensive vascular network that transports blood containing essential nutrients, facilitates gas exchange, removes metabolic waste products, and transfers critical signaling molecules, which are crucial for regulating cellular processes, tissue homeostasis, and physiological responses [1]. During embryonic development, vasculogenesis, which is the de novo formation of blood vessels from angioblasts, differentiates into endothelial cells that subsequently assemble into a primitive vascular network. Upon establishment of this initial plexus, angiogenesis, characterized by the sprouting of new vessels from preexisting ones, takes over to expand the vascular system [2]. While vasculogenesis is essential during early embryonic development, angiogenesis persists throughout life, playing a critical role in processes such as tissue growth, repair, and especially tumorigenesis [3].

Tumor growth remains restricted, typically not exceeding 2–3 mm in diameter during the avascular phase, and tumors rely on passive diffusion to acquire oxygen and essential nutrients from adjacent tissues. However, as tumor mass increases, diffusion alone becomes insufficient to sustain the escalating metabolic demands,

A. Sinha · P. K. Kujur
Cancer Biology Laboratory, School of Life Sciences, Jawaharlal Nehru University, New Delhi, India

R. P. Singh (✉)
Cancer Biology Laboratory, School of Life Sciences, Jawaharlal Nehru University, New Delhi, India

Department of Pharmaceutical Sciences, Skaggs School of Pharmacy and Pharmaceutical Sciences, University of Colorado, Anschutz Medical Campus, Aurora, CO, USA

Gautam Buddha University, Greater Noida, Uttar Pradesh, India
e-mail: rana_singh@mail.jnu.ac.in

A. S. Ramteke et al. (eds.), *Molecular Targets in Cancer Therapy*,
https://doi.org/10.1007/978-981-96-7336-0_6

leading to the development of a hypoxic, ischemic, and acidic microenvironment. This hypoxic microenvironment induces the secretion of pro-angiogenic factors like VEGF while simultaneously downregulating angiogenesis inhibitors, which initiate the "angiogenic switch," a process first described by Judah Folkman in the 1970s as a critical driver of tumor growth and metastasis [4, 5]. This shift in the balance of pro- and antiangiogenic factors activates endothelial cells and promotes neovascularization, allowing the tumor to access the circulatory system and sustain further proliferation. The angiogenic switch, essential for sustained tumor growth and metastasis, is not a single event but a dynamic process influenced by various factors within the tumor microenvironment, including immune cells, stromal cells, and the extracellular matrix [6, 7]. Here, we highlight the potential of angiogenesis as a preventative as well as therapeutic target in the control and management of tumor growth.

2 Traditional Medicine and Angiogenesis

Antiangiogenic therapy aims to inhibit the formation of new blood vessels within tumors, thereby depriving them of the resources needed for growth and spread. The concept of targeting angiogenesis for cancer treatment emerged in the mid-twentieth century with pioneering work by Judah Folkman. Since then, significant progress has been made in understanding the molecular mechanisms driving tumor angiogenesis, leading to the development of various antiangiogenic agents. Traditional medicine has been utilized for centuries in the treatment of various ailments, and in recent years, there has been increasing interest in its potential to modulate angiogenesis, the process of new blood vessel formation from preexisting vasculature. Several traditional medicinal systems, Chinese, Korean, Native American, African, Unani, and Ayurveda, have been investigated for their ability to either promote or inhibit angiogenesis, depending on the therapeutic context. Here, phytochemicals with antiangiogenic potentials against cancer are discussed (Fig. 1 and Table 1).

3 Phytochemicals Having Antiangiogenic Activity Against Cancer

3.1 Berberine

Berberine is a naturally occurring compound in goldenseal, Oregon grape, tree turmeric, and other plants. It has been shown to possess anticancer, anti-inflammatory, and antiangiogenesis properties against gastric cancer, colorectal cancer, colon cancer, liver cancer, breast cancer, and others. Berberine has an antiangiogenic effect on gastric cancer, triggering HIF-1α degradation via a proteasomal proteolytic pathway in the SC-M1 cell line [8]. It showed antiangiogenic effect in an orthotopic model of human hepatocellular carcinoma MHCC-97L via downregulation of HIF-1α and reduced production of VEGF, resulting in tumor growth inhibition [9].

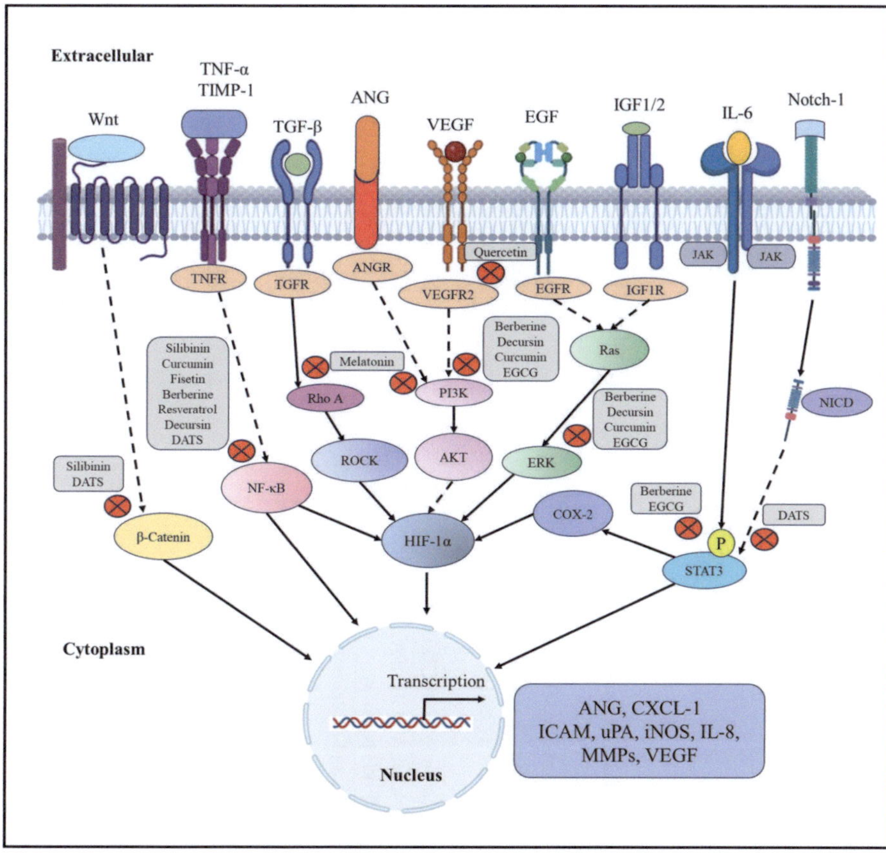

Fig. 1 Angiogenesis signaling pathways targeted by phytochemicals

Berberine caused downregulation of MMP-9 mRNA expression and suppressed tumor development in azoxymethane (AOM) and dextran sulfate sodium (DSS)-induced colorectal cancer in mice [10]. It inhibited VEGFR2/ERK pathway in its antiangiogenic effect on glioblastoma xenografts [11]. Berberine also showed in vivo antiangiogenic activity in B16F-10 melanoma in C57BL/6 αmice by inhibiting HIF, VEGF, COX-2, NO, and NF-κB. In a recent study, berberine reduced the mRNA and protein expression of HIF-1α in MDA-MB-231, MCF-7, and 4T1 cells and inhibited the growth of breast cancer [12].

3.2 Resveratrol

Resveratrol is a natural polyphenolic flavonoid found in red wine, mulberries, peanuts, and rhubarb. It possesses anticarcinogenic, antiviral, neuroprotective, anti-inflammatory, and antioxidant properties. Resveratrol inhibits angiogenesis in

Table 1 Phytochemicals targeting tumor angiogenesis in various cancers

S. no.	Phytochemicals	Cancer type	Molecular mechanism of action	In vitro study in cells	In vivo study in animal model	References
1	Curcumin 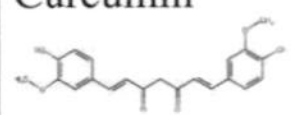	Breast cancer	Inhibits mRNA expression of VEGF and bFGF	MDA-MB-231	–	[52]
			Suppresses VEGF expression via NF-κB inhibition	MDA-MB-231	Xenograft	[69]
			Suppresses CXCL1, 2 via NF-κB inhibition	MDA-MB-231	–	[57]
			Deregulates cyclin D1, PECAM-1, and p65 via NF-κB pathway	MDA-MB-231	Xenograft	[70]
			Suppresses VEGFR2/3 expression	MDA-MB-231	Xenograft	[71]
			Reduces secretion of VEGF	T47-D	–	[72]
		Prostate cancer	Suppresses MMP-2 and MMP-9 expression	DU145	Xenograft	[55]
			Decreases CXCL1 and -2 via NF-κB inhibition.	PC-3	Xenograft	[58]
			Reduces MMP-9 activity and downregulates cellular matriptase	PC-3	Xenograft	[56]
			Decreases expression level of CD31	LNCaP	Xenograft	[73]
			Inhibits the activation, VEGF, uPA, MMP-2, MMP-9 via of NF-κB pathway	–	LNCaP Xenografts	[74]
			Suppresses mRNA and proteins expression of VEGF	PC-3	–	[75]
		Colon cancer	Inhibits COX-2 expression and PGE-2 production	HIMECs	–	[76]
			Inhibits NF-κB, uPA and MMP-9 via AMPK activation	SW480	–	[60]

		Liver cancer	Upregulates Bax and p53 and downregulates Bcl-2	HepG2 Huh 7 THLE-2	–	[77]
			Reduces tumor-induced over-expression of COX-2 and serum VEGF	–	HepG2 Xenograft	[78]
			Down-regulates transcriptional activity of HIF-1α and VEGF expression	HepG2	–	[79]
			Downregulates VEGF, CD31, and α-smooth muscle actin expression levels	Huh-7 MHCC-97H HepG2	HepG2 Xenograft	[80]
			Decreases VEGF, IL-6, and SDF-1 expression via inhibiting HIF-1α	HepG2	–	[81]
			Reduces mRNA expression of MMP-9, -2 and VEGF-A	–	Dalton's lymphoma bearing mice	[82]
			Reduces MMP-9 production through downregulation of STAT3 and AKT signaling pathways	HCC Cell	–	[83]
		Fibrosarcoma	Inhibits VEGF production and modulate VEGFR2 signaling pathway	HUVEC	T241-VEGF tumor-bearing mice	[84]

(continued)

Table 1 (continued)

S. no.	Phytochemicals	Cancer type	Molecular mechanism of action	In vitro study in cells	In vivo study in animal model	References
		Lung cancer	Inhibits MMP-2 and -9 via MEKK3, p-ERK signaling pathways	A549	–	[85]
			Reduces both transgen-hVEGF and endogenous VEGF overexpression	–	Transgenic mice carrying A_{165} (*hVEGF-A_{165}*) gene	[86]
			Suppresses the protein expression of MMP-2, MMP-7, VEGF through JAK-STAT3 signaling pathway	NCI-H446	–	[87]
			Decreases mRNA expression level of CD31 and CD105, biomarkers of endothelium	NCI-H460	NCI-H460 Ectopic xenograft	[88]
			Decreases VEGF expression via PI3K/Akt/mTOR signaling pathways	HUVEC	A549 xenograft	[89]
			Downregulates mRNA expression of Notch and HIF-1 and suppression of VEGF and NF-κB expression	A549	Xenograft	[90]
		Pancreatic cancer	Downregulates NF-κB and NF-κB-regulated genes	MIA PaCa-2	Xenograft	[53]
		Chronic myelogenous leukemia	Reduces HIF-1α activity via the miR-22/IPO7 axis	K562	–	[61]

2	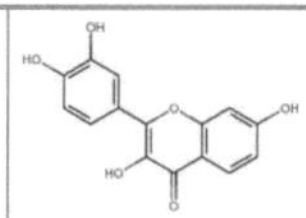 Fisetin	Breast cancer	Downregulates MMP-9 expression via PKC/ ROS/ MAPK pathways	MCF-7	–	[91]
			Decreases MMP-2 and MMP-9 production mediated by heme oxygenase-1	4T1 JC	–	[49]
		Lung cancer	Downregulates NF-κB-regulated genes, such as MMP-9, ICAM-1, and VEGF	H1299	–	[92]
			Downregulates the expressions of MMP-2 and u-PA at both the protein and mRNA levels	A549	–	[93]
			Decreases the expression of eNOS, VEGF, iNOS, MMP-2 and-9	A549	–	[50]
		Prostate cancer	Decreases the expression of eNOS, VEGF, iNOS, MMP-2 and-9	DU145	–	[50]
		Cervical cancer	Inhibits mRNA and protein level of uPA expression	SiHa CaSki	–	[51]
		Pancreatic cancer	Inhibits NF-κB mediated protein expression and activity of MMP-9	AsPC-1	–	[94]
		Human osteosarcoma	Reduces NF-κB and uPA protein levels as well as MMP-2 and MMP-9 protein expression	U-2 OS	–	[46]
		Diabetic retinopathy	Downregulates VEGF expression	Human retinal microvascular endothelial cells (HRMECs)	–	[48]
		Fibrosarcoma	Inhibits VEGF, MMP-1, MMP-3, MMP-7, and MMP-9	HT-1080 HUVECs	–	[95]

(continued)

Table 1 (continued)

S. no.	Phytochemicals	Cancer type	Molecular mechanism of action	In vitro study in cells	In vivo study in animal model	References
3	Decursin 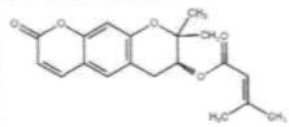	Preclinical angiogenesis model	Suppresses MMP-2 and -9 activities	HUVEC ex vivo rat aortic ring	–	[41]
		Colon cancer	MMP-2 and MMP-9 expressions were downregulated by JNK/PI3K and ERK/JNK signaling pathway	CT-26	–	[43]
		Breast cancer	Inhibits MMP-9 expression through suppression of PI3K, ERK and NF-κB activation	MDA-MB-231	–	[96]
		Fibrosarcoma	Inhibits TNFα-induced expression of pro-inflammatory mediators, such as MMP-9 and IL-8	HT1080	–	[97]
4	Silibinin 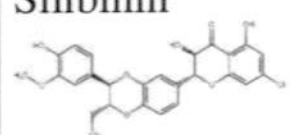	Colorectal cancer	Downregulates the expression of HIF-1α, VEGF, COX-2, Ang-2, and Ang-4 as well as the expression of MMP-2, MMP-9, CCR-2, and CXCR-4	CT26	–	[63]
			Downregulates NOS, COX, HIF-1α, and VEGF expression	–	HT29 xenograft	[64]
			Decreases VEGF and iNOS level through suppressing β-catenin signaling pathway	LoVo	SW480 cells xenograft	[65]
		Prostate cancer	Decreases HIF-1α expression with strong reduction in hypoxia-induced NADPH oxidase (NOX) activity	LNCaP	22Rv1	[67]
			Decreases VEGF, and VEGFR2 expression along with MMP-2 and MMP-9 protein expression	–	Transgenic adenocarcinoma of the mouse prostate (TRAMP) model	[98]

		Breast cancer	Decreases TGF-β2-induced MMP-2, and MMP-9 expression levels	Triple negative breast cancer (TNBC)	–	[68]
		Lung cancer	Decreases expression of VEGF and of iNOS and COX-2	–	Urethane-induced lung tumors	[99]
			Inhibits HIF-1α, NF κB, and STAT-3 activation, and upregulation of the angiopoietin-2 (Ang-2) and Ang-receptor tyrosine kinase (Tie-2)	–	Urethane-induced lung tumors	[100]
			Decreases VEGFR2 level, and STAT3 and NF-κB activation through iNOS regulation	–	Urethane-induced lung tumors	[101]
		Skin cancer	Decreases HIF-1α and VEGF protein level via modulation of NF-κB transcription factor	–	Female SKH-1 hairless mice	[102]
5	Berberine 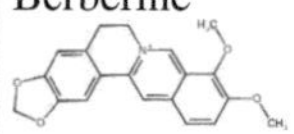	Gastric cancer	Triggers HIF-1α degradation via proteasomal proteolytic pathway	SC-M1	–	[8]
		Glioblastoma	Declines mRNA expression level of CD31 and; inhibits the phosphorylation of VEGFR2 and ERK	HUVEC	-U87 ectopic xenograft -U87orthotopic xenograft	[11]
		Tongue cancer	Downregulates u-PA, MMP-2 and -9 expressions through the FAK, IKK, and NF-κB mediated pathways	SCC-4	–	[103]
		Breast cancer	Decreases VEGF-induced fibronectin expression through the inhibition of the PI3K/AKT pathway	MCF7 T47D	–	[104]
			Reduces mRNA and protein expression of HIF-1α	MDA-MB-231 MCF-7 4T1	4T1/Luc mice	[12]
			Decreases MMP-1 and MMP-9 mRNA and protein expressions through suppressing PKC-α phosphorylation	MCF-7	–	[105]

(continued)

Table 1 (continued)

S. no.	Phytochemicals	Cancer type	Molecular mechanism of action	In vitro study in cells	In vivo study in animal model	References
		Liver cancer	Downregulates HIF-1α and reduced production of VEGF	MHCC-97L SK-Hep1	Orthotopic model of HCC (MHCC-97L	[9]
			Downregulates the expression COX-2, NF-κB, uPA, and MMP-9	Bel-7402 SMMC-7721	–	[106]
		Skin cancer	Inhibits the expression of HIF, VEGF, COX-2, NO, and NF-κB	Rat aortic ring	B16F-10 melanoma cells in C57BL/6 mice	[107]
		Colorectal cancer	Decreases the expression of vascular endothelial growth factor (VEGF)	SW480	–	[108]
			Downregulates the expression levels of COX-2 via JAK2/STAT3 signaling pathway	SW620 LoVo	Xenograft	[109]
			Downregulates MMP-9 mRNA expression	–	AOM-DSS CRC model	[10]
6	Resveratrol 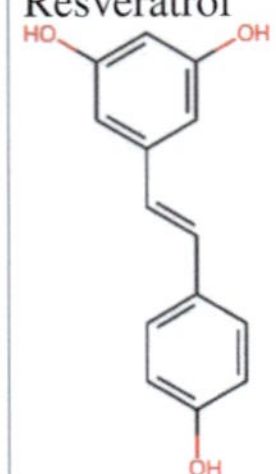	Colon cancer	Reduces hemoglobin percentage	–	COLO205-luc Xenograft	[110]
		Colorectal cancer	Reduces VEGF level	Caco2 HCT116	–	[14]
		Oral cancer	Decreases the expression of CD133 and VEGF-A	H-357 SCC-9	Xenograft	[17]
		Breast cancer	Reduces extracellular levels of VEGF	MDA-MB-231	Xenograft	[13]
		Brain cancer	Suppresses the expression of VEGF mRNA	RT-2	Xenograft	[111]
		Lung cancer	Inhibits Heme oxygenase-1 (HO-1) mediated MMP-9 and MMP-2 expression	A549	–	[15]
		Liver cancer	Suppresses HSC-induced IL-6 and VEGF expression and ROS production	HepG2	–	[16]
			Inhibits VEGF protein and mRNA expression through NF-κB	HepG2		[112]

7	Diallyl trisulfide (DATS) 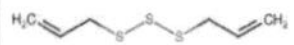	Brain cancer	Decreases the expression and activities of MMP-2, MMP-9, and VEGF by suppressing Wnt/β-catenin signaling	C6 U87	–	[113]
			Inhibits the expression of VEGF	–	U87MG orthotopic xenograft	[21]
		Colon cancer	Inhibits protein levels of MMP-2, MMP-7 and -9, and VEGF secretion	HT-29	Murine xenograft	[20]
		Prostate cancer	Inhibits the expression of VEGF, IL-6, HIF-1α, MMP-2, MMP-7 and MMP-9, through NF-κB	–	Orthotopic model	[18]
		Bone cancer	Decreases the mRNA and protein expression of VEGF, MMP-2, and MMP-9 via inactivation of Notch-1 activity	U2OS SaOS-2 MG-63	–	[19]
		Breast cancer	Inhibits mRNA and protein expressions of HIF-1α target genes MMP-2 and VEGF	MDA-MB-231	–	[114]
8	Melatonin 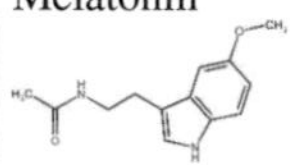	Gastric cancer	Reduces mRNA and protein levels of HIF-1α and VEGF	SGC-7901	Xenograft	[28]
			Inhibits VEGF and angiogenin secretion	SGC-7901	Chick embryo tumor xenograft	[115]
		Prostate cancer	Attenuates the mRNA level of HIF-1α, HIF-2α, and VEGF and also suppressed VEGF at the protein level and VEGF production	PC-3	–	[116]

(continued)

Table 1 (continued)

S. no.	Phytochemicals	Cancer type	Molecular mechanism of action	In vitro study in cells	In vivo study in animal model	References
		Breast cancer	Reduces the expression of VEGFR2 and VEGFR3	MDA-MB-231	Xenograft	[117]
			Reduces VEGF-A protein expression	MDA-MB-231 MCF-7	–	[25]
			Decreases protein expression of IGF-IR, HIF-1α, and VEGF via upregulation of miR-152-3p	MDA-MB-468	Xenograft	[23]
			Decreases the gene and protein expression of IGF-1R and VEGF via regulation of miR-148a-3p	MDA-MB-231	Xenograft	[24]
			Decreases the protein level VEGF-A, VEGFR1, and VEGFR2	4T1-derived 4TLM	Xenograft	[118]
			Reduces protein expression of VEGF-C, VEGFR2, and VEGFR3, MMP9, and Angiogenin	MCF-7	–	[119]
			Reduces protein expression of VEGFR2 and Angiogenin	MDA-MB-231	–	
		Leukemia	Reduces IL-6 secretion and VEGF mRNA expression	NB4	–	[27]
		Bone cancer	Decreases VEGFA mRNA and protein expression via upregulating the expression of miR-424-5p	SaOS2 MG63	–	[26]
		Oral cancer	Inhibits the expression of the genes HIF-1α, VEGF, and ROCK-1	SCC9	–	[120]
		Brain cancer	Downregulates VEGF mRNA expression and secretion	SH-SY5Y	–	[29]
		Kidney cancer	Inhibition of HIF-1α expression	–	Renal adenocarcinoma cell (RENCA) Xenograft	[121]
		Colon cancer	Inhibits both mRNA and protein levels of VEGF expression and downregulates HIF-1α protein	HCT116	–	[122]

9	(−)-epigallocatechin-3-gallate (EGCG) 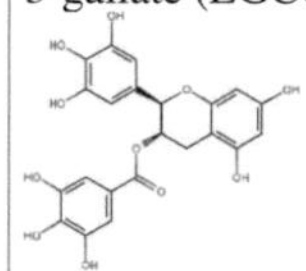	Gastric cancer	Decreases VEGF secretion and inhibits VEGF mRNA expression	SGC-7901	Xenograft	[31]
			Inhibits VEGF expression via suppressing STAT3 Pathway	AGS	–	[123]
		Cervical cancer	Upregulates TIMP-1 expression and downregulation of MMP-9 expression	HeLa	–	[34]
		Lung cancer	Inhibits mRNA and protein expression of HIF-1α and VEGF	A549	Xenograft	[124]
			Increases endostatin expression and suppresses VEGF expression	A549	Xenograft	[35]
			Downregulates HIF-1α, VEGF, COX-2 via PI3K/Akt and ERK1/2 signaling pathways and upregulates p53 and β-catenin protein levels	A549	Xenograft	[32]
		Breast cancer	Inhibits CD44, VEGF, Ki-67, and MMP-2 expression	–	DMBA -induced mammary cancer	[33]
		Colorectal cancer	Decreases expression levels of VEGF, HIF-1, IGF-1, IGF-2, and EGF via inhibition of ERK and Akt proteins	SW837	Xenografts	[125]
		Pancreatic cancer	Inhibits VEGF, CD31, MMP-2, MMP-7, MMP-9, and MMP-12	–	AsPC-1 Xenograft	[30]

(continued)

Table 1 (continued)

S. no.	Phytochemicals	Cancer type	Molecular mechanism of action	In vitro study in cells	In vivo study in animal model	References
10	Quercetin 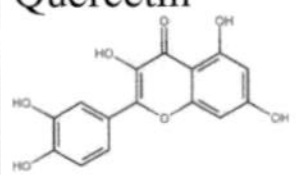	Esophageal cancer	Decreases protein levels of VEGF-A, MMP9, and MMP2	Eca109	–	[37]
		Prostate cancer	Increases thrombospondin-1 (TSP-1) protein and mRNA expression	PC-3	Xenograft	[126]
			Downregulates mRNA expression of Osteopontin (OPN) and VEGF isoforms	PC-3 LNCaP	–	[127]
			Inhibits VEGFR activation/ phosphorylation	PC-3	–	[128]
		Breast cancer	Downregulates mRNA expression and protein levels of VEGF and VEGFR2	MCF-7	Xenograft	[39]
			Suppresses HIF-1α and VEGF gene transcription	MCF-7	–	[129]
		Intraocular cancer	Downregulates the level of VEGF at the mRNA and protein levels	RB Y79	–	[130]
		Bone cancer	Downregulates mRNA and protein expression levels of HIF-1α, VEGF, MMP2, and MMP9	HOS	–	[38]
		Liver cancer	Reduces VEGFR2 mRNA and protein expression	Huh-7 SK-HEP-1	Xenograft	[40]

human breast cancer by reducing extracellular VEGF levels [13]. VEGF levels in colorectal cancer cells, Caco2 and HCT116, along with angiogenesis are reduced by resveratrol [14]. It inhibits matrix metalloproteases, including MMP-2 and MMP-9, and suppresses heme oxygenase-1 in the A549 lung cancer cell line [15]. Resveratrol exerts an antiangiogenic effect in human hepatocellular carcinoma by suppressing IL-6 and VEGF expression and ROS production [16]. It reduces the expression of VEGF mRNA in RT-2 glioma cells and suppresses the growth of brain cancer. Recently, resveratrol was found to inhibit oral cancer by decreasing the expression of CD133 and VEGF-A in H-357 and SCC-9 cell lines [17].

3.3 Diallyl Trisulfide (DATS)

Diallyl trisulfide (DATS) is usually found in garlic and is an effective compound against cancer cells. DATS inhibits orthotopic prostate tumor growth by inhibiting the expression of VEGF, IL-6, HIF-1α, MMP-2, MMP-7, and MMP-9 through the NF-κB pathway [18]. It also inhibits the proliferation, invasion, and angiogenesis of osteosarcoma by decreasing the mRNA and protein expression of VEGF, MMP-2, and MMP-9 via inactivation of Notch-1 activity [19]. DATS reduced breast cancer metastasis by inhibiting the mRNA and protein expressions of HIF-1α target genes such as MMP-2 and VEGF. It significantly inhibited invasion, migration, and angiogenesis of human colon cancer HT-29 cells via inhibition of protein levels of MMP-2, MMP-7, and MMP-9, and VEGF secretion [20]. The U87MG orthotopic xenograft model, treated with DATS, resulted in the inhibition of VEGF expression, preventing tumor progression in human glioblastoma [21].

3.4 Melatonin

Melatonin is known as a mammalian neurohormone; however, it is also present in plants and works as an antioxidant and signaling molecule. It has been found in apple, barley, bean, cucumber, grapes, maize, potato, rice, and tomato [22]. The anticancer effects of melatonin have been widely studied in various types of cancers. Several studies have reported the anticancer effects of melatonin on breast cancer. It suppresses the expression of angiogenesis-related proteins, HIF-1α, and VEGF via upregulation of miR-152-3p [23]. Similarly, it also suppresses the gene and protein expression of IGF-1R and VEGF via regulation of miR-148a-3p in MDA-MB-231 breast cancer cells [24]. It also had an anti-tumor effect on the breast cancer cell lines, MDA-MB-231 and MCF-7, because it decreases the expression of the VEGF-A protein [25]. It also had an antitumor effect on the breast cancer cell lines, MDA-MB-231 and MCF-7 involving decreased expression of the VEGF-A protein [26]. Antiangiogenic and anti-inflammatory effects of melatonin in leukemia through reduced VEGF mRNA expression and interleukin-6 secretion in NB4 cells have been observed [27]. A recent study also reported the anticancer effect of melatonin on gastric cancer via inhibition of the VEGF and angiogenin secretion in the SGC-7901 cell line [28]. Melatonin also downregulated VEGF mRNA expression and its secretion in SH-SY5Y ovarian cancer cells [29]. Thus, melatonin possesses anticancer activity against various types of cancer.

3.5 (−)-Epigallocatechin-3-gallate (EGCG)

(−)-Epigallocatechin-3-gallate (EGCG) is a compound mainly found in green tea. This polyphenolic compound is believed to have several health benefits associated with the consumption of green tea. Various studies have reported the cytotoxic effects of EGCG in many cancer cells. EGCG inhibited metastasis and angiogenesis of human pancreatic cancer cells by inhibiting VEGF, CD31, MMP-2, MMP-7, MMP-9, and MMP-12 in an AsPC-1 xenograft animal model [30]. EGCG inhibits the growth of gastric cancer by decreasing VEGF secretion and inhibiting VEGF mRNA expression [31]. Another study documented that EGCG suppressed HIF-1α and VEGF protein expression induced by nicotine in A549 xenografts of nude mice [32]. EGCG has also been found to inhibit chemically induced mammary cancer in rats by inhibiting CD44, VEGF, Ki-67, and MMP-2 expression [33]. EGCG effectively inhibited the migration of HeLa cells by modulating the expression of MMP-9 and TIMP-1 [34]. According to another study, EGCG treatment of A549 lung cancer cells suppressed angiogenesis by increasing endostatin expression and decreasing VEGF expression [35]. EGCG exerted growth-inhibitory effects on colorectal cancer cells by decreasing the expression levels of VEGF, HIF-1 α, IGF-1, IGF-2, and EGF and via inhibition of ERK and AKT signaling. Overall, EGCG has a strong potential for suppressing the growth and progression of cancer.

3.6 Quercetin

Quercetin is a flavanol and is found mostly in onions, grapes, broccoli, cherries, berries, and citrus fruits. According to previous studies, quercetin provides a variety of benefits, including anti-inflammatory, antioxidant, and anticancer properties [36].

A study demonstrated the antiangiogenic effect of quercetin on esophageal cancer cells, where it significantly reduced the expression of VEGF-A, MMP-2, and MMP-9 [37]. It was shown that quercetin treatment of prostate cancer PC-3 and LNCaP cells causes downregulation of mRNA expression of osteopontin (OPN) and VEGF isoforms, which are known to promote angiogenesis and drug resistance. The osteosarcoma invasion, migration, and angiogenesis were inhibited by the downregulation of HIF-1α, VEGF, MMP-2, and MMP-9 mRNA and protein expression levels in HOS cells by quercetin [38]. It is also found to inhibit angiogenesis in MCF-7 breast cancer cells and xenograft models, which was associated with the downregulation of VEGF and VEGFR2 expression [39]. Recently, quercetin was found to impair angiogenesis in Huh-7 and SK-HEP-1 hepatocellular carcinoma (HCC) cells by reducing VEGFR2 mRNA and protein expression [40]. Thus, quercetin could serve as a cancer chemopreventive compound through dietary intervention.

3.7 Decursin

Decursin is a pyranocoumarin compound isolated from the roots of a Korean herb *Angelica gigas*. It has been shown to possess anticancer effects in a variety of models including prostate cancer. In a study conducted by our group, we found that decursin inhibits antiangiogenic activities by suppressing proliferation and inducing G1 phase cell cycle arrest in human umbilical vein endothelial cells (HUVECs). Capillary tube formation, migration, and invasion, which are primary properties of endothelial cells, were also decreased. Mechanistically, it involved the inhibition of MMP-2 and MMP-9 activities. The in vitro results were further validated by its effect on ex vivo rat aortic ring assay, in which decursin could inhibit the sprouting of newer vessels [41]. In another study, decursin significantly attenuated lung tumor growth in mice via inhibition of microvessel density and the expression of the VEGFR2-signaling molecule as well as MMP-2 [42]. The antimetastatic activity of decursin in colon cancer was found to be due to the downregulation of MMP-2 and MMP-9 expression by JNK/PI3K and ERK/JNK pathways [43]. Further, decursin inhibited 12-O-tetradecanoylphorbol-13-acetate (TPA)-induced MMP-9 expression in MCF-7 cells through the suppression of NF-κB pathways [44]. Recently, it was found that decursin promotes the proteasomal degradation of HIF-1α in lung and colorectal cancer cells under hypoxia [45]. Thus, decursin has a strong antiangiogenic potential against various cancers.

3.8 Fisetin

Fisetin, a natural flavonol, is commonly found in strawberries and other fruits and vegetables, and has shown strong efficacy against many cancers. It inhibits the mobility and migration of U-2 OS cells by reducing NF-κB and uPA protein levels and inhibiting MMP-2 and MMP-9 [46]. In a recent study, fisetin significantly inhibited VEGF and epithelial cell adhesion molecule (EpCAM) via upregulation of AKAP12 in a colorectal cancer patient-derived organoid xenograft model (CRC PDOX) model [47]· Fisetin has been also found to prevent angiogenesis in diabetic retinopathy (DR) by downregulating VEGF expression [48]). It also inhibits cell migration by decreasing MMP-2 and MMP-9 production mediated by heme oxygenase-1 in breast cancer 4T1 and JC cells [49]. A study from our laboratory demonstrated that fisetin exhibits strong antiangiogenic potential by inhibiting various angiogenic attributes in human umbilical vein endothelial cells (HUVECs). It also decreased the expression of pro-angiogenic mediators, including eNOS, VEGF, iNOS, MMP-2, and MMP-9 in A549 lung cancer and DU145 prostate cancer cells [50]. Fisetin has been also found to inhibit the invasion and migration of cervical cancer cells through the suppression of the expression of urokinase plasminogen activator (uPA) and NF-κB signaling pathway [51]. Thus, fisetin has shown a promising potential in angioprevention of cancer.

3.9 Curcumin

A derivative of plant *Curcuma longa*, commonly known as turmeric, curcumin is known to possess anti-inflammatory, antimutagenic, and anticarcinogenic properties. Its antiangiogenic activity is reported in HUVECs. It has been shown to downregulate transcript levels of VEGF and bFGF in triple-negative breast cancer cell line MDA-MB-231 cells [52]. Curcumin also inhibited tumor growth and angiogenesis in breast cancer mouse model through the modulation of NF-κB pathway [53]. It suppresses the synthesis of many MMPs through the downregulation of AP-1 and NF-κB. Treatment of breast cancer cells with curcumin led to significantly lower levels of MMP-1 and MMP-2 mRNA and protein [54]. Curcumin reduces MMP-2 and MMP-9 expression and suppresses the growth and invasion potential of tumor cells in DU145 cells in both culture and xenograft experiments [55]. Similarly, curcumin downregulated MMP-9 activity in cell and xenograft models to prevent prostate cancer tumor growth and metastasis [56]. Curcumin has been found to decrease CXCL1 and CXCL2 in breast and prostate cancer cells via inhibition of NF-κB. It directly targets the IκBα pathway, which results in the inhibition of NF-κB [57, 58]. Curcumin targets the FGF angiogenic signaling pathway and inhibits the expression of MMP-6 as well as neovascularization in the mouse cornea [59]. Curcumin has a potential role in suppressing metastasis in colon cancer cells via AMPK activation and inhibition of NF-κB, uPA, and MMP-9 [60]. Curcumin increases miR-22 in chronic myelogenous leukemia cells, which directly targets IPO7, suggesting that curcumin alters HIF-1 activity via the miR-22/IPO7 axis [61]. Thus, curcumin might play a significant role in dietary cancer chemoprevention.

3.10 Silibinin

Silibinin, a flavanolignan compound, is isolated from the seeds of a plant *Silybum marianum* and is used since ancient times for the treatment of liver disorders. Silymarin/silibinin, a component of the milk thistle plant and artichoke, inhibits VEGF secretion in cancer cells and disrupts tube formation by endothelial cells. It has been reported to inhibit secreted as well as cellular MMP-2 levels and to suppress capillary formation in human umbilical vein endothelial cells in Matrigel assay [62]. Silibinin was found to significantly downregulate the expression of angiogenesis and metastasis-promoting proteins, including HIF-1α, VEGF, COX-2, Ang-2, and Ang-4, as well as the expression of MMP-2, and MMP-9 in mouse colon cells [63]. It also decreases the expression of VEGF, iNOS, NOS3, COX-1, COX-2, and HIF-1α in human colorectal cancer [64]. Velmurugan et al. reported the preventive effect of silibinin against colorectal cancer, which targeted angiogenesis regulators and decreased the protein level of VEGF and iNOS through suppression of the β-catenin signaling pathway in the CRC SW480 xenograft model [65]. Silibinin treatment decreased HIF-1α protein in ovarian clear-cell carcinoma cells, HAC-2,

OVISE, and RMG-1, and eIF4E2 and RPS6 mRNA in HAC-2 and RMG-1 cells [66]. The efficacy of silibinin to target cell growth and angiogenesis in prostate cancer cells was also analyzed under hypoxic conditions. The result indicated that silibinin strongly decreased hypoxia-induced HIF-1α expression and NADPH oxidase (NOX) activity [67]. Silibinin also showed anticancer effects on the metastatic potential of TNBC cells, which was associated with decreased TGF-β2-induced MMP-2 and MMP-9 expression levels [68]. Thus, silibinin possesses a strong antiangiogenic activity against various cancers.

4 Medicinal Plants with Anticancer Activities

Medicinal plants have been used for thousands of years in Asian and African folk remedies, and many plants are used for their health advantages in Western countries. These plants contain biologically active compounds that can inhibit the growth of tumors and have antiangiogenic properties. These plants have benefits over chemical drugs and can have synergistic effects, influencing different aspects of disease pathology simultaneously [131]. Grape seed polyphenols are commercially marketed as grape seed extract (GSE), which approximately contain 75–95% procyanidins. GSE is known to have antitumor activity against breast, lung, skin, and gastrointestinal cancers. It has been reported to inhibit the growth of human prostate carcinoma cells in culture and nude mice xenograft models. GSE inhibited tumor angiogenesis via downregulation of VEGF and upregulation of IGFBP-3 in tumor cells [132]. GSE could also inhibit tumor angiogenesis in breast cancer cells in mice. Its antiangiogenic activity involves the inhibition of kinase activity of VEGFR2 and MAPK downstream signaling [133]. Grape seed proanthocyanidins also exhibit antiangiogenic effects in colorectal and liver cancers via inhibition of VEGFR expression [134] (Fig. 2).

Oak is a shrub whose different parts have been used as a traditional drug. It also inhibits tumor growth and has an antiangiogenic effect. Oak extract reduces the expression of angiogenic factors, such as MMP-9, and inhibits VEGF secretion in HUVECs [135]. Anthocyanin-rich extract from black rice (AEBR) has shown beneficial effects against breast cancer and suppresses tumor growth and angiogenesis in MDA-MB-453 cell xenograft by reducing the expression of angiogenesis factors MMP-2, MMP-9, and uPA in tumor tissue [131]. Wheatgrass is the young grass of the common wheat plant. It has a high content of vitamins, minerals, and amino acids, which play an important role in its anticancer properties. Studies have demonstrated that wheatgrass inhibits metastasis and angiogenesis in liver cancer by decreasing the expression of VEGF, MMP-9, and COX-2 via the PI3K/AKT pathway [136]. Wheatgrass extract was also found to suppress HIF-1α expression in A549 human lung adenocarcinoma cells [137]. These reports suggest that many such plants have angiopreventive properties against various cancers (Table 2).

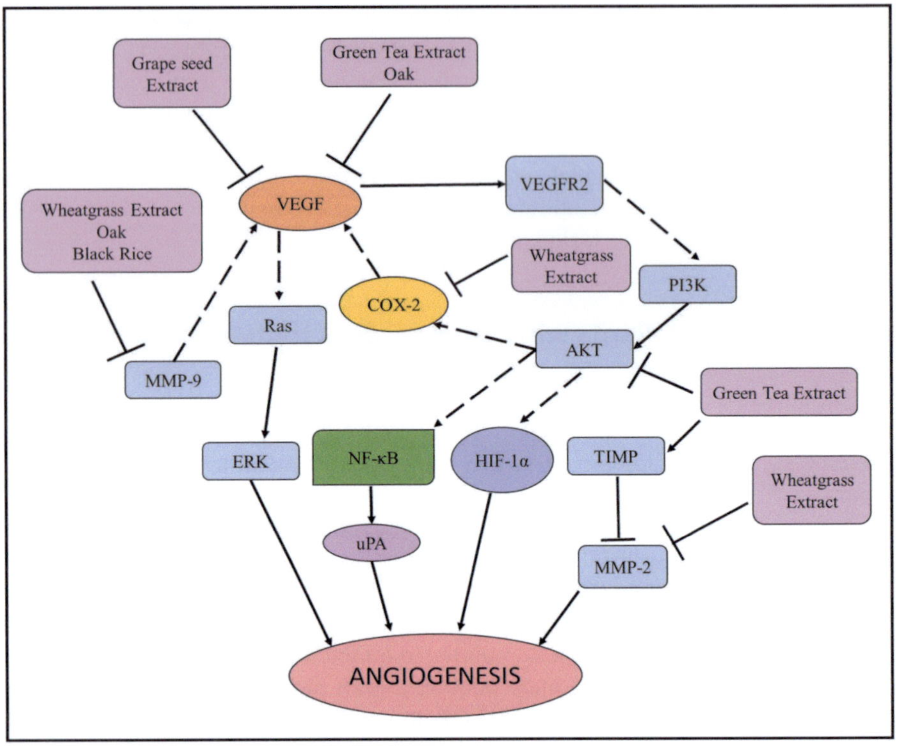

Fig. 2 Angiogenic pathways modulated by plant extracts

5 Challenges and Future Directions

Targeting angiogenesis has emerged as a promising strategy for cancer therapy, with several antiangiogenic agents already in clinical use and others in development. Despite the promising preclinical and clinical data, the clinical success of antiangiogenic therapies has been limited. Tumors often develop resistance to these treatments through mechanisms, such as upregulation of alternative angiogenic pathways, FGFs, or angiopoietins, to bypass the effects of VEGF inhibition or recruitment of alternative cell types that can contribute to blood vessel formation. Additionally, some tumors may become more invasive or aggressive due to the normalization of the blood vessels after angiogenesis inhibition, making them more susceptible to metastasis. Moreover, the toxicity and side effects of angiogenesis inhibitors, such as hypertension, proteinuria, bleeding, and impaired wound healing, may limit the long-term use of these therapies and necessitate careful patient management. The tumor microenvironment plays a pivotal role in the effectiveness of angiogenesis inhibitors. The heterogeneous nature of the tumor vasculature, including areas of hypoxia and necrosis, can complicate the delivery of

Table 2 Medicinal plant extract targeting angiogenesis in cancer

S. No.	Medicinal plant extract	Type of cancer	Molecular mechanism	In vitro study (cell line)	In vivo study model	References
1	Grape seed extract	Prostate cancer	Downregulates VEGF and upregulates IGFBP-3 protein expression	–	DU145 xenograft	[132]
		Breast cancer	Inhibits kinase activity of VEGFR2	MDA-MB-231	Xenograft	[133]
2	Grape seed proanthocyanidins	Colon cancer	Suppresses the expression of VEGF and Ang1	SW620	–	[134]
		Lung cancer	Reduces the levels of VEGF protein	–	A549 and H1299 xenografts	[138]
		Liver cancer	Decreases mRNA and protein expression of VEGF	–	H22 xenograft	[139]
3	*Ficus Carica*	Breast cancer	Decreases mRNA expression of MMP2	MDA-MB-231	–	[140]
4	Black rice	Breast cancer	Reduces expression of MMP-2, MMP-9, and uPA	MDA-MB-453	–	[131]
5	Wheatgrass extract	Liver cancer	Decreases the protein expression of VEGF, MMP-9, and COX-2 and increases TIMP	Hep-2	–	[136]
6	Green tea extract	Breast cancer	Inhibits expression of VEGFR and MMP-9	MDA-MB-231	Xenograft	[141]

antiangiogenic agents. Furthermore, the immune cells present within the tumor microenvironment can influence the response to angiogenesis inhibition.

To improve the efficacy of antiangiogenic therapies, ongoing research is focused on understanding the dynamic and heterogeneous nature of the tumor vasculature. Innovative strategies are being developed to normalize the abnormal blood vessels in tumors, thereby enhancing drug delivery and improving the response to other therapeutic agents. Additionally, biomarkers to predict the response to angiogenesis inhibition are under investigation, with the aim of personalizing treatments to individual patients. Ongoing research into combination therapies, biomarker identification, and novel drug delivery systems will be essential in enhancing the clinical outcomes of angiogenesis-targeted therapies. Despite the challenges, the continued exploration of angiogenesis inhibition holds significant potential for improving cancer treatment and outcomes.

Acknowledgments The original works of authors in the article were supported by DST-PURSE, DST-FIST, UGC-RN, DBT-Builder-SLS, and UGC-DRS, New Delhi.

References

1. Liu ZL, et al. Angiogenic signaling pathways and anti-angiogenic therapy for cancer. Signal Transduct Target Ther. 2023;8:198. https://doi.org/10.1038/s41392-023-01256-w.
2. Risau W, Flamme I. Vasculogenesis. Annu Rev Cell Dev Biol. 1995;11:73–91. https://doi.org/10.1146/annurev.cb.11.110195.000445.
3. Rivard A, Isner JM. Angiogenesis and Vasculogenesis in treatment of cardiovascular disease. Mol Med. 1998;4:429–40. https://doi.org/10.2119/molmed.1998.4.429.
4. Folkman J. Tumor angiogenesis: therapeutic implications. N Engl J Med. 1971;285:1182–6. https://doi.org/10.1056/NEJM197110142851104.
5. Folkman J, et al. Isolation of a tumor factor responsible for angiogenesis. J Exp Med. 1971;133:275–88. https://doi.org/10.1084/jem.133.2.275.
6. Baeriswyl V, Christofori G. The angiogenic switch in carcinogenesis. Semin Cancer Biol. 2009;19:329–37. https://doi.org/10.1016/j.semcancer.2009.07.004.
7. Bergers G, Benjamin LE. Tumorigenesis and the angiogenic switch. Nat Rev Cancer. 2003;3:401–10. https://doi.org/10.1038/nrc1093.
8. Lin S, et al. Berberine inhibits HIF-1alpha expression *via* enhanced proteolysis. Mol Pharmacol. 2004;66:612–9. https://doi.org/10.1124/mol.104.001890.
9. Tsang CM, et al. Berberine suppresses Id-1 expression and inhibits the growth and development of lung metastases in hepatocellular carcinoma. Biochim Biophys Acta. 2015;1852:541–51. https://doi.org/10.1016/j.bbadis.2015.01.001.
10. Chen H, et al. Berberine inhibits intestinal carcinogenesis by suppressing intestinal pro-inflammatory genes and oncogenic factors through modulating gut microbiota. BMC Cancer. 2022;22:566. https://doi.org/10.1186/s12885-022-09493-3.
11. Jin F, et al. Berberine inhibits angiogenesis in glioblastoma xenografts by targeting the VEGFR2/ERK pathway. Pharm Biol. 2018;56:665–71. https://doi.org/10.1080/13880209.2017.1398836.
12. Sun Y, et al. Berberine inhibits breast carcinoma proliferation and metastasis under hypoxic microenvironment involving gut microbiota and endogenous metabolites. Pharmacol Res. 2023;193:106817. https://doi.org/10.1016/j.phrs.2023.106817.

13. Garvin S, et al. Resveratrol induces apoptosis and inhibits angiogenesis in human breast cancer xenografts *in vivo*. Cancer Lett. 2006;231:113–22. https://doi.org/10.1016/j.canlet.2005.11.011.
14. Fouad M, et al. Resveratrol inhibits proliferation, angiogenesis and induces apoptosis in colon cancer cells. Hum Exp Toxicol. 2013;32:1067–80. https://doi.org/10.1177/0960327113487276.
15. Liu P, et al. Resveratrol inhibits human lung adenocarcinoma cell metastasis by suppressing heme oxygenase 1-mediated nuclear factor-κB pathway and subsequently downregulating expression of matrix metalloproteinases. Mol Nutr Food Res. 2010;54:54. https://doi.org/10.1002/mnfr.200900211.
16. Yan Y, et al. Resveratrol inhibits hepatocellular carcinoma progression driven by hepatic stellate cells by targeting Gli-1. Mol Cell Biochem. 2017;434:17–24. https://doi.org/10.1007/s11010-017-3109-3.
17. Pradhan R, et al. Resveratrol nanoparticle attenuates metastasis and angiogenesis by deregulating inflammatory cytokines through inhibition of CAFs in oral cancer by CXCL-12/IL-6-dependent pathway. J Nutr Biochem. 2023;113:109257. https://doi.org/10.1016/j.jnutbio.2023.109257.
18. Shankar S, et al. Diallyl trisulfide increases the effectiveness of TRAIL and inhibits prostate cancer growth in an orthotopic model: molecular mechanisms. Mol Cancer Ther. 2008;7:2328–38. https://doi.org/10.1158/1535-7163.MCT-07-2379.
19. Li Y, et al. Diallyl trisulfide inhibits proliferation, invasion and angiogenesis of osteosarcoma cells by switching on suppressor microRNAs and inactivating of Notch-1 signaling. Carcinogenesis. 2013;34:1601–10. https://doi.org/10.1093/carcin/bgt100.
20. Lai K, et al. Diallyl trisulfide inhibits migration, invasion and angiogenesis of human colon cancer HT-29 cells and umbilical vein endothelial cells, and suppresses murine xenograft tumour growth. J Cell Mol Med. 2015;19:474–84. https://doi.org/10.1111/jcmm.12438.
21. Das A, et al. Single agent efficacy of the HDAC inhibitor DATS in preclinical models of glioblastoma. Cancer Chemother Pharmacol. 2018;82:945–52. https://doi.org/10.1007/s00280-018-3740-1.
22. Salehi B, et al. Melatonin in medicinal and food plants: occurrence, bioavailability, and health potential for humans. Cells. 2019;8:681. https://doi.org/10.3390/cells8060681.
23. Marques JHM, et al. Melatonin restrains angiogenic factors in triple-negative breast cancer by targeting miR-152-3p: *in vivo* and *in vitro* studies. Life Sci. 2018;208:131–8. https://doi.org/10.1016/j.lfs.2018.04.026.
24. Lacerda JZ, et al. Therapeutic potential of Melatonin in the regulation of MiR-148a-3p and Angiogenic factors in breast cancer. MicroRNA. 2019;8:237–47. https://doi.org/10.2174/2211536608666190520112520.
25. Gelaleti GB, et al. Efficacy of melatonin, IL-25 and siIL-17B in tumorigenesis-associated properties of breast cancer cell lines. Life Sci. 2017;183:98–109. https://doi.org/10.1016/j.lfs.2017.06.005.
26. Vimalraj S, et al. Melatonin regulates tumor angiogenesis *via* miR-424-5p/VEGFA signaling pathway in osteosarcoma. Life Sci. 2020;256:118011. https://doi.org/10.1016/j.lfs.2020.118011.
27. Amirzargar MR, et al. Angiogenesis, coagulation, and fibrinolytic markers in acute promyelocytic leukemia (NB4): an evaluation of melatonin effects. J Pineal Res. 2023;75:75. https://doi.org/10.1111/jpi.13005.
28. Wang R-X, et al. Melatonin downregulates nuclear receptor RZR/RORγ expression causing growth-inhibitory and anti-angiogenesis activity in human gastric cancer cells *in vitro* and *in vivo*. Oncol Lett. 2016;12:897–903. https://doi.org/10.3892/ol.2016.4719.
29. González A, et al. Melatonin inhibits angiogenesis in SH-SY5Y human neuroblastoma cells by downregulation of VEGF. Oncol Rep. 2017;37:2433–40. https://doi.org/10.3892/or.2017.5579.
30. Shankar S. EGCG inhibits growth, invasion, angiogenesis and metastasis of pancreatic cancer. Front Biosci. 2008;13:440. https://doi.org/10.2741/2815.

31. Zhu B-H. (−)-Epigallocatechin-3-gallate inhibits growth of gastric cancer by reducing VEGF production and angiogenesis. World J Gastroenterol. 2007;13:1162. https://doi.org/10.3748/wjg.v13.i8.1162.
32. Shi J. Epigallocatechin-3-gallate inhibits nicotine-induced migration and invasion by the suppression of angiogenesis and epithelial-mesenchymal transition in non-small cell lung cancer cells. Oncol Rep. 2015;33:2972–80. https://doi.org/10.3892/or.2015.3889.
33. Abd El-Rahman SS. Epigallocatechin-3-Gallate: the prospective targeting of cancer stem cells and preventing metastasis of chemically-induced mammary cancer in rats. Am J Med Sci. 2017;354:54–63. https://doi.org/10.1016/j.amjms.2017.03.001.
34. Sharma C. (−)-Epigallocatechin-3-Gallate induces apoptosis and inhibits invasion and migration of human cervical cancer cells. Asian Pacific J Cancer Prev. 2012;13:4815–22. https://doi.org/10.7314/APJCP.2012.13.9.4815.
35. Sakamoto Y. Effects of Epigallocatechin-3-gallate (EGCG) on A549 lung cancer tumor growth and angiogenesis. Biosci Biotechnol Biochem. 2013;77:1799–803. https://doi.org/10.1271/bbb.120882.
36. Anand DA. Overviews of biological importance of quercetin: a bioactive flavonoid. Pharmacogn Rev. 2016;10:84. https://doi.org/10.4103/0973-7847.194044.
37. Liu Y. Quercetin inhibits invasion and angiogenesis of esophageal cancer cells. Pathol Res Pract. 2021;222:153455. https://doi.org/10.1016/j.prp.2021.153455.
38. Lan H. Quercetin inhibits cell migration and invasion in human osteosarcoma cells. Cell Physiol Biochem. 2017;43:553–67. https://doi.org/10.1159/000480528.
39. Zhao X. Quercetin inhibits angiogenesis by targeting calcineurin in the xenograft model of human breast cancer. Eur J Pharmacol. 2016;781:60–8. https://doi.org/10.1016/j.ejphar.2016.03.063.
40. Xiong W. Quercetin inhibits endothelial & hepatocellular carcinoma cell crosstalk *via* reducing extracellular vesicle-mediated VEGFR2 mRNA transfer. Mol Carcinog. 2024;63:2254–68. https://doi.org/10.1002/mc.23807.
41. Bhat TA. Inhibition of angiogenic attributes by decursin in endothelial cells and *ex vivo* rat aortic ring angiogenesis model. Indian J Exp Biol. 2011;49:848–56.
42. Jung MH. Decursin and decursinol angelate inhibit VEGF-induced angiogenesis *via* suppression of the VEGFR-2-signaling pathway. Carcinogenesis. 2009;30:655–61. https://doi.org/10.1093/carcin/bgp039.
43. Son SH. Decursin and decursinol from Angelica gigas inhibit the lung metastasis of murine colon carcinoma. Phytother Res. 2011;25:959–64. https://doi.org/10.1002/ptr.3372.
44. Kim J-M. Decursin prevents TPA-induced invasion through suppression of PKCα/p38/NF-κB-dependent MMP-9 expression in MCF-7 human breast carcinoma cells. Int J Oncol. 2014;44:1607–13. https://doi.org/10.3892/ijo.2014.2327.
45. Ge Y. Decursin promotes HIF-1α proteasomal degradation and immune responses in hypoxic tumour microenvironment. Phytomedicine. 2020;78:153318. https://doi.org/10.1016/j.phymed.2020.153318.
46. Chen J-K. Fisetin suppresses human osteosarcoma U-2 OS cell migration and invasion *via* affecting FAK, uPA and NF-κB signaling pathway *in vitro*. In Vivo. 2019;33:801–10. https://doi.org/10.21873/invivo.11542.
47. Kim N. Fisetin induces the upregulation of AKAP12 mRNA and anti-angiogenesis in a patient-derived organoid xenograft model. Biomed Pharmacother. 2023;167:115613. https://doi.org/10.1016/j.biopha.2023.115613.
48. Lai M. Fisetin prevents angiogenesis in diabetic retinopathy by downregulating VEGF. J Ophthalmol. 2023;2023:7951928. https://doi.org/10.1155/2023/7951928.
49. Tsai C-F. Fisetin inhibits cell migration *via* inducing HO-1 and reducing MMPs expression in breast cancer cell lines. Food Chem Toxicol. 2018;120:528–35. https://doi.org/10.1016/j.fct.2018.07.059.
50. Bhat TA. Fisetin inhibits various attributes of angiogenesis *in vitro* and *in vivo*—implications for angioprevention. Carcinogenesis. 2012;33:385–93. https://doi.org/10.1093/carcin/bgr282.

51. Chou R-H. Fisetin inhibits migration and invasion of human cervical cancer cells by down-regulating urokinase plasminogen activator expression through suppressing the p38 MAPK-dependent NF-κB signaling pathway. PLoS One. 2013;8:e71983. https://doi.org/10.1371/journal.pone.0071983.
52. Di G. Analysis of anti-proliferation of curcumin on human breast cancer cells and its mechanism. Zhonghua Yi Xue Za Zhi. 2003;83:1764–8.
53. Bimonte S. Curcumin inhibits tumor growth and angiogenesis in an orthotopic mouse model of human pancreatic cancer. Biomed Res Int. 2013;2013:810423. https://doi.org/10.1155/2013/810423.
54. Bachmeier BE. The role of curcumin in prevention and management of metastatic disease. Int J Mol Sci. 2018;19:1716. https://doi.org/10.3390/ijms19061716.
55. Hong JH. The effects of curcumin on the invasiveness of prostate cancer *in vitro* and *in vivo*. Prostate Cancer Prostatic Dis. 2006;9:147–52. https://doi.org/10.1038/sj.pcan.4500856.
56. Cheng T-S. Curcumin-targeting pericellular serine protease matriptase role in suppression of prostate cancer cell invasion, tumor growth, and metastasis. Cancer Prev Res (Phila). 2013;6:495–505. https://doi.org/10.1158/1940-6207.CAPR-12-0293-T.
57. Bachmeier BE. Curcumin downregulates the inflammatory cytokines CXCL1 and −2 in breast cancer cells *via* NFkappaB. Carcinogenesis. 2008;29:779–89. https://doi.org/10.1093/carcin/bgm248.
58. Killian PH. Curcumin inhibits prostate cancer metastasis *in vivo* by targeting the inflammatory cytokines CXCL1 and −2. Carcinogenesis. 2012;33:2507–19. https://doi.org/10.1093/carcin/bgs312.
59. Mohan R. Curcuminoids inhibit the Angiogenic response stimulated by fibroblast growth Factor-2, including expression of matrix metalloproteinase Gelatinase B. J Biol Chem. 2000;275:10405–12. https://doi.org/10.1074/jbc.275.14.10405.
60. Tong W. Curcumin suppresses colon cancer cell invasion *via* AMPK-induced inhibition of NF-κB, uPA activator and MMP9. Oncol Lett. 2016;12:4139–46. https://doi.org/10.3892/ol.2016.5148.
61. Monteleone F, et al. SWATH-MS based quantitative proteomics analysis reveals that curcumin alters the metabolic enzyme profile of CML cells by affecting the activity of miR-22/IPO7/HIF-1α axis. J Exp Clin Cancer Res. 2018;37:170. https://doi.org/10.1186/s13046-018-0843-y.
62. Jiang C, et al. Anti-Angiogenic potential of a cancer Chemopreventive flavonoid antioxidant, Silymarin: inhibition of key attributes of vascular endothelial cells and angiogenic cytokine secretion by cancer epithelial cells. Biochem Biophys Res Commun. 2000;276:371–8. https://doi.org/10.1006/bbrc.2000.3474.
63. Sameri S, et al. Targeting the hallmarks of cancer: the effects of silibinin on proliferation, cell death, angiogenesis, and migration in colorectal cancer. BMC Compl Med Ther. 2021;21:160. https://doi.org/10.1186/s12906-021-03330-1.
64. Singh RP, et al. Silibinin inhibits colorectal cancer growth by inhibiting tumor cell proliferation and angiogenesis. Cancer Res. 2008;68:2043–50.
65. Velmurugan B, et al. Silibinin exerts sustained growth suppressive effect against human colon carcinoma SW480 xenograft by targeting multiple signaling molecules. Pharm Res. 2010;27:2085–97. https://doi.org/10.1007/s11095-010-0207-6.
66. Miyazawa M, et al. Hypoxia-inducible factor-1α suppression in ovarian clear-cell carcinoma cells by Silibinin Administration. Anticancer Res. 2020;40:6791–8. https://doi.org/10.21873/anticanres.14702.
67. Deep G, et al. Silibinin inhibits hypoxia-induced HIF-1α-mediated signaling, angiogenesis and lipogenesis in prostate cancer cells: *in vitro* evidence and *in vivo* functional imaging and metabolomics. Mol Carcinog. 2017;56:833–48. https://doi.org/10.1002/mc.22537.
68. Kim S, et al. Silibinin inhibits triple negative breast cancer cell motility by suppressing TGF-β2 expression. Tumour Biol. 2016;37:11397–407. https://doi.org/10.1007/s13277-016-5000-7.

69. Chakraborty G, et al. Curcumin suppresses breast tumor angiogenesis by abrogating osteopontin-induced VEGF expression. Mol Med Rep. 2008;1:641–6. https://doi.org/10.3892/mmr_00000005.
70. Bimonte S, et al. Dissecting the role of Curcumin in tumour growth and angiogenesis in Mouse Model of human breast cancer. Biomed Res Int. 2015;2015:1–7. https://doi.org/10.1155/2015/878134.
71. Ferreira LC, et al. Effect of Curcumin on pro-angiogenic factors in the xenograft model of breast cancer. Anti Cancer Agents Med Chem. 2015;15:1285–96. https://doi.org/10.2174/1871520615666150520093644.
72. Carroll CE, et al. Curcumin inhibits MPA-induced secretion of VEGF from T47-D human breast cancer cells. Menopause. 2008;15:570–4. https://doi.org/10.1097/gme.0b013e31814fae5d.
73. Dorai T, et al. Therapeutic potential of curcumin in human prostate cancer. III. Curcumin inhibits proliferation, induces apoptosis, and inhibits angiogenesis of LNCaP prostate cancer cells *in vivo*. Prostate. 2001;47:293–303. https://doi.org/10.1002/pros.1074.
74. Shankar S, et al. Curcumin sensitizes TRAIL-resistant xenografts: molecular mechanisms of apoptosis, metastasis and angiogenesis. Mol Cancer. 2008;7:16. https://doi.org/10.1186/1476-4598-7-16.
75. Deng G, et al. Curcumin inhibits the expression of vascular endothelial growth factor and androgen-independent prostate cancer cell line PC-3 *in vitro*. Zhonghua Nan Ke Xue. 2008;14:116–21.
76. Binion DG, et al. Curcumin inhibits VEGF-mediated angiogenesis in human intestinal microvascular endothelial cells through COX-2 and MAPK inhibition. Gut. 2008;57:1509–17. https://doi.org/10.1136/gut.2008.152496.
77. Harakeh S, et al. Novel curcumin nanoformulation induces apoptosis, and reduces migration and angiogenesis in liver cancer cells. Artif Cells Nanomed Biotechnol. 2023;51:361–70. https://doi.org/10.1080/21691401.2023.2238756.
78. Yoysungnoen P, et al. Effects of curcumin on tumor angiogenesis and biomarkers, COX-2 and VEGF, in hepatocellular carcinoma cell-implanted nude mice. Clin Hemorheol Microcirc. 2006;34:109–15.
79. Bae M-K, et al. Curcumin inhibits hypoxia-induced angiogenesis *via* down-regulation of HIF-1. Oncol Rep. 2006;15:1557–62.
80. Tian S, et al. Curcumin inhibits the growth of liver cancer by impairing myeloid-derived suppressor cells in murine tumor tissues. Oncol Lett. 2021;21:286. https://doi.org/10.3892/ol.2021.12547.
81. Shao S, et al. Curcumin suppresses hepatic stellate cell-induced Hepatocarcinoma angiogenesis and invasion through downregulating CTGF. Oxidative Med Cell Longev. 2019;2019:8148510. https://doi.org/10.1155/2019/8148510.
82. Das L, et al. Long term effect of curcumin in regulation of glycolytic pathway and angiogenesis *via* modulation of stress activated genes in prevention of cancer. PLoS One. 2014;9:e99583. https://doi.org/10.1371/journal.pone.0099583.
83. Chiablaem K, et al. Curcumin suppresses vasculogenic mimicry capacity of hepatocellular carcinoma cells through STAT3 and PI3K/AKT inhibition. Anticancer Res. 2014;34:1857–64.
84. Fu Z, et al. Curcumin inhibits angiogenesis and improves defective hematopoiesis induced by tumor-derived VEGF in tumor model through modulating VEGF-VEGFR2 signaling pathway. Oncotarget. 2015;6:19469–82. https://doi.org/10.18632/oncotarget.3625.
85. Lin S-S, et al. Curcumin inhibits the migration and invasion of human A549 lung cancer cells through the inhibition of matrix metalloproteinase-2 and -9 and Vascular Endothelial Growth Factor (VEGF). Cancer Lett. 2009;285:127–33. https://doi.org/10.1016/j.canlet.2009.04.037.
86. Tung Y-T, et al. Curcumin reduces pulmonary tumorigenesis in vascular endothelial growth factor (VEGF)-overexpressing transgenic mice. Mol Nutr Food Res. 2011;55:1036–43. https://doi.org/10.1002/mnfr.201000654.
87. Yang C-L, et al. Curcumin blocks small cell lung cancer cells migration, invasion, angiogenesis, cell cycle and neoplasia through Janus kinase-STAT3 signalling pathway. PLoS One. 2012;7:e37960. https://doi.org/10.1371/journal.pone.0037960.

88. Xu X, et al. Curcumin inhibits human non-small cell lung cancer xenografts by targeting STAT3 pathway. Am J Transl Res. 2017;9:3633–41.
89. Jiao D, et al. Curcumin inhibited HGF-induced EMT and angiogenesis through regulating c-Met dependent PI3K/Akt/mTOR signaling pathways in lung cancer. Mol Ther Oncolytics. 2016;3:16018. https://doi.org/10.1038/mto.2016.18.
90. Li X, et al. Anticancer effects of curcumin on nude mice bearing lung cancer A549 cell subsets SP and NSP cells. Oncol Lett. 2018;16:6756–62. https://doi.org/10.3892/ol.2018.9488.
91. Noh E-M, et al. Fisetin regulates TPA-induced breast cell invasion by suppressing matrix metalloproteinase-9 activation *via* the PKC/ROS/MAPK pathways. Eur J Pharmacol. 2015;764:79–86. https://doi.org/10.1016/j.ejphar.2015.06.038.
92. Sung B, et al. Fisetin, an inhibitor of cyclin-dependent kinase 6, down-regulates nuclear factor-kappaB-regulated cell proliferation, antiapoptotic and metastatic gene products through the suppression of TAK-1 and receptor-interacting protein-regulated IkappaBalpha kinase activation. Mol Pharmacol. 2007;71:1703–14. https://doi.org/10.1124/mol.107.034512.
93. Liao Y-C, et al. Involvement of the ERK signaling pathway in fisetin reduces invasion and migration in the human lung cancer cell line A549. J Agric Food Chem. 2009;57:8933–41. https://doi.org/10.1021/jf902630w.
94. Murtaza I, et al. Fisetin, a natural flavonoid, targets chemoresistant human pancreatic cancer AsPC-1 cells through DR3-mediated inhibition of NF-kappaB. Int J Cancer. 2009;125:2465–73. https://doi.org/10.1002/ijc.24628.
95. Park JH, et al. Fisetin inhibits matrix metalloproteinases and reduces tumor cell invasiveness and endothelial cell tube formation. Nutr Cancer. 2013;65:1192–9. https://doi.org/10.1080/01635581.2013.828090.
96. Kim W-J, et al. Decursinol angelate blocks transmigration and inflammatory activation of cancer cells through inhibition of PI3K, ERK and NF-kappaB activation. Cancer Lett. 2010;296:35–42. https://doi.org/10.1016/j.canlet.2010.03.012.
97. Lee S-H, et al. A novel derivative of decursin, CSL-32, blocks migration and production of inflammatory mediators and modulates PI3K and NF-κB activities in HT1080 cells. Cell Biol Int. 2012;36:683–8. https://doi.org/10.1042/CBI20110257.
98. Singh RP, et al. Silibinin inhibits established prostate tumor growth, progression, invasion, and metastasis and suppresses tumor angiogenesis and epithelial-mesenchymal transition in transgenic adenocarcinoma of the mouse prostate model mice. Clin Cancer Res. 2008;14:7773–80. https://doi.org/10.1158/1078-0432.CCR-08-1309.
99. Singh RP, et al. Effect of silibinin on the growth and progression of primary lung tumors in mice. J Natl Cancer Inst. 2006;98:846–55. https://doi.org/10.1093/jnci/djj231.
100. Tyagi A, et al. Growth inhibition and regression of lung tumors by silibinin: modulation of angiogenesis by macrophage-associated cytokines and nuclear factor-κB and signal transducers and activators of transcription 3. Cancer Prev Res. 2009;2:74–83. https://doi.org/10.1158/1940-6207.CAPR-08-0095.
101. Ramasamy K, et al. Silibinin prevents lung tumorigenesis in wild-type but not in iNOS−/− mice: potential of real-time micro-CT in lung cancer chemoprevention studies. Clin Cancer Res. 2011;17:753–61. https://doi.org/10.1158/1078-0432.CCR-10-2290.
102. Gu M, et al. Silibinin inhibits inflammatory and angiogenic attributes in photocarcinogenesis in SKH-1 hairless mice. Cancer Res. 2007;67:3483–91. https://doi.org/10.1158/0008-5472.CAN-06-3955.
103. Ho Y-T, et al. Berberine suppresses *in vitro* migration and invasion of human SCC-4 tongue squamous cancer cells through the inhibitions of FAK, IKK, NF-kappaB, u-PA and MMP-2 and -9. Cancer Lett. 2009;279:155–62. https://doi.org/10.1016/j.canlet.2009.01.033.
104. Kim S, et al. Berberine suppresses TPA-induced fibronectin expression through the inhibition of VEGF secretion in breast cancer cells. Cell Physiol Biochem. 2013;32:1541–50. https://doi.org/10.1159/000356591.
105. Kim S, et al. Berberine suppresses the TPA-induced MMP-1 and MMP-9 expressions through the inhibition of PKC-α in breast cancer cells. J Surg Res. 2012;176:e21–9. https://doi.org/10.1016/j.jss.2011.11.1041.

106. Wang X, et al. Up-regulation of PAI-1 and down-regulation of uPA are involved in suppression of invasiveness and motility of hepatocellular carcinoma cells by a natural compound berberine. Int J Mol Sci. 2016;17:577. https://doi.org/10.3390/ijms17040577.
107. Hamsa TP, et al. Antiangiogenic activity of berberine is mediated through the downregulation of hypoxia-inducible factor-1, VEGF, and proinflammatory mediators. Drug Chem Toxicol. 2012;35:57–70. https://doi.org/10.3109/01480545.2011.589437.
108. Chidambara Murthy KN, et al. The natural alkaloid berberine targets multiple pathways to induce cell death in cultured human colon cancer cells. Eur J Pharmacol. 2012;688:14–21. https://doi.org/10.1016/j.ejphar.2012.05.004.
109. Liu X, et al. Berberine inhibits invasion and metastasis of colorectal cancer cells *via* COX-2/PGE2 mediated JAK2/STAT3 signaling pathway. PLoS One. 2015;10:e0123478. https://doi.org/10.1371/journal.pone.0123478.
110. Sudha T, et al. Resveratrol and its nanoformulation attenuate growth and the angiogenesis of xenograft and orthotopic colon cancer models. Molecules. 2020;25:1412. https://doi.org/10.3390/molecules25061412.
111. Tseng S-H, et al. Resveratrol suppresses the angiogenesis and tumor growth of gliomas in rats. Clin Cancer Res. 2004;10:2190–202. https://doi.org/10.1158/1078-0432.CCR-03-0105.
112. Yu H-B, et al. Resveratrol inhibits VEGF expression of human hepatocellular carcinoma cells through a NF-kappa B-mediated mechanism. Hepato-Gastroenterology. 2010;57:1241–6.
113. Tao Q, et al. Diallyl trisulfide inhibits proliferation, invasion and angiogenesis of glioma cells by inactivating Wnt/β-catenin signaling. Cell Tissue Res. 2017;370:379–90. https://doi.org/10.1007/s00441-017-2678-9.
114. Wei Z, et al. Diallyl trisulfides, a natural histone deacetylase inhibitor, attenuate HIF-1α synthesis, and decreases breast cancer metastasis. Mol Carcinog. 2017;56:2317–31. https://doi.org/10.1002/mc.22686.
115. Wang R, et al. Activity of melatonin against gastric cancer growth in a Chick embryo tumor xenograft model. Cancer Manag Res. 2021;13:8803–8. https://doi.org/10.2147/CMAR.S329728.
116. Sohn EJ, et al. Upregulation of miRNA3195 and miRNA374b mediates the anti-angiogenic properties of melatonin in hypoxic PC-3 prostate cancer cells. J Cancer. 2015;6:19–28. https://doi.org/10.7150/jca.9591.
117. Jardim-Perassi BV, et al. Effect of melatonin on tumor growth and angiogenesis in xenograft model of breast cancer. PLoS One. 2014;9:e85311. https://doi.org/10.1371/journal.pone.0085311.
118. Karadas AK, et al. Melatonin decreases metastasis, primary tumor growth and angiogenesis in a mice model of breast cancer. Hum Exp Toxicol. 2021;40:1545–57. https://doi.org/10.1177/09603271211002883.
119. Jardim-Perassi BV. Melatonin regulates angiogenic factors under hypoxia in breast cancer cell lines. Anti Cancer Agents Med Chem. 2016;16:347–58. https://doi.org/10.2174/1871520615666150511094201.
120. Goncalves N. Molecular markers of angiogenesis and metastasis in lines of oral carcinoma after treatment with melatonin. Anti Cancer Agents Med Chem. 2014;14:1302–11. https://doi.org/10.2174/1871520614666140812110246.
121. Kim K. Melatonin suppresses tumor progression by reducing angiogenesis stimulated by HIF-1 in a mouse tumor model. J Pineal Res. 2013;54:264–70. https://doi.org/10.1111/j.1600-079X.2012.01030.x.
122. Park SY. Melatonin suppresses tumor angiogenesis by inhibiting HIF-1α stabilization under hypoxia. J Pineal Res. 2010;48:178–84. https://doi.org/10.1111/j.1600-079X.2009.00742.x.
123. Zhu BH. (−)-Epigallocatechin-3-gallate inhibits VEGF expression induced by IL-6 *via* Stat3 in gastric cancer. World J Gastroenterol. 2011;17:2315. https://doi.org/10.3748/wjg.v17.i18.2315.
124. Li X. Epigallocatechin-3-Gallate inhibits IGF-I-stimulated lung cancer angiogenesis through downregulation of HIF-1α and VEGF expression. Lifestyle Genom. 2013;6:169–78. https://doi.org/10.1159/000354402.

125. Shimizu M. (−)-Epigallocatechin gallate inhibits growth and activation of the VEGF/VEGFR axis in human colorectal cancer cells. Chem Biol Interact. 2010;185:247–52. https://doi.org/10.1016/j.cbi.2010.03.036.
126. Yang F. Quercetin inhibits angiogenesis through thrombospondin-1 upregulation to antagonize human prostate cancer PC-3 cell growth *in vitro* and *in vivo*. Oncol Rep. 2016;35:1602–10. https://doi.org/10.3892/or.2015.4481.
127. Mirzaei A. Quercetin can be a more reliable treatment for metastatic prostate cancer than the localized disease: an *in vitro* study. J Cell Mol Med. 2023;27:1725–34. https://doi.org/10.1111/jcmm.17783.
128. Pratheeshkumar P. Quercetin inhibits angiogenesis mediated human prostate tumor growth by targeting VEGFR-2 regulated AKT/mTOR/P70S6K signaling pathways. PLoS One. 2012;7:e47516. https://doi.org/10.1371/journal.pone.0047516.
129. Oh SJ. Inhibition of angiogenesis by quercetin in tamoxifen-resistant breast cancer cells. Food Chem Toxicol. 2010;48:3227–34. https://doi.org/10.1016/j.fct.2010.08.028.
130. Song W. Quercetin inhibits angiogenesis-mediated human retinoblastoma growth by targeting vascular endothelial growth factor receptor. Oncol Lett. 2017;14:3343–8. https://doi.org/10.3892/ol.2017.6623.
131. Hoseinkhani Z. Medicinal plants extracts with antiangiogenic activity: where is the link? Adv Pharm Bull. 2020;10:370–8. https://doi.org/10.34172/apb.2020.045.
132. Singh RP. Grape seed extract inhibits advanced human prostate tumor growth and angiogenesis and upregulates insulin-like growth factor binding protein-3. Int J Cancer. 2004;108:733–40. https://doi.org/10.1002/ijc.11620.
133. Wen W. Grape seed extract inhibits angiogenesis *via* suppression of the vascular endothelial growth factor receptor signaling pathway. Cancer Prev Res. 2008;1:554–61. https://doi.org/10.1158/1940-6207.CAPR-08-0040.
134. Huang S. Grape seed proanthocyanidins inhibit colon cancer-induced angiogenesis through suppressing the expression of VEGF and Angl. Int J Mol Med. 2012;30:1410–6. https://doi.org/10.3892/ijmm.2012.1147.
135. Yarani R. *In vitro* inhibition of angiogenesis by hydroalcoholic extract of oak (Quercus infectoria) acorn shell *via* suppressing VEGF, MMP-2, and MMP-9 secretion. Pharm Biol. 2013;51:361–8. https://doi.org/10.3109/13880209.2012.729147.
136. Shakya G. Inhibition of metastasis and angiogenesis in Hep-2 cells by wheatgrass extract - an *in vitro* and *in silico* approach. Toxicol Mech Methods. 2018;28:205–18. https://doi.org/10.1080/15376516.2017.1388460.
137. Sim JH. Wheatgrass extract ameliorates hypoxia-induced mucin gene expression in A549 cells. Pharmacogn Mag. 2017;13:7–12. https://doi.org/10.4103/0973-1296.197660.
138. Akhtar S. Grape seed Proanthocyanidins inhibit the growth of human non-small cell lung cancer xenografts by targeting insulin-like growth factor binding Protein-3, tumor cell proliferation, and angiogenic factors. Clin Cancer Res. 2009;15:821–31. https://doi.org/10.1158/1078-0432.CCR-08-1901.
139. Feng LL. Effect of grape Procyanidins on tumor angiogenesis in liver cancer xenograft models. Asian Pac J Cancer Prev. 2014;15:737–41. https://doi.org/10.7314/APJCP.2014.15.2.737.
140. Zhang Y. Extracts and components of Ficus carica leaves suppress survival, cell cycle, and migration of triple-negative breast cancer MDA-MB-231 cells. Onco Targets Ther. 2018;11:4377–86. https://doi.org/10.2147/OTT.S171601.
141. Leong H. Green tea catechins inhibit angiogenesis through suppression of STAT3 activation. Breast Cancer Res Treat. 2009;117:505–15. https://doi.org/10.1007/s10549-008-0196-x.

Inflammation as a Target in Cancer Therapy

I. S. Aswathy, Sabu Vidya, R. S. Arya, Abhirami Harjith, Amy S. Mathew, K. S. Janasevakan, M. Caroline Diana Sherly, Noble T. Baby, and Chandrashekaran Guruvayoorappan

1 Introduction

Inflammation is an essential immune response initiated in response to either an infection or a tissue injury, and it will become pathological if it persists for a longer time, which, in turn, leads to chronic inflammatory conditions. Chronic inflammation leads to the development of various chronic diseases, such as cardiovascular disease, diabetes, cancer, autoimmune disorders, and other neurological diseases [1]. In 1863, Rudolf Ludwig Carl Virchow first recognized a functional link between inflammation and cancer by indicating the presence of leukocytes within the tumors. He hypothesized that inflammatory responses are one of the predisposing conditions responsible for tumor progression, referring it as "Out of the 'normal' inflammatory hyperplasia," which is often linked to alterations in the physiological healing process [2, 3]. Chronic inflammation plays a crucial role in tumor development by creating a tumor microenvironment (TME) [4]. The innate and adaptive immune cells such as monocytes and T lymphocytes directly interact with the inflammatory tumor microenvironment to produce inflammatory mediators in response to signals that are generated by the tumor [5, 6].

Inflammatory mediators are chemicals that play a major role in the progression of chronic inflammatory responses via acting on various target cells and associated organs in response to an external stimulus or irritant. Mast cells, neutrophils, eosinophils, macrophages, lymphocytes, endothelial cells, and platelets are the major cell types that usually generate inflammatory responses. Similarly, lipid mediators (prostaglandins and leukotrienes), polypeptide mediators, cytokines (interleukins

I. S. Aswathy · S. Vidya · R. S. Arya · A. Harjith · A. S. Mathew · K. S. Janasevakan · M. Caroline Diana Sherly · N. T. Baby · C. Guruvayoorappan (✉)
Laboratory of Immunopharmacology and Experimental Therapeutics, Division of Cancer Research, Regional Cancer Centre (Research Centre, University of Kerala), Thiruvananthapuram, Kerala, India

A. S. Ramteke et al. (eds.), *Molecular Targets in Cancer Therapy*,
https://doi.org/10.1007/978-981-96-7336-0_7

and tissue necrosis factor), and vasoactive amines (histamine and serotonin) are some of the cell-derived mediators involved in the development of inflammatory responses [7]. These inflammatory mediators then activate various signaling pathways (NF-kB, PI3K/AKT, JAK/STAT, mTOR) that participate in the development of chronic inflammation via activating transcription factors such as NF-κB, p53, HIF-1α, PPAR-γ, Nrf2, and AP-1. Upon activation, these transcription factors make changes in the expression of oncogenes, thereby enhancing the survival of cancer cells [8].

Surgery, chemotherapy, immunotherapy, radiotherapy, and hormone therapy are the conventional treatment methods used for cancer treatment. Among the different conventional or primary cancer treatment modalities, chemotherapy is the most commonly used conventional method widely employed for treating cancer. Chemotherapy helps reduce the size of the tumor, cancer-associated symptoms, and extend the lifespan of patients by modulating the immune system only if it is applied along with other conventional treatment methods such as surgery or radiation [9–11]. Even though chemotherapy is the most common method employed to combat cancer progression it is considered as a form of cytotoxic therapy since it interferes with basic cellular functions such as cell proliferation, maintenance, metastasis, angiogenesis, and apoptosis, thereby affecting not only cancerous cells but also normal cells. Chemotherapy is associated with a lot of complications that adversely affect the internal organs in variable intensities. These adverse effects immediately affect skin, hair, bone marrow, gastrointestinal tract, and kidneys and gradually spread toward the brain, heart, and lungs. During chemotherapy, patients will suffer from multiple drug resistance due to genetic factors, increased drug efflux, increased metabolism of xenobiotics, growth factors, and enhanced DNA repair. All these will reduce the curative effect of chemotherapeutic drugs and generate issues related to cancer treatment. Moreover, its low specificity, toxicity, resistivity, and hypersensitivity reactions make it necessary to search for an alternative that enhances the antitumor potential and reduces the side effects [12–14].

All the above reports highlight the importance of targeting inflammation, which plays a pivotal role in the development of cancer. Since inflammatory response plays a major role in cancer progression by promoting the growth of tumor cells, therapeutic agents capable of interacting with the key factors that are actively participating in the initiation of chronic inflammation will be a promising tool for controlling cancer. Today, numerous avenues have been opened in the field of cancer research for its treatment via controlling chronic inflammatory responses. This chapter will discuss the various treatment modalities that can be used to treat cancer via targeting chronic inflammation: the immunomodulatory effect via targeting checkpoint inhibitors, modulating macrophage activities; role of nutraceuticals, dietary supplements and targeting the microbiome on inflammation; combination therapies over conventional treatment strategies; nanomedicine and epigenetic modulation; participation of cancer stem cells as therapeutic targets, personalized medicine, and emerging therapies to target inflammation in cancer therapy.

2 Targeting Inflammatory Pathways, Inflammatory Cytokines, and Chemokines to Enhance the Effect of Therapy

The development of cancer and chronic inflammation is tightly linked to the 5-LOX pathway. There is evidence that 5-LOX products may have a part in the early phases of colorectal cancer (CRC) and 5-LOX stimulated by IL-13 and IL-4 is extensively seen in monocytes and neutrophils [15]. LOX inhibition results in decreased fibronectin assembly and collagen cross-linking, increased drug penetration, and downregulated ITGA5/FN1 expression. These effects lead to the activation of apoptosis, the suppression of FAK/Src signaling, and the resensitization to chemotherapy [16]. In vitro, LOX inhibition with β-aminopropionitrile (BAPN) dramatically lowers collagen deposition and increases drug penetration in tumor heterospheroids [17]. Acetyl-CoA carboxylase1 (ACC1) enzyme activity and lipid synthesis were inhibited as a result of AMP-activated protein kinase (AMPK) being activated by Alox12 depletion, according to mechanism studies [18]. Additionally, the study by Hou et al. revealed that the conventional anti-fibrosis medication silibinin suppressed non-small cell lung carcinoma (NSCLC) cell migration in a way that was reliant on EGFR and LOX [19].

The development of inflammation and cancer is largely dependent on the activation of NF-κB [20] and is responsible for enhanced resistance to treatment in glioblastoma (GBM) [21, 22]. The regulation of NF-κB signaling and chemoresistance is mediated via the hMTR4-PDIA3P1-miR 125/124-TRAF6 regulatory axis, which might be used in anticancer treatment [23]. Co-treatment using NF-κB inhibitor BAY 11–7082 with temozolomide (TMZ) reduced GBM viability, inhibited the NF-κB pathway, and increased apoptosis, all of which had a substantial impact on cell growth [24]. The findings by Buhrmann et al. highlight the potential of Calebin A for treating CRC by blocking the NF-κB associated inflammatory pathway, hence suppressing the multicellular tumor microenvironment, which promotes CRC cell invasion and malignancy in HCT116 cells in vitro [25]. C-miR146a, when repeatedly administered intravenously, prevented the spread of disseminated HL-60 leukemia by suppressing the expression of NF-κB target genes, which highlight the possibility of treating inflammatory and myeloproliferative diseases using myeloid cell-targeted miR-146a mimics [26]. Punicalagin has been found to inhibit the growth of cervical cancer cells and promote cell death by reducing NF-kappa B activity, showing it might serve as a potential candidate for anticancer drug development, demonstrating antagonistic effects on NF-κB signaling pathways to hinder the progression of cancer cells [27].

In a hypoxic microenvironment, study by Zhang et al. demonstrated an HIF-1a/IL-1b signaling loop between cancer cells and tumor-associated macrophages, which led to metastasis and the epithelial-mesenchymal transition of cancer cells. More significantly, their findings point to the possible use of an anti-inflammatory approach in the treatment of hepatocellular carcinoma (HCC) [28] as in the "inflammatory foci" of tissues, HER2-negative cells may have an advantage in growing due to the overexpression of HIF1-α [29]. Arsenite-induced glycolysis was enhanced by monocarboxylate transporter-4 (MCT-4), mediated by hypoxia inducible factor-1α (HIF-1α) and the last byproduct of glycolysis, lactate, was discharged into the extracellular space. Thus, arsenite-induced liver carcinogenesis was facilitated by the generation of pro-inflammatory cytokines, which were enhanced by the acidic microenvironment [30]. By reducing tumor development and boosting CD8+ T-cell memory and inflammatory capacity, HIF1α inhibitors also mimic the effects of HIF1α ablation. Moreover, combining a Treg inhibitor with an HIF1α inhibitor can significantly decrease the growth of tumors [31].

It has been demonstrated that IL-6 activates signaling pathways, the most well-studied of which are the Janus kinase (JAK) and STAT3 pathways, which cause tumor growth [32]. Iron overload may promote IL-6 synthesis, which may encourage the malignant transformation of breast cancer cells in an inflammatory environment rich in paracrine or autocrine IL-6 [33]. Consequently, co-targeting the IL-6 and IL-8 signaling pathways might be a useful therapy strategy for ovarian cancer [34]. Xiao et al. confirmed that IL-6 activates the JAK2/STAT3/Sox4/NLRP3 pathway, which has an oncogenic role in the evolution of oral squamous cell carcinoma (OSCC) [35] and an in vivo investigation showed that administering NVP-BEZ235, a novel dual-targeted ATP-competitive PI3K/mTOR inhibitor and anti-IL-6 Ab together more successfully decreased the HCC tumor burden [36]. Furthermore, berberine BBR inhibited the in vivo development of tumors xenografted with MKN-45. BBR's protective benefits against gastric cancer are linked to its ability to disrupt the IL-6/JAK2/STAT3 signaling pathway [37].

The pro-inflammatory cytokine TNF-α is mostly released by macrophages and is involved in the pathophysiology of malignant and inflammatory autoimmune disorders [38]. It has been shown that elevated TNF-α in the OSCC tumor microenvironment promotes invasion by first supporting the pro-invasive and pro-inflammatory characteristics of OSCC cells and then via its paracrine pathway, which help in the activation and recruitment of inflammatory cells [39].

A number of inflammatory illnesses and the development of tumors are linked to the chemokine receptor CXCR2 and its ligands. The CXCLs/CXCR2 axis draws neutrophils to inflammatory areas, which is important in the cancer microenvironment. Through the activation of NF-κB/P300, CXCL1 promotes the development of colon cancer, and CXCL1-based treatment offers a potentially innovative way to stop the development of colon cancer [40]; also in pancreatic cancers (PCs), the CXCR2/1 antagonist has shown anticancer and antimetastatic properties [41]. It has been demonstrated that obstructing CXCL1-CXCR2 slows the growth of tumors and lowers the TME's macrophage recruitment [42]. CXCL12 expression in CAFs might be blocked by rocuronium bromide (RB) via repressing the PI3K/AKT/mTOR signaling pathway and autophagy, which would impair the CXCL12-mediated development of EC tumors [43]; additionally, Liu et al. proposed that miR-27b-5p may have inhibited the growth of ovarian cancer by targeting CXCL1 [44].

Cancer therapy-induced inflammation typically gives remaining cancer cells with resistance to subsequent rounds of treatment, hastening cancer progression. As a result, integrating anti-inflammatory approaches with cancer treatment has increased anticancer efficacy in a number of clinical scenarios and in vivo experiments [45].

3 Anti-inflammatory Drugs in Cancer Prevention and Therapy

A wide range of biologicals with different treatment strategies have been developed for inflammation, including anticytokine therapies, B lymphocyte depletion, or blockage of lymphocyte trafficking. The history of anti-inflammatory drugs begins with the discovery of aspirin, and is still the world's most commonly used drug. The ability of aspirin to prevent the production of thromboxanes and prostaglandins, by inhibiting COX-1 and COX-2, has made it the most commonly used drug [46].

Several epidemiological studies have demonstrated that long-term use of nonsteroidal anti-inflammatory drugs (NSAIDs), such as aspirin, celecoxib, diclofenac, diflunisal, ibuprofen, indomethacin, naproxen, oxaprozin, piroxicam, salsalate, sulindac, and tolmetin, significantly reduces the chances of malignancy and cancer [47]. Anti-inflammatory drugs were found to be useful adjuvants for conventional therapeutic strategies and also could decrease the side effects of chemotherapeutics. Patients with metastatic prostate cancer, who received celecoxib and docetaxel, concurrently experienced less bone marrow damage, while those who received celecoxib with capecitabine or FOLFIRI (folinic acid, fluorouracil, and irinotecan) experienced fewer episodes of diarrhea [48, 49]. Aspirin and other anti-inflammatory medications can serve as antithrombotic medications, reduce arterial thrombosis by allowing chemotherapeutics to pass through, and improve prognosis. By suppressing the expression and activity of matrix metalloproteinases (MMPs) 3 and 9 as well as VEGF, anti-inflammatory medications such COX-2 inhibitors or dexamethasone decreased neurotoxicity and stabilized the blood–brain barrier [50].

Another potential mechanism of anti-inflammatory agents in cancer is its ability to modify the pharmacokinetics of the chemodrugs. Glucocorticoids changed the spleen and bone marrow's uptake of gemcitabine or carboplatin and increased the quantity of medications that reached the tumor in mice models, but they did not significantly alter plasma pharmacokinetics [51]. Glucocorticoids can alter the pharmacokinetics of traditional medications by enhancing their anticancer benefits and reducing their adverse effects. Furthermore, anti-inflammatory drugs can alter the way chemotherapeutic drugs are metabolized.

Chemosensitization by anti-inflammatory medications is likely another factor contributing to the combination of anti-inflammatory and chemotherapeutic treatments' potent anticancer effects. Anti-inflammatory drugs can also remodel the tumor immune environment to enhance immune checkpoint blockade efficacy. It was shown that, in xenograft mouse models, therapeutically targeting the COX2/PGE_2/EP2-4 pathway with anti-inflammatory drugs synergized with ICB in mouse cancer models [52].

A list of compounds including 8-anilinoimidazopyridines, thieno pyrimidinone, etc., or derivatives like 5-substituted quinazolinone derivatives, or a combination of drugs like curcumin and at least one NSAID possessing both anticancer and anti-inflammatory activities have been patented [47].

4 Nanomedicine in Cancer Therapy Targeting Inflammation

Nanomedicine offers innovative approaches by delivering anti-inflammatory agents precisely to the tumor microenvironment [53]. The approaches utilize nanoparticles and nanoscale systems to diagnose, treat, and monitor diseases thus regulating inflammation. These nanomedicines can potentially disrupt the tumor support system and enhance the efficacy of cancer remedies. Its most noteworthy potential lies in improving precision in drug delivery and reducing side effects, particularly in cancer. Nanomedicine utilizing nanoscale materials for diagnostic imaging and therapy holds particular promise since they have distinct properties such as TME penetration to deliver the drugs selectively [54]. Targeted delivery improves treatment efficiency by reducing systemic toxicity, a complex issue in conventional chemotherapy.

4.1 Current Landscape of Nanomedicine in Cancer Therapy

Targeted drug delivery of nanoparticles (NPs) like lipid-based carriers, polymeric NPs, and dendrimers can be engineered to recognize cancer-specific receptors and to navigate the body defenses prolonging circulation time and avoiding immune clearance. Advances include developing ligand-modified nanoparticles that specifically target overexpressed receptors on cancer cells. Studies also underscore the efficacy of ligand-targeted NPs that release drugs only in the tumor microenvironment (TME) where unique conditions, e.g., low pH or high enzyme levels can trigger drug release enhancing healing and reducing systemic toxicity [55]. Nanomedicine has also enabled combination therapies in which drug delivery can be integrated with immunomodulatory agents; these approaches have given promise in boosting antitumor immune responses and overcoming resistance in challenging cancers such as non-small cell lung cancer.

The potential for enhancing immune system activity against tumors by sustained drug release and immune checkpoint inhibitors is a focus in ongoing clinical trials [56]. Metabolism-targeted nanomedicine is a flourishing area in nanomedicine that includes metabolism in TME that supports rapid cell division and survival. Metabolism-targeted nanomedicine can interfere with pathways that supply nutrients, energy, and growth such as glucose, glutamine, and lipid metabolism [57]. This approach exploits the unique metabolic dependencies of tumors potentially broadening the effectiveness of therapies across multiple cancer types.

4.2 Nanoparticle Delivery System for Cancer Therapy

Nanoparticles, the most widely studied component of nanomedicine, range in size from 1 to 100 nm and can be engineered to carry therapeutic agents including peptides, proteins, and nucleic acids. Nanoparticles as drug delivery systems have gained significant attention in recent years due to their ability to overcome many limitations associated with traditional drug delivery methods [58]. These systems are designed to deliver the drugs specifically to targeted regions increasing drug bioavailability, reducing damage, and enhancing therapeutic outcomes. Nanoparticles offer several advantages over conventional drug delivery systems such as improved solubility of hydrophobic drugs, enhanced circulation time, and reduced systemic toxicity. These characteristics are important where inflammation-associated mechanisms can often undermine therapeutic efficacy. Several types of nanoparticles have been explored for drug delivery including lipid-based nanoparticles called liposomes consisting of spherical vesicles with an aqueous core. They are biocompatible and therefore used for the delivery of both hydrophobic and hydrophilic drugs. One of the liposomal formulations of doxorubicin called Doxil is the earliest FDA-approved nanoparticle-based drug used in cancer therapy [59]. Gold and silver nanoparticles that are classified as metallic nanoparticles have gained significant importance due to their ability to carry therapeutic agents and tunable optical properties. Gold nanoparticles are being explored in cancer therapy as they can be modified for photothermal therapy [60]. Dendrimers are highly branched tree-like structures that offer multiple attachment sites for drugs and targeting ligands. Their size and surface chemistry can be precisely controlled making them ideal for targeted drug delivery [61]. Another set of lipid-based nanoparticles called solid lipid nanoparticles (SLNs) and nanostructured lipid carriers (NLCs) offer controlled drug release and improved stability of drug formulations [62]. They are increasingly being used in the delivery of poorly soluble drugs because they have high encapsulation efficiency and loading capacity (Table 1).

Table 1 Types of nanoparticles for cancer therapy and their characteristics

Nanoparticle	Size range (nm)	Advantages	Therapeutic targets
Liposomes	50–200	Biodegradable, low toxicity	Inflammatory cytokines, tumor cells
PLGA NPs	100–300	Long circulation time	Anti-inflammatory agents
Gold NPs	5–100	Nondegradable, multifunctional	Targeted cancer cells
Dendrimers	10–50	Customizable surface	Cytokine inhibitors
Nanospheres	50–300	Biocompatible	Both hydrophilic and lipophilic drugs can be administered
Solid lipid nanoparticles	10–1000	Biocompatible	Targeted cancer cells

4.3 Targeting Inflammatory Pathways Using Nanoparticles

The ability of nanocarriers to deliver anti-inflammatory drugs directly to inflamed cancer tissues is pivotal for improving therapeutic outcomes. Anti-inflammatory drugs such as nonsteroidal anti-inflammatory drugs (NSAIDs) and corticosteroids have been shown to reduce tumor-promoting inflammation [63]. However, their systemic toxicity limits their clinical use. Nanocarriers can address these challenges by delivering drugs specifically to the tumor, sparing healthy tissues. They can also be designed for the simultaneous release of chemotherapeutic agents and anti-inflammatory drugs.

This approach targets both inflammation and cancer cells simultaneously enhancing therapeutic ability. Recent studies have explored the use of nanoparticles for the delivery of traditional small-molecule drugs and biomolecules such as siRNA, mRNA, and proteins and thus offer tremendous possibilities as drug delivery systems [64].

Nanoparticle-based drugs have already reached the market and challenges such as biocompatibility, immune evasion, and large-scale production still need to be addressed. Continued research into nanoparticles, design surface functionalization, and understanding their interactions with biological systems will likely improve their clinical utility. The future of nanoparticle drug delivery lies in personalized medicine where therapies are customized based on genetic and molecular characteristics of individual patients.

5 Personalized Immunotherapy for Inflammation in Cancer

Over the past few decades, new technologies have helped to improve cancer treatment methods. Immunotherapy is one of the emerging cancer treatment techniques, which helps to increase the efficacy of cancer treatment and offers protection against tumor recurrence by boosting the body's immune system to recognize and eliminate cancer cells [65]. Effectiveness of the immunotherapy depends on individual immune responses. Individual inflammatory or genetic profiles, which are responsible for the development of a tumor microenvironment (TME) and tumor progression differ with each patient. Therefore, traditional therapies have some limitations, which include chances of drug resistance due to inappropriate drug and dosage selection. The heterogenic nature of the tumor also leads to poor treatment responses and drug resistance [66]. Personalized immunotherapy is based on the individual genetic or inflammatory profiling. It will help with the selection of the right drug and correct dosage for the patients. Personalized therapy also helps with real-time monitoring of individual treatment response, thereby allowing clinicians to change treatment strategies accordingly for the better outcome of the patients [67].

Emerging genetic profiling technologies like next generation sequencing (NGS) can provide a complete genetic mutational landscape. It will help to identify driver gene mutations, which are responsible for altered pathways leading to chronic inflammation. The identified individual gene variations can be used to design personalized targeted therapies [68]. Gene variations responsible for tumor microenvironment formation may vary with each individual. Therefore, drug selection also varies accordingly for each person. Blood-based NGS is useful for the real-time monitoring of genes responsible for inflammation like cytokine genes such as IL-1, IL-6, TNF, etc., and also transcription factors like NFκB and STAT3 during the entire treatment period in a serial manner. Serial blood analysis within every 3/6 months will help to identify any new genetic variation or chances of recurrence of the disease [69]. Neoantigens are mutation-derived cancer-specific antigens present in cancer cells and are a target for many immunotherapy strategies [70]. The unique nature of the neoantigens is an attractive-personalized immunotherapy target. The gene mutation responsible for neoantigen formation may vary with individual patients. Commonly mutated genes in various cancers are TP53, BRAF, KRAS, etc. [71]. Therefore, biomarker-specific drug selection is essential for curative treatment. Potential personalized neoantigen vaccines provide long immune responses, which give protection from disease recurrence. RNA sequencing (RNA Seq) analysis also provides a complete picture of gene expression variations related to inflammation in cancer. It will help to differentiate the genes, which regulate chronic inflammation that leads to cancer from acute inflammation that leads to allergic reactions. There is a difference in genes, which are involved in metastatic tumor progression from primary tumor. RNA Seq analysis from both the primary and metastatic site of the same individual will help to differentiate genes, which are regulating primary tumor progression from metastatic tumor progression [72]. Personalized drug selection based on individual gene expression analysis has great advantages. Recent studies demonstrated that targeted biomarker-based treatment is a much better outcome than non-personalized treatment strategies [73].

6 Anti-inflammatory Nutraceuticals and Dietary Interventions

Diet has a critical role in the regulation of numerous metabolic activities. Thus, nutrition and its components can contribute to a state of well-being by lowering the risks associated with certain illnesses and enhancing the quality of life [74]. Plant-based remedies have been employed to treat inflammatory diseases and ailments since ancient times [75] and also dietary phytochemicals may help prevent cancer by inhibiting COX-2, among other ways [76]. Bioavailable nutraceuticals comprise microencapsulated bioactive chemicals such as phenolic compounds, flavonoids, carotenoids, antioxidant molecules, and other plant metabolites. Encapsulating nutraceuticals allows them to keep their qualities and interact with the gastrointestinal system while also enhancing their solubility and bioavailability [77–79]. For instance, resveratrol and its derivatives are structurally interesting possibilities for the creation of anti-inflammatory drugs. The limited bioavailability of resveratrol is thought to be one of the major barriers to translating the great success stories of resveratrol in the laboratory into encouraging outcomes in clinical trials [80]. By suppressing PD-L1 via the IL-6/JAK2/STAT3 axis, a nano-formulated resveratrol (Res-NP) attenuates carcinogenesis in vitro, in ovo, ex vivo, and in vivo model systems. It also deregulates the link between activated macrophage-M1 and cancer-associated fibroblasts (CAFs) [81]. Curcumin can reduce the proliferation and differentiation of cancer stem cells by inhibiting inflammatory cytokines leading to the reduction in tumor development and repopulation following radiation in addition to death signaling pathways [82] and it is also known to suppress cancer stem cell development by inhibiting transcription factors like activator protein-1, NF-κB, STAT-3, phosphodiesterases, cytokines like IL-1, IL-6, and IL-8, as well as chemokine receptors like CXCR1 and CXCR2 [83–86]. Curcumin inhibited the CAF-mediated activation of the JAK/STAT3 signaling pathway in gastric cancer (GC) cells, according to experimental models, which clearly implies that GC cell's CAF-induced chemoresistance is reversed by inhibiting the JAK/STAT3 signaling pathway [87]. Dietary ω3-polyunsaturated fatty acids (PUFA) have attracted interest in recent years as possible anti-inflammatory and anticancer medications, particularly in the large intestine, where pro-inflammatory circumstances cause colon cancer. Epidemiological, experimental, and clinical evidence suggests that ω3-PUFA may play a role in various stages of CRC management, including antineoplastic activity against human CRC cells, improving radiation and chemotherapy efficacy, alleviating cancer-associated secondary complications, and preventing CRC recurrence [88]. And thus, understanding the particular impacts of dietary components can help prevent and treat a variety of ailments, including cancer [89].

7 Combination Therapies: Targeting Inflammation Alongside Conventional Cancer Treatment

The inflammatory response has a great impact on enhancing the development of cancer in various parts of the body via activating signaling cascades. Thus, therapeutic agents that are capable of targeting inflammatory signaling cascades will be a suitable remedy to inhibit cancer progression by enhancing the efficiency of treatment therapies. Medicinal plants are enriched with biologically active compounds, which are secondary metabolites with pharmacological properties. They exert multi-target action while participating in the signaling cascade and also help to nullify the adverse effects of drugs via neutralizing the effects imparted by the drug and in turn ameliorate treatment strategies. Since the bioactive compounds possess antioxidant and other pharmacological effects, they have been used since ancient times for treating chronic inflammatory diseases such as cardiovascular disease, diabetes, and cancer due to their curative effects [90]. According to previous reports, phytocompounds with anti-inflammatory properties have the potential to treat cancer progression by preventing tumor growth and metastasis via inhibiting angiogenesis. Hence, the application of bioactive compounds having anti-inflammatory potential will be an effective method, which can be used as an alternative treatment strategy to overcome the adverse effects exerted by conventional treatment therapies [91, 92].

In the case of a complex disease such as cancer, targeted monotherapy alone leads to the development of drug resistance by actively surviving tumor cells and this has become a serious challenge for cancer treatment. However, combination therapy has the potential to tackle the issue due to the synergistic action exerted by the combination of two or more therapeutic agents [93, 94]. Combination therapy describes the combination of bioactive compounds with conventional therapy for treating cancer. The combination of these multiple therapies has demonstrated its therapeutic efficacy in various pathological conditions through targeting both multiple pathways and also the different components present in the same pathway thereby enhance the quality of life of affected persons to target both multiple pathways and also the different components present in the same pathway [95, 96]. Combining two or more pharmacological agents having therapeutic potential for targeting cancer cells exerts a positive impact on treating cancer. The strategy behind the application of combination therapy by using more than one drug is that it offers several advantages over monotherapy such as it helps to delay the establishment of resistance, reduces the duration of treatment, and allows minimum dosage of drugs. This will not only enhance the therapeutic efficiency of chemotherapeutic agents but also reduce their toxicity [97]. Combination therapy can achieve favorable outcomes such as additive, synergistic, or potentiation effects. If the therapeutic effect attained is equal to the sum of the individual therapeutic agents, the combination is said to be additive. Synergistic therapy is a type of combination therapy that helps to create a therapeutic effect greater than the effect exerted by a total of each component. If one medicine enhances the therapeutic efficacy of another by regulating its pharmacokinetic profile, it is said to be potentiation [98].

Mechanisms of synergistic effects involved in combination therapy are mainly:

(i) Multi-target effect—Two or more drugs that will synergistically act on multiple targets on both cellular and functional constituents.
(ii) Modulation of pharmacokinetic and physicochemical properties—Synergy will help to modulate the most important physicochemical property such as solubility of the drugs in order to enhance its bioavailability.
(iii) Impairment of resistant mechanism—Natural derivatives present in combination therapy antagonize the drug resistance mechanism through its pharmacological activities and thereby enhance the therapeutic benefits to patients.
(iv) Elimination or neutralization potential—Bioactive components have the potential to reduce the toxicity exerted by its associated drugs [99].

Drug–herbal combination therapy is a promising option for treating cancer because of its synergistic anticancer impact, reduced toxicity, and broader clinical use compared to monotherapy. This combined therapy enhances the patient's quality of life by raising the survival rate and dramatically lowering the resistance of cancer cells by increasing their sensitivity [100]. Combining an anticancer medication with a natural compound improves the overall therapeutic effect against cancer cells by working in concert to target various signaling pathways that contribute to the development of cancer. This will help to exert cytotoxicity on cancer cells while protecting healthy, normal cells. Since combination therapy provides a novel strategy that enhances the efficacy of treatment by improving the special qualities of natural compounds and mitigating the negative effects of chemotherapeutic medications, this treatment modality is widely recognized for treating cancer. This paves the way for developing an effective and personalized treatment approach in future for treating the chronic inflammatory disease, cancer [101, 102]. Hence, combination therapy can be recommended as a standard first-line treatment strategy over conventional therapies due to its enhanced therapeutic potential and reduced side effects.

8 Microbiome and Inflammation-Targeted Therapies

The human body hosts a vast number of microbial cells, carrying more genes than the human genome, with up to thousands of bacterial species present at any time. The microbiome and its interactions with the immune, endocrine, and nervous systems are linked to diseases like cancer, inflammatory bowel disease, and depression. Research shows a significant relationship between the microbiome and inflammation, contributing to tumor formation and increased cancer risk [45, 103]. The microbiome, vital for maintaining the balance within the body, is crucial for the functioning of almost all human physiological processes. When the dynamic and complex ecosystem of microbial community becomes imbalanced, it is termed dysbiosis. Various serious health issues, such as colorectal cancer (CRC), irritable bowel syndrome (IBS), inflammatory bowel disease (IBD), etc., have been associated with dysbiosis [104].

Traditional therapies of cancer include chemotherapy, and surgery, whereas manipulation of the microbiome like fecal-based transplantation, probiotics, and microbiome-centered therapies as a treatment for cancer is a novel approach where intestinal microbiota modulation, improvement of intestinal barrier immunomodulation, etc. is focused [105, 106].

The impact of probiotic bacteria on cancer prevention by controlling the oxidative status is complex as it involves many factors like activating antioxidant enzymatic pathways, regulating inflammatory pathways, changing microbiota-derived metabolites, etc. Lactic acid bacteria (LAB) found in yoghurt and fermented milks have been found to have therapeutic effects against diseases like cancer, infection, and inflammation. They also have immunomodulating and immunostimulant properties. Consumption of LAB and yoghurt can decrease cancer occurrence and prevent intestinal carcinogenesis [107]. For instance, in a study conducted by [151] has shown that a probiotic cocktail containing *Lactobacillus* spp., and *Bifidobacterium* spp., downregulated JAK genes and STAT genes, reducing inflammation-associated diseases like inflammatory bowel disease. A study utilizing BALB/c mice demonstrated that a cyclical yoghurt diet could inhibit colorectal carcinoma development. The inflammatory response preceding tumor formation was reduced with yoghurt consumption. Yoghurt consumption lowered pro-inflammatory cytokine levels, suggesting distinct immune mechanisms compared to indomethacin, ultimately downregulating immune responses while exerting their anti-inflammatory antitumor effects [108].

Prebiotics, nondigestible food components, offer health benefits by modulating the host's gut microbiota or the probiotics. They are classified by the number of monomers bound together, with oligosaccharides being the most preferred. Common prebiotics include xylooligosaccharides, fructooligosaccharides, isomaltooligosaccharides, transgalactooligosaccharides, galactooligosaccharides, etc. [109]. Synbiotics—a combination of probiotics and prebiotics found in food, medicine, and supplements improve microbial function and survival in the gastrointestinal tract. Fermented milk, for instance, has been shown to lower colonic tumors by controlling inflammation and gut health [110]. Postbiotics which are obtained from inactivated probiotics or their metabolites, have been shown to exhibit selective cytotoxicity and have the potential to be incorporated in supplements as well as in adjunctive therapies [111].

Fecal microbiota transplantation (FMT) is a technique by which the intestinal microbial diversity is restored by transplanting gut microbiota from healthy donors to patients, often sourced from a centralized stool bank [112]. A study conducted by Yu et al. [113] explored FMT on colorectal cancer progression in mice with induced microbial dysbiosis. FMT was shown to increase immune cells like CD8+ T, CD49b+ NK, Foxp3+ Treg cells, infiltrating into the intestine and downregulating inflammatory cytokines like IL1a, IL6, etc.

The ability of microbiome manipulation to lower inflammation and improve therapeutic results has drawn a lot of interest in the context of cancer treatment in the recent times. Table 2 lists some of the strategies of microbiome manipulation along with clinical trials.

9 Epigenetic Modulation of Inflammation in Cancer

Chromatin shape is critical for transcriptional regulation and has an impact on crucial cellular functions like as growth, differentiation, DNA repair, and apoptosis. Epigenetic changes, such as chromatin remodeling, histone acetylation, methylation, and phosphorylation, and higher order chromosomal connections, affect chromatin accessibility to transcription factors (TFs) and other transcriptional machinery in response to a variety of stimuli. Dysregulation of these systems can activate inflammatory signaling pathways, which contribute to cancer formation and dissemination [126]. In addition to controlling the expression of inflammatory genes, epigenetic modifications are also impacted by inflammatory signaling pathways that are triggered by pro-inflammatory cytokines. The epigenetic landscape may be altered by this reciprocal interaction, which could promote carcinogenesis by upregulating pro-inflammatory genes and suppressing tumor suppressor genes [127, 128]. NF-κB responds to pro-inflammatory signals like tumor necrosis factor-alpha (TNFα) and lipopolysaccharide (LPS) by translocating to the nucleus, binding chromatin, and inducing structural remodeling to regulate downstream transcriptional outputs [129]. Pro-inflammatory gene transcription is facilitated by the recruitment

Table 2 Microbiome manipulation strategies to reduce inflammation in cancer

Strategy	Mechanism of action	Effect on inflammation	Clinical trials	Examples
Probiotics	Introduction of beneficial microorganisms	Downregulating pro-inflammatory cytokine levels like TNF-α, IL-6, etc. and upregulating the production of anti-inflammatory cytokines like IL-4, IL-10, etc. [114]	(i) NCT00936572 Reducing pathogens and modulates local immunity [115] (ii) NCT03782428 Reduced pro-inflammatory cytokines-TNF-α, IL-6, IL-10, IL-12, IL-17A, IL-17C, and IL-22 [116]	*Lactobacillus rhamnosus* GG, *Lactobacillus acidophilus*, *Lactobacillus delbrueckii* sub sp. *bulgaricus*, *Bifidobacterium lactis etc*
Prebiotics	Nondigestible fibers that promote the growth of probiotics	Promotes microbiomes that produce anti-inflammatory cytokines [117]	(i) ChiCTR-IOR-17013269 Improves serum immunologic indicators before operation in CRC patients [118]	Xylooligosaccharides, Fructooligosaccharides, Isomaltooligosaccharides, Transgalactooligosaccharides, Galactooligosaccharides etc. [109]
Synbiotics	Combination of probiotics and prebiotics	Reduction of inflammation by gut microbiome restoration [119]	(i) IRCT2017092023861N6 Levels of TGF-β, IL-10, and adiponectin in the serum in women with breast cancer related lymphedema [120]	Synbiotic combination of probiotic *Lactobacillus gasseri* 505 and prebiotic *Cudrania tricuspidata* [110]
Postbiotics	Use of metabolic by-products of probiotics	Reduction of acute inflammatory response, gut barrier disruption, activation of pathways of innate immunity [121]	(i) ONC/OSS-19/2021 Improving the management of abemaciclib-induced diarrhea leads to less severe symptoms, fewer instances of dose reductions, and shorter periods of discomfort [122]	HT-CECT-7347 [121] PostbiotiX [122]
Fecal microbiota transplantation	Stool transfer from a healthy donor to rebalance microbiota.	Upregulation of anti-inflammatory cytokines and downregulation of pro-inflammatory cytokines [113, 123]	(i) NL5829 Improved response and survival in patients with metastatic gastrointestinal cancer [124]	Cancer immunotherapy trials [124] and gastrointestinal disorders [125]

of epigenetic regulators by NF-κB, including BRD4 and CBP/p300 acetyl transferases [130, 131].

Helicobacter pylori-induced gastric cancer is a classic example of how persistent inflammation can alter the host's epigenome. Pro-inflammatory genes like TNFα and IL-1β are transcriptionally driven by inflammatory pathways such PI3K/Akt, Wnt/β-catenin, and NF-κB, which are activated by *H. pylori*. These genes' expression profiles are changed by epigenetic changes, such as abnormal DNA methylation, in their promoter regions [132, 133]. Notably, these methylation modifications are associated with inflammation rather than infection itself. Immunosuppressant therapy reduced inflammation-associated methylation alterations in a gerbil model while having no significant effect on bacterial colonization [134, 135]. Epstein-Barr virus infections affect DNA methylation, silencing tumor suppressor genes like LOX and p16Ink4a and increasing the activity of pro-inflammatory genes like IL-8 and TNFα, leading to oncogenesis [136]. Nitric oxide produced by macrophages during *H. pylori* infection has been recognized as an important modulator of DNA methylation alterations in immune-related genes, such as PTPRR and MAL, confirming the link between infection-driven inflammation and epigenetic reprogramming [137]. While bacterial or viral infections are the primary drivers of inflammation, noninfectious exposures including allergens, dust, and chemicals can also cause chronic inflammatory states [138]. For example, smoking, a lifestyle-related inflammatory stimulus, causes broad epigenetic changes, including aberrant DNA and histone modifications, as well as microRNA dysregulation, which promote lung carcinogenesis [139]. Nicotine treatment in mice was found to affect the balance of DNA methylation and demethylation, altering histone acetylation and gene expression to promote lung cancer growth [140]. Hormonal therapies, such as sex steroids, have also been linked in epigenetic reprogramming, which influences methylation and inflammatory signaling in prostate cancer patients [141].

10 Macrophage Polarization and Its Role in Cancer: Insight into Therapeutic Strategies

Macrophages are malleable, allowing them to switch between M1 (classically activated) and M2 (alternatively activated) states based on the signals they encounter. They can be present in both human and animal malignancies. M1 macrophages are responsible for inflammation and immunological activation, whereas M2 macrophages aid in tissue healing and immune repression. In cancer, M2 polarization frequently promotes tumor development, but M1 polarization can increase antitumor immunity. Modulating macrophage polarization, particularly by targeting specific signaling pathways or using nanocarriers, appears to be a potential technique for treating inflammatory illnesses as well as cancer [142]. NF-κB and other signaling pathways have a role in macrophage polarization. In cancer therapy, M2-related signaling can be disrupted while M1-related pathways are enhanced. Macrophages in the tumor microenvironment (TME) release cytokines, chemokines, growth factors, and matrix metalloproteinases, with immune checkpoint receptor expression being especially high in M2 macrophages. These macrophages may contribute to tumor development, angiogenesis, and immunological suppression [3].

11 Inflammation and Cancer Stem Cells: Interplay and Therapeutic Targeting

Cancer stem cells (CSCs) are a small subset of cells within the tumor tissue. Dick and his colleagues were the first to report CSCs in acute myeloid leukemia (AML) [143]. Like normal stem cells, these CSCs exhibit self-renewal and proliferative potential. They are involved in tumorigenesis, tumor growth, metastasis, and recurrence. They also differentiate into different cell types within the tumor tissue, contributing to its heterogeneity [144, 145]. Conventional cancer treatment methods target only the tumor tissue and not the CSCs as they are highly resistant which, in turn, leads to metastasis and relapse. During tumor progression, to adapt and interact with the tumor microenvironment more CSCs are induced from differentiated cells [145]. CSCs exhibit enhanced resistance to hypoxic conditions, induce angiogenesis, and escape immune surveillance by lowering tumor-specific antigens' expression. They also gain resistance by elevating anti-inflammatory cytokines and growth factors [146]. Just like stem cells positioned in normal organs require interactions with the microenvironment to maintain their stemness, interactions with the tumor environment are necessary for CSC survival [147, 148]. The importance of tumor microenvironment in CSC survival opens the way for newer strategies to control tumor progression and metastasis.

The inflammatory microenvironment is essential for the CSC niche. Greater insight into the interplay between CSCs and inflammatory microenvironments opened the door for the design of more efficient therapeutic strategies against tumorigenesis and its progression [144]. Increased generation of inflammation mediators such as lipid peroxidation products (LPPs), reactive oxygen (ROS), and nitrogen (RNS) species under chronic inflammatory conditions causes' damage to stem cell DNA and their cumulative effect converts the normal stem cells to CSCs [149, 150]. Hence, the occurrence of LPPs, RNS, ROS, inflammatory cytokines, and chemokines in higher concentrations in the microenvironment of normal stem cells promotes the formation of CSCs.

12 Conclusion

In this chapter, we discussed different aspects of inflammation, its impact on cancer development, and specific areas focusing on the inflammatory milieu for target-oriented drug delivery and cancer therapy. Inflammation-related biological processes had greater influence in the different stages of cancer development as explained in this chapter and therefore, inflammation-oriented therapy can be an effective tool in cancer therapy. The advantages of combination therapy over conventional therapy were recommended as a primary treatment source because of the synergistic effect exerted by it. Utilizing anti-inflammatory agents from natural sources for chemopreventive activities in vivo trials was also discussed. The role played by dietary components in cancer therapy nowadays cannot be ignored. Personalized treatment strategies can be adopted to track responsible inducers in inflammation leading to cancer. As stated earlier, this approach could open new avenues in treatment modalities based on patients' genetic and molecular profiles. The influence of immune check point inhibitors in cancer therapy was also highlighted; at the same time, more research outcomes are essential for optimization. Even though precision in drug delivery and reduced toxicity were the advantages of nano-based drug delivery in cancer therapy, several advancements can be made in this field that decide unmet clinical needs. The treatment parameters also depend on various specific biomarkers on which cancer stem cells depend thereby considering it as an effective tool of targeting; despite promising preclinical and observational proof showing the use of anti-inflammatory drugs for cancer therapy or recurrence prevention, randomized controlled trials have not shown significant improvements in survival or recurrence risk reduction. Future efforts should be focused on addressing limitations like advancing our understanding of the complexity of the interplay of various immune pathways, the intricacy of tumor microenvironment, and developing synergistic therapies.

Acknowledgments The authors thank Dr. Rekha A Nair, Director, Regional Cancer Centre (RCC), and Dr. Lakshmi S, Head, Division of Cancer Research, RCC, for providing the necessary requirements and assistance.

Declaration The authors declare that there are no conflicts of interest.

References

1. Prasad S, et al. Age-associated chronic diseases require age-old medicine: role of chronic inflammation. Prev Med (Baltim). 2012;54:S29–37.
2. Korniluk A, et al. From inflammation to cancer. Irish J Med Sci. 2017;186:57–62.
3. Balkwill F, Mantovani A. Inflammation and cancer: back to Virchow? Lancet. 2001;357:539–45.
4. Wen Y, et al. Chronic inflammation, cancer development and immunotherapy. Front Pharmacol. 2022;13:1040163.
5. Wang J, et al. Crosstalk between cancer and immune cells: role of tumor-associated macrophages in the tumor microenvironment. Cancer Med. 2019;8:4709–21.
6. Sahai E, et al. A framework for advancing our understanding of cancer-associated fibroblasts. Nat Rev Cancer. 2020;20:174–86.
7. Rang HP, et al. Local hormones, inflammation and immune reactions. Pharmacology. 2007;5:217–43.
8. Dharshini LCP, et al. Regulatory components of oxidative stress and inflammation and their complex interplay in carcinogenesis. Appl Biochem Biotechnol. 2023;195:2893–916.
9. Kris MG, et al. Chemotherapy for lung cancers: here to stay. Am Soc Clin Oncol Educ B. 2014;34:e375–80.
10. Amjad E, et al. Systems biology comprehensive analysis on breast cancer for identification of key gene modules and genes associated with TNM-based clinical stages. Sci Rep. 2020;10:10816.

11. Gahtori R, et al. Anticancer plant-derivatives: deciphering their oncopreventive and therapeutic potential in molecular terms. Future J Pharm Sci. 2023;9:14.
12. Nikolaou M, et al. The challenge of drug resistance in cancer treatment: a current overview. Clin Exp Metastasis. 2018;35:309–18.
13. Schirrmacher V. From chemotherapy to biological therapy: a review of novel concepts to reduce the side effects of systemic cancer treatment. Int J Oncol. 2019;54:407–19.
14. Bukowski K, et al. Mechanisms of multidrug resistance in cancer chemotherapy. Int J Mol Sci. 2020;21:3233.
15. Mariani F, et al. Inflammatory pathways in the early steps of colorectal cancer development. World J Gastroenterol. 2014;20:9716–31.
16. Saatci O, et al. Targeting lysyl oxidase (LOX) overcomes chemotherapy resistance in triple negative breast cancer. Nat Commun. 2020;11:2416.
17. Zhu S, et al. Combination therapy of lox inhibitor and stimuli-responsive drug for mechanochemically synergistic breast cancer treatment. Adv Healthc Mater. 2023;12:2300103.
18. Huang Z, et al. ALOX12 inhibition sensitizes breast cancer to chemotherapy via AMPK activation and inhibition of lipid synthesis. Biochem Biophys Res Commun. 2019;514:24–30.
19. Hou X, et al. Silibinin inhibits NSCLC metastasis by targeting the EGFR/LOX pathway. Front Pharmacol. 2018;9:21.
20. Khan H, et al. Targeting NF-κB signaling pathway in cancer by dietary polyphenols. Crit Rev Food Sci Nutr. 2020;60:2790–800.
21. Kapoor GS, et al. Distinct domains in the SHP-2 phosphatase differentially regulate epidermal growth factor receptor/NF-κB activation through Gab1 in glioblastoma cells. Mol Cell Biol. 2004;24:823–36.
22. Yamini B. NF-κB, mesenchymal differentiation and glioblastoma. Cells. 2018;7:125.
23. Xie C. A hMTR4-PDIA3P1-miR-125/124-TRAF6 regulatory axis and its function in NF kappa B signaling and chemoresistance. Hepatology. 2020;71:1660–77.
24. Avci NG, et al. NF-κB inhibitor with Temozolomide results in significant apoptosis in glioblastoma via the NF-κB(p65) and actin cytoskeleton regulatory pathways. Sci Rep. 2020;10:13352.
25. Buhrmann C, et al. Targeting NF-κB signaling by calebin A, a compound of turmeric, in multicellular tumor microenvironment: potential role of apoptosis induction in CRC cells. Biomedicines. 2020. https://doi.org/10.3390/BIOMEDICINES8080236.
26. Su YL, et al. Myeloid cell-targeted miR-146a mimic inhibits NF-κB-driven inflammation and leukemia progression *in vivo*. Blood. 2020;135:167–80.
27. Zhang L, et al. Punicalagin promotes the apoptosis in human cervical cancer (ME-180) cells through mitochondrial pathway and by inhibiting the NF-kB signaling pathway. Saudi J Biol Sci. 2020;27:1100–6.
28. Zhang J, et al. Hypoxia-inducible factor-1α/interleukin-1β signaling enhances hepatoma epithelial-mesenchymal transition through macrophages in a hypoxic-inflammatory microenvironment. Hepatology. 2018;67:1872–89.
29. Piasecka D, et al. Upregulation of HIF1-α via an NF-κB/COX2 pathway confers proliferative dominance of HER2-negative ductal carcinoma in situ cells in response to inflammatory stimuli. Neoplasia (United States). 2020;22:576–89.
30. Luo F, et al. Enhanced glycolysis, regulated by HIF-1α via MCT-4, promotes inflammation in arsenite-induced carcinogenesis. Carcinogenesis. 2017;38:615–26.
31. Wei P, et al. Combination therapy of HIFα inhibitors and Treg depletion strengthen the anti-tumor immunity in mice. Eur J Immunol. 2023;53:2250182.
32. Browning L, et al. IL-6 and ovarian cancer: inflammatory cytokines in promotion of metastasis. Cancer Manag Res. 2018;10:6685–93.
33. Cheng M, et al. Iron promotes breast cancer cell migration via IL-6/JAK2/STAT3 signaling pathways in a paracrine or autocrine IL-6-rich inflammatory environment. J Inorg Biochem. 2020;210:111159.
34. Zhang R, et al. Combined inhibition of IL-6 and IL-8 pathways suppresses ovarian cancer cell viability and migration and tumor growth. Int J Oncol. 2022;60:1–9.

35. Xiao L, et al. Interleukin-6 mediated inflammasome activation promotes oral squamous cell carcinoma progression via JAK2/STAT3/Sox4/NLRP3 signaling pathway. J Exp Clin Cancer Res. 2022;41:166.
36. Wang Y, et al. The synergistic antitumor effect of IL-6 neutralization with NVP-BEZ235 in hepatocellular carcinoma. Cell Death Dis. 2022;13:146.
37. Xu M, et al. Berberine inhibits gastric cancer development and progression by regulating the JAK2/STAT3 pathway and downregulating IL-6. Life Sci. 2022;290:120266.
38. Schwabe RF, et al. TNF-α-induced liver injury: role of IKK, JNK, and ROS pathways. Am J Physiol Liver Physiol. 2006;290:G583–9.
39. Brierly G, et al. Tumour necrosis factor alpha (TNF-α) and oral squamous cell carcinoma. Cancers (Basel). 2023;15:1841.
40. Zhuo C, et al. CXCL1 promotes colon cancer progression through activation of NF-κB/P300 signaling pathway. Biol Direct. 2022;17:34.
41. Prajapati DR, et al. Small molecule antagonist of CXCR2 and CXCR1 inhibits tumor growth, angiogenesis, and metastasis in pancreatic cancer. Cancer Lett. 2023;563:216185.
42. Zhao H, et al. Blocking the CXCL1-CXCR2 axis enhances the effects of doxorubicin in HCC by remodelling the tumour microenvironment via the NF-κB/IL-1β/CXCL1 signalling pathway. Cell Death Discov. 2023;9:120.
43. Li J, et al. Rocuronium bromide suppresses esophageal cancer via blocking the secretion of C-X-C motif chemokine ligand 12 from cancer associated fibroblasts. J Transl Med. 2023;21:248.
44. Liu CH, et al. Tumor-suppressor miRNA-27b-5p regulates the growth and metastatic behaviors of ovarian carcinoma cells by targeting CXCL1. J Ovarian Res. 2020;13:92.
45. Zhao H, et al. Inflammation and tumor progression: signaling pathways and targeted intervention. Signal Transduct Target Ther. 2021;6:263.
46. Dinarello CA. Anti-inflammatory agents: present and future. Cell. 2010;140:935–50.
47. Khandia R, et al. Interplay between inflammation and cancer. Adv Protein Chem Struct Biol. 2020;119:199–245.
48. Javle MM, et al. Celecoxib and mucosal protection: translation from an animal model to a phase I clinical trial of celecoxib, irinotecan, and 5-fluorouracil. Clin Cancer Res. 2007;13:965–71.
49. Albouy B, et al. Preliminary results of the Prostacox phase II trial in hormonal refractory prostate cancer. BJU Int. 2007;100:770–4.
50. Candelario-Jalil E, et al. Cyclooxygenase inhibition limits blood-brain barrier disruption following intracerebral injection of tumor necrosis factor-α in the rat. J Pharmacol Exp Ther. 2007;323:488–98.
51. Wang H, et al. Pretreatment with dexamethasone increases antitumor activity of carboplatin and gemcitabine in mice bearing human cancer xenografts: *in vivo* activity, pharmacokinetics, and clinical implications for cancer chemotherapy. Clin Cancer Res. 2004;10:1633–44.
52. Pelly VS, et al. Anti-inflammatory drugs remodel the tumor immune environment to enhance immune checkpoint blockade efficacy. Cancer Discov. 2021;11:2602–19.
53. Nirmala MJ, et al. Cancer nanomedicine: a review of nano-therapeutics and challenges ahead. RSC Adv. 2023;13:8606–29.
54. Mundekkad D, et al. Nanoparticles in clinical translation for cancer therapy. Int J Mol Sci. 2022;23:1685.
55. Yan S, et al. Different targeting ligands-mediated drug delivery systems for tumor therapy. Pharmaceutics. 2024;16:248.
56. Shao K, et al. Nanoparticle-based immunotherapy for cancer. ACS Nano. 2015;9:16–30.
57. Jin J, et al. Targeting glutamine metabolism as a therapeutic strategy for cancer. Exp Mol Med. 2023;55:706–15.
58. Patra JK, et al. Nano based drug delivery systems: recent developments and future prospects. J Nanobiotechnol. 2018;16:1–33.
59. Allen TM, Cullis PR. Liposomal drug delivery systems: from concept to clinical applications. Adv Drug Deliv Rev. 2013;65:36–48.

60. Jain S, et al. Gold nanoparticles as novel agents for cancer therapy. Br J Radiol. 2012;85:101–13.
61. Kesharwani P, et al. Dendrimer as nanocarrier for drug delivery. Prog Polym Sci. 2014;39:268–307.
62. Müller RH, et al. Solid lipid nanoparticles (SLN) for controlled drug delivery-a review of the state of the art. Eur J Pharm Biopharm. 2000;50:161–77.
63. Thun MJ, et al. Nonsteroidal anti-inflammatory drugs as anticancer agents: mechanistic, pharmacologic, and clinical issues. J Natl Cancer Inst. 2002;94:252–66.
64. Paunovska K, et al. Drug delivery systems for RNA therapeutics. Nat Rev Genet. 2022;23:265–80.
65. Esfahani K, et al. A review of cancer immunotherapy: from the past, to the present, to the future. Curr Oncol. 2020;27:87–97.
66. Pérez-Ruiz E, et al. Cancer immunotherapy resistance based on immune checkpoints inhibitors: targets, biomarkers, and remedies. Drug Resist Updat. 2020;53:100718.
67. Kiyotani K, et al. Personalized immunotherapy in cancer precision medicine. Cancer Biol Med. 2021;18:955.
68. Maciejko L, et al. Cancer immunotherapy and personalized medicine: emerging technologies and biomarker-based approaches. J Mol Biomark Diagn. 2017;8:350.
69. Groisberg R, et al. Immunotherapy and next-generation sequencing guided therapy for precision oncology: what have we learnt and what does the future hold? Expert Rev Precis Med Drug Dev. 2018;3:205–13.
70. Guan H, et al. Tumor neoantigens: novel strategies for application of cancer immunotherapy. Oncol Res. 2023;31:437.
71. Duan Z, et al. Targeting the cancer neoantigens p53 and KRAS with TCR mimic antibodies. Antib Ther. 2021;4:208.
72. Lau D, et al. RNA sequencing of the tumor microenvironment in precision cancer immunotherapy. Trends Cancer. 2019;5:149–56.
73. Schwaederle M, et al. Association of biomarker-based treatment strategies with response rates and progression-free survival in refractory malignant neoplasms: a meta-analysis. JAMA Oncol. 2016;2:1452–9.
74. Dinu M, et al. Consumption of ultra-processed foods is inversely associated with adherence to the Mediterranean diet: a cross-sectional study. Nutrients. 2022;14:2073.
75. Mueller M, et al. Anti-inflammatory activity of extracts from fruits, herbs and spices. Food Chem. 2010;122:987–96.
76. Liu RH. Potential synergy of phytochemicals in cancer prevention: mechanism of action. J Nutr. 2004;134:3479S–85S.
77. Zou L, et al. Designing excipient emulsions to increase nutraceutical bioavailability: emulsifier type influences curcumin stability and bioaccessibility by altering gastrointestinal fate. Food Funct. 2015;6:2475–86.
78. Liu X, et al. Enhancement of nutraceutical bioavailability using excipient nanoemulsions: role of lipid digestion products on bioaccessibility of carotenoids and phenolics from mangoes. J Food Sci. 2016;81:N754–61.
79. Gonçalves RFS, et al. Advances in nutraceutical delivery systems: from formulation design for bioavailability enhancement to efficacy and safety evaluation. Trends Food Sci Technol. 2018;78:270–91.
80. Prakash V, et al. Resveratrol as a promising nutraceutical: implications in gut microbiota modulation, inflammatory disorders, and colorectal cancer. Int J Mol Sci. 2024. https://doi.org/10.3390/ijms25063370.
81. Pradhan R, et al. Nano formulated resveratrol inhibits PD-L1 in oral cancer cells by deregulating the association between tumor associated macrophages and cancer associated fibroblasts through IL-6/JAK2/STAT3 signaling axis. J Nutr Biochem. 2024;125:109568.
82. Song B, et al. Suppressive effects of fisetin on mice T lymphocytes *in vitro* and *in vivo*. J Surg Res. 2013;185:399–409.

83. Chen C, et al. C086, a novel analog of curcumin, induces growth inhibition and down-regulation of NFκB in colon cancer cells and xenograft tumors. Cancer Biol Ther. 2011;12:797–807.
84. Abusnina A, et al. Tumour growth inhibition and anti-angiogenic effects using curcumin correspond to combined PDE2 and PDE4 inhibition. Thromb Haemost. 2015;114:319–28.
85. Sordillo PP, Helson L. Curcumin and cancer stem cells: curcumin has asymmetrical effects on cancer and normal stem cells. Anticancer Res. 2015;35:599–614.
86. Teymouri M, et al. Biological and pharmacological evaluation of dimethoxycurcumin: a metabolically stable curcumin analogue with a promising therapeutic potential. J Cell Physiol. 2018;233:124–40.
87. Ham IH, et al. Curcumin inhibits the cancer-associated fibroblast-derived chemoresistance of gastric cancer through the suppression of the JAK/STAT3 signaling pathway. Int J Oncol. 2022;61:1–12.
88. Miccadei S, et al. ω3 Polyunsaturated fatty acids as immunomodulators in colorectal cancer: new potential role in adjuvant therapies. Front Immunol. 2016;7:486.
89. Caponio GR, et al. Nutraceuticals: focus on anti-inflammatory, anti-cancer, antioxidant properties in gastrointestinal tract. Antioxidants. 2022. https://doi.org/10.3390/antiox11071274.
90. Petrovska BB. Historical review of medicinal plants' usage. Pharmacogn Rev. 2012;6:1–5.
91. Hoseinkhani Z, et al. Medicinal plants extracts with antiangiogenic activity: where is the link? Adv Pharm Bull. 2020;10:370.
92. Wang S-T, et al. Tea polyphenols and their chemo preventive and therapeutic effects on colorectal cancer. World J Gastroenterol. 2020;26:562.
93. Gillies RJ, et al. Evolutionary dynamics of carcinogenesis and why targeted therapy does not work. Nat Rev Cancer. 2012;12:487–93.
94. Gupta B, et al. Solid matrix-based lipid nanoplatforms as carriers for combinational therapeutics in cancer. J Pharm Investig. 2017;47:461–73.
95. Gadde S. Multi-drug delivery nanocarriers for combination therapy. MedChemComm. 2015;6:1916–29.
96. Foucquier J, Guedj M. Analysis of drug combinations: current methodological landscape. Pharmacol Res Perspect. 2015;3:e00149.
97. Marimani M. Combination therapy against multidrug resistance. In: Combination therapy against multidrug resistance. Elsevier; 2020. p. 39–64.
98. Hu Q, et al. Recent advances of cocktail chemotherapy by combination drug delivery systems. Adv Drug Deliv Rev. 2016;98:19–34.
99. Pezzani R, et al. Synergistic effects of plant derivatives and conventional chemotherapeutic agents: an update on the cancer perspective. Medicina (B Aires). 2019;55:110.
100. Wu J, et al. Exploration of the use of natural compounds in combination with chemotherapy drugs for tumor treatment. Molecules. 2023;28:1022.
101. Kearney T, et al. Pepper spray injury severity: ten-year case experience of a poison control system. Prehospital Emerg Care. 2014;18:381–6.
102. Zhong WL, et al. Antitumor effect of natural product molecules against lung cancer. In: A Global scientific vision-prevention, diagnosis, and treatment of lung cancer. IntechOpen; 2017. https://doi.org/10.5772/67241.
103. Gilbert JA, et al. Current understanding of the human microbiome. Nat Med. 2018;24:392–400.
104. Dixit K, et al. Restoration of dysbiotic human gut microbiome for homeostasis. Life Sci. 2021;278:119622.
105. Bajaj JS, et al. Promises of microbiome-based therapies. J Hepatol. 2022;76:1379–91.
106. Śliżewska K, et al. The role of probiotics in cancer prevention. Cancers (Basel). 2021. https://doi.org/10.3390/cancers13010020.
107. de LeBlanc A, et al. The application of probiotic fermented milks in cancer and intestinal inflammation. Proc Nutr Soc. 2010;69:421–8.
108. Moreno D, de Leblanc A, et al. Inflammatory immune response. Eur J Inflamm. 2004;2:21–31.
109. Mahdavi M, et al. Preventing colorectal cancer through prebiotics. Microorganisms. 2021;9:1325.

110. Oh NS, et al. Cancer-protective effect of a synbiotic combination between Lactobacillus gasseri 505 and a Cudrania tricuspidata leaf extract on colitis-associated colorectal cancer. Gut Microbes. 2020;12:1785803.
111. Chuah L-O, et al. Postbiotic metabolites produced by Lactobacillus plantarum strains exert selective cytotoxicity effects on cancer cells. BMC Complement Altern Med. 2019;19:1–12.
112. Chen D, et al. Fecal microbiota transplantation in cancer management: current status and perspectives. Int J Cancer. 2019;145:2021–31.
113. Yu H, et al. Fecal microbiota transplantation inhibits colorectal cancer progression: reversing intestinal microbial dysbiosis to enhance anti-cancer immune responses. Front Microbiol. 2023;14:1126808.
114. Azad MAK, et al. Immunomodulatory effects of probiotics on cytokine profiles. Biomed Res Int. 2018;2018:8063647.
115. Gianotti L, et al. A randomized double-blind trial on perioperative administration of probiotics in colorectal cancer patients. World J Gastroenterol WJG. 2010;16:167.
116. Zaharuddin L, et al. A randomized double-blind placebo-controlled trial of probiotics in post-surgical colorectal cancer. BMC Gastroenterol. 2019;19:1–8.
117. Bilal M, et al. Dietary component-induced inflammation and its amelioration by prebiotics, probiotics, and synbiotics. Front Nutr. 2022;9:931458.
118. Xie X, et al. Effects of prebiotics on immunologic indicators and intestinal microbiota structure in perioperative colorectal cancer patients. Nutrition. 2019;61:132–42.
119. Mohanty D, et al. Prebiotics and synbiotics: recent concepts in nutrition. Food Biosci. 2018;26:152–60.
120. Saneei Totmaj A, et al. The effects of synbiotic supplementation on serum anti-inflammatory factors in the survivors of breast cancer with lymphedema following a low calorie diet: a randomized, double-blind, clinical trial. Nutr Cancer. 2022;74:869–81.
121. Martorell P, et al. Heat-treated Bifidobacterium longum CECT-7347: a whole-cell postbiotic with antioxidant, anti-inflammatory, and gut-barrier protection properties. Antioxidants. 2021;10:536.
122. De Sanctis R, et al. Optimizing abemaciclib-induced diarrhea management in patients with breast cancer: a pragmatic 2-group study using a postbiotic microbiota stabilizer. Oncologist. 2024;29:e1113.
123. Burrello C, et al. Therapeutic faecal microbiota transplantation controls intestinal inflammation through IL10 secretion by immune cells. Nat Commun. 2018;9:5184.
124. de Clercq NC, et al. Fecal microbiota transplantation from overweight or obese donors in cachectic patients with advanced gastroesophageal cancer: a randomized, double-blind, placebo-controlled, phase II study. Clin Cancer Res. 2021;27:3784–92.
125. Wang J-W, et al. Fecal microbiota transplantation: review and update. J Formos Med Assoc. 2019;118:S23–31.
126. Simon S, Labarriere N. PD-1 expression on tumor-specific T cells: friend or foe for immunotherapy? Onco Targets Ther. 2018;7:e1364828.
127. Bhatt D, Ghosh S. Regulation of the NF-κB-mediated transcription of inflammatory genes. Front Immunol. 2014;5:71.
128. Bannister AJ, Kouzarides T. Regulation of chromatin by histone modifications. Cell Res. 2011;21:381–95.
129. Bernardo G, et al. The emerging role of the microbiota in breast cancer progression. Cells. 2023;12:1945.
130. Brown JD, et al. NF-κB directs dynamic super enhancer formation in inflammation and atherogenesis. Mol Cell. 2014;56:219–31.
131. Amrutkar RD, et al. Precision targeting of BET proteins-navigating disease pathways, inhibitor insights, and shaping therapeutic frontiers: a comprehensive review. Curr Drug Targets. 2024;26:147.
132. Shanmugam MK, Sethi G. Role of epigenetics in inflammation-associated diseases. Epigenet Dev Dis. 2012:627–57.

133. Sah DK, et al. Reactive oxygen species and H. pylori infection: a comprehensive review of their roles in gastric cancer development. Antioxidants. 2023;12:1712.
134. He Z, et al. Cellular plasticity and fate determination in gastric carcinogenesis. Iscience. 2024;27:109465.
135. Hattori N, Ushijima T. Epigenetic impact of infection on carcinogenesis: mechanisms and applications. Genome Med. 2016;8:1–13.
136. Toller IM. Molecular mechanisms of inflammation-induced cancer: T-cells as targets for prevention and cure of Heliobacter-related gastric preneoplasia. 2010.
137. Maccani JZJ, Maccani MA. Altered placental DNA methylation patterns associated with maternal smoking: current perspectives. Adv Genomics Genet. 2015;2015:205–14.
138. Zhang L, et al. Targeting ion channels for the treatment of lung cancer. Biochim Biophys Acta (BBA) Rev Cancer. 1876;2021:188629.
139. Zong D, et al. The role of cigarette smoke-induced epigenetic alterations in inflammation. Epigenetics Chromatin. 2019;12:1–25.
140. Morel D, et al. Combining epigenetic drugs with other therapies for solid tumours—past lessons and future promise. Nat Rev Clin Oncol. 2020;17:91–107.
141. Uz U, Eskiizmir G. Association between interleukin-6 and head and neck squamous cell carcinoma: a systematic review. Clin Exp Otorhinolaryngol. 2021;14:50–60.
142. Bhat AA, et al. Cytokine-and chemokine-induced inflammatory colorectal tumor microenvironment: emerging avenue for targeted therapy. Cancer Commun. 2022;42:689–715.
143. Bonnet D, Dick JE. Human acute myeloid leukaemia is organized as a hierarchy that originates from a primitive hematopoietic cell. Nat Med. 1997;3:730–7.
144. Zhang S, et al. Interplay between inflammatory tumor microenvironment and cancer stem cells. Oncol Lett. 2018;16:679–86.
145. Chu X, Tian W, et al. Cancer stem cells: advances in knowledge and implications for cancer therapy. Signal Transduct Target Ther. 2024;9:170.
146. Albini A, et al. Cancer stem cells and the tumor microenvironment: interplay in tumor heterogeneity. Connect Tissue Res. 2015;56:414–25.
147. Voog J, Jones DL. Stem cells and the niche: a dynamic duo. Cell Stem Cell. 2010;6:103–15.
148. Kise K, et al. Tumor microenvironment for cancer stem cells. Adv Drug Deliv Rev. 2016;99:197–205.
149. Okada F. Inflammation and free radicals in tumor development and progression. Redox Rep. 2002;7:357–68.
150. Ohnishi S, et al. DNA damage in inflammation-related carcinogenesis and cancer stem cells. Oxidative Med Cell Longev. 2013;2013:387014.
151. Aghamohammad, S., et al. The effects of the probiotic cocktail on modulation of the NF-kB and JAK/STAT signaling pathways involved in the inflammatory response in bowel disease model. BMC Immunol. 2022; 23:8.

Epithelial–Mesenchymal Transition: Insights into Molecular Pathways and Cancer Therapeutics

Mantu Lal and Ashu Bhan Tiku

1 Introduction

Epithelial–mesenchymal transition (EMT) is a developmental process that plays a significant role in cancer progression also. Epithelial cells can undergo this process under certain circumstances, during which they acquire migratory and invasive characteristics and lose their cell polarity and cell–cell adhesions to become mesenchymal in nature. Cell polarity is the innate asymmetry observed in cells, whether in their composition, arrangement, or form. Hence, the spatial variations that are observed in a cell's shape, structure, and function collectively fall under the umbrella of cell polarity. All types of developing, migratory, and epithelial cells depend on cell polarity to operate. Cell polarity is controlled through cells' interactions with their neighboring cells as well as the extracellular matrix (ECM) [1, 2]. EMT can take place in polarized epithelial cells, which normally use their basal surface to communicate with the basement membrane. As a result, they can acquire the traits of a mesenchymal cell, including greater invasiveness, improved migration potential, increased resistance to apoptosis, and significantly higher ECM component production. The breakdown of the basement membrane underneath an epithelial layer and the emergence of a mesenchymal cell capable of migrating out of it are indicators that an epithelial layer transition has completed [3].

This phenomenon was observed for the first time by Elizabeth Hay, who had been studying the process of embryonic development and gastrulation in chickens [4]. The underlying plasticity of the epithelial phenotype is brought forth by EMT via the demonstration of the capacity of epithelial cells to partially or fully transform into mesenchymal cells and back [5]. A lot of critical internal events happen in

M. Lal · A. B. Tiku (✉)
Radiation and Cancer Therapeutics Lab, School of Life Sciences, Jawaharlal Nehru University, New Delhi, India
e-mail: abtiku@mail.jnu.ac.in

A. S. Ramteke et al. (eds.), *Molecular Targets in Cancer Therapy*,
https://doi.org/10.1007/978-981-96-7336-0_8

epithelial cells during EMT: (1) loss of cell junctions and apical-basal polarity, (2) alteration of the signaling pathways that determine cell shape, and (3) reprograming of gene expression. All these events synergistically lead to an increase in the motility of the cell and signify the presence of an invasive phenotype [6, 7]. Therefore, EMT is the "differentiation switch" that induces the transformation of polarized epithelial cells into contractile and motile mesenchymal cells [8].

Most adult tissues and organs are produced by a sequence of processes that induce both EMT and its opposite process, mesenchymal–epithelial transition (MET). This phenomenon shows the complexity by which EMT signaling pathways work, as they can trigger EMT in developmental, normal, as well as transformed environments.

EMT is a crucial stage in the development of primary tumors into metastatic ones. Although up to 90% of cancer-related deaths are caused by metastasis yet this process is still one of the most mysterious pathological events in the development of cancer [9]. Several signaling pathways, such as NF-κB, Notch, Wnt, RTK/Ras, TGFβ, and Hedgehog, influence EMT regulation along the course of tumor development. One of the key mechanisms of EMT is the repression of E-cadherin by transcriptional regulators such as Twist and Snail [10]. However, there are still many unknowns in the gene regulatory network underlying EMT. Since EMT is a fundamental process involved in embryogenesis, organ development, wound healing, fibrosis, and cancer, unraveling the underlying complex gene regulatory mechanisms is crucial for developing effective EMT-targeted cancer therapies [11].

2 Major Steps Involved in EMT Development in Cancer

In metastatic tumors, the steps of epithelial–mesenchymal transition (EMT) involve a series of molecular and cellular changes. Changes in the expression of important cell–surface proteins, cytoskeletal proteins, extracellular proteins, transcription factors, and in particular microRNAs are involved in the induction and drive of EMT. Many of these dysregulated proteins and microRNAs can therefore be used as biomarkers to study and monitor EMT. The various steps involved are as follows.

2.1 Loss of Cell Polarity and Cell–Cell Adhesion

Epithelial cells typically exhibit a polarized structure and are connected by tight junctions. During EMT, these cells lose their polarity, and the connections between them are diminished. Hence, the process of EMT finally leads to outcomes such as downregulation of epithelial marker proteins and reduction of adherent junction proteins, including cytokeratin, E-cadherin, β-catenin, and γ-catenin, which are essential for cell–cell adhesion [7, 12].

It has been observed that the induction of EMT is particularly driven by the loss of E-cadherin either through downregulation or delocalization. The repression of this molecule, which is a critical factor for maintaining epithelial integrity, is seen most often in cells undergoing EMT. For example, in a study, it was found that immune-deficient mice that received cell lines lacking E-cadherin exhibited higher tumorigenicity and metastasis [13]. It has also been observed that O-glycosylation can hinder the transport of E-cadherin to the plasma membrane, which can impede its function [14]. On the other hand, however, E-cadherin has pro-metastatic functions as well, indicating that maintaining certain epithelial characteristics may aid in the development of carcinomas [15–17].

2.2 Gain of Mesenchymal Markers

As cells lose their epithelial characteristics, they begin to express markers typical of mesenchymal cells, such as N-cadherin, P-cadherin, intermediate filament vimentin, and fibronectin. This process is regulated by various transcription factors and signaling pathways [18]. These markers are associated with increased motility and invasive capabilities as they affect both the morphological and molecular plasticity of cells and are noted to be upregulated in cells with mesenchymal characteristics [19–21].

2.3 Increased Motility and Invasiveness

The cells acquire the ability to move through the ECM, which is a hallmark of mesenchymal cells. This newly acquired feature occurs because of the mesenchymal markers gained that were mentioned in the previous step. The increased motility facilitates the migration of cells away from the original epithelium of their origin.

Other than internal cellular factors, many other external factors, such as stromal cells from blood vessels, lymphatic endothelial cells, cancer-associated fibroblasts (CAFs), and inflammatory cells that originate from the bone marrow, often play a role in regulating the movement and invasion of cancer cells, as well as guiding their path through the ECM [22, 23].

2.4 Extracellular Matrix Remodeling

The role of the ECM, growth factors such as TGF-β in polarity determination, and similar other molecules engaged in extracellular modeling during cancer progression and metastasis signifies the importance of the microenvironment in the pathogenesis process [24]. The ECM's mechanical properties and composition are key regulators of EMT, and in turn, EMT drives changes in the ECM [25]. Cells secrete matrix metalloproteinases (MMPs) and other enzymes that degrade the surrounding ECM, facilitating migration and invasion [26, 27]. MMPs release proteolytic laminin and collagen type IV fragments from the basement membrane, which influence cell adhesion and migration [28–31].

2.5 Epigenetic Modifications

Dysregulation of gene expression is a major driver of EMT and plays a role in the development and progression of several cancers. Changes in DNA methylation and histone modifications are important for a stable transition from an epithelial to a mesenchymal phenotype. Epigenetic mechanisms frequently control the expression of key EMT transcription factors (EMT-TFs), including Snail, Twist, and ZEB1. This can lead to an increase in mesenchymal markers such as vimentin and N-cadherin while simultaneously decreasing epithelial markers such as E-cadherin [32, 33]. Hence, epigenetic modifications also play a crucial role in the process of EMT, particularly through the regulation of gene expression.

2.6 Molecular Reprogramming

Transcription factors such as Snail, Zeb1/Zeb2, and Twist play pivotal roles in driving the EMT process. As established earlier, the epithelial marker, E-cadherin, is repressed during EMT. When E-cadherin is downregulated, β-catenin is released. This protein then moves to the nucleus and activates transcription factors like ZEB, Twist, Snail, and Slug, which can then go on to suppress the expression of E-cadherin in a positive feedback loop.

These factors repress epithelial gene expression and activate mesenchymal gene expression through interactions with various epigenetic regulators and chromatin remodeling machinery. Cell migration, tissue morphogenesis, and the development of cancer are all aided by the stimulation of these transcription factors [34, 35].

Since multiple molecular pathways cross talk with individual facets of EMT, driving cancer spread and helping the tumor gain invasive and metastatic abilities, a comprehensive knowledge of the steps involved in EMT is essential for the development of therapeutic approaches to inhibit EMT and halt the progression of cancer.

3 Various Molecular Pathways Involved in EMT

In the past, signaling pathways were examined separately and thought of as linear, noninteracting channels; however, studies in the newly developing field of systems biology have highlighted the intricate network of signals throughout development as well as tumorigenesis. Thus, a greater awareness of the significance of pathway cross talk has resulted from these investigations [24].

It has been demonstrated that TGF-β activates all ERK, p38 MAPK, and JNK MAPKs in a variety of cell types [36, 37] like hematopoietic stem cells [38], breast cancer [39], prostate cancer [40], melanoma [41], etc. These pathways interact in complex networks that promote the EMT process [42]. The development of EMT depends on both Smad and non-Smad pathways that are induced by TGFβ. The Wnt, SHH, and TGF-β signaling pathways work together and interact with each other via key transcription factors like Smad to control EMT, facilitating metastasis in the process. The reason for this is that various signals, such as Notch, Integrin, TNF-α, and EGF, can work in tandem with TGF-β to drive invasion and metastasis [8, 43, 44].

TGF-β is one of the main elements controlling EMT because of its vast influence over the signaling that regulates the process. It has been noted that it is critically important for the continued maintenance of the state of EMT because when this pathway is inhibited via the use of TGF-β antagonists, a reversal is seen and the opposite phenomenon, MET, is induced instead. This was seen in the case of a mouse mammary carcinoma cell line JygMC(A), which has the natural tendency to metastasize to the liver and the lungs. However, when TGF-β was inhibited in the cells using adenoviruses carrying Smad7 cDNA (an antagonist of TGF-β signaling), the cells lost their spontaneous metastatic ability in a xenograft in vivo model [45]. Similarly, knocking down Smad4 results in the loss of EMT and retention of E-cadherin in breast cancer cells [46], pancreatic cancer cells [47], and mouse hepatocytes in vitro [48]. In a very interesting study from 2012, Gu et al. demonstrated that gastric cancer-derived exosomes could induce mesenchymal stem cells (MSCs) to be differentiated into carcinoma-associated fibroblasts (CAFs) [37]. They postulated that this phenomenon occurred due to an exosome-dependent transfer of TGF-β [49]. In another study focusing on breast cancer, it was noted that the accumulation of cholesterol in the endoplasmic reticulum could trigger ER stress, which in turn inhibited EMT through modulation of the TGF-β/Smad signaling pathway [50]. Expression of N-cadherin in pancreatic cancer was correlated with neural invasion and was influenced by growth factors such as FGF-2 and TGF-β [51].

NF-κB is another important signaling pathway found to play an essential role in EMT. TGF-β can function via Smad-independent pathways and activate TAK1 (TGF-β-activated kinase 1), which is a MAP kinase (MAPK). TAK1 can then go on to activate NF-κB by phosphorylating the IKK complex upon TGF-β induction; this interaction highlights the interplay between TGF-β and NF-κB pathways [52, 53]. Additionally, the NF-κB/EMT axis mediates drug resistance in tumor cells, and it has been observed that NF-κB inhibition can lead to a decrease in the metastatic abilities of a tumor in many cancer types [52, 54].

These signaling pathways, along with other numerous molecular changes, cause the activation of one or more transcriptional repressors of the E-cadherin gene, like the Snail proteins (Snail1, Snail2), ZEB1, ZEB2 (Sip1), E47, and the Twist proteins. All of these repressors act upon *CDH1,* which is the gene that encodes for E-cadherin [55]. Twist, in particular, has been found to be a crucial regulator of EMT, as its expression is connected to the activation of NF-κB signaling pathways, which amplifies the overall EMT process [56]. The activation of these transcriptional repressors, thus, directly ties into the first step of EMT, that is, the repression of E-cadherin and β-catenin, leading to loss of cell–cell adhesion and changes in cell polarity.

Two of the major signals implicated in EMT are cytoskeletal remodeling by Rho GTPase signaling and Rho-regulated actin motility. Alterations in these primarily change the processes of binding, polymerization, and stability of the actin-based cytoskeleton [57, 58]. It has been found that TGF-β-induced EMT leads to changes in the orientation of the tubulin-based microtubule assembly in MCF-7, A549, and HaCaT cell lines. Before EMT, the microtubules were placed along the cell edges while after it, they became perpendicular to the cell membrane. As a result of this change, the area covered by the microtubule assembly increased and all the cell lines studied had a larger surface area after EMT [59]. It has also been observed that depletion of cytokeratin 18 (CK18) can induce partial EMT and stemness by affecting EpCAM expression in breast cancer [60]. Figure 1 depicts these changes at the cellular level.

Vimentin is an important mesenchymal marker, as it has been consistently found to be upregulated across various cancers. It represents a good biomarker for identifying pathological EMT and can be a potential therapeutic target. This protein can increase cell "stiffness" with its intricate filamentous network, which leads to good cancer cell survival within the crowded conditions of a solid tumor. Thus, vimentin provides a viscous yet elastic framework that safeguards the nucleus and other important organelles by maintaining their position and structure [61]. Since vimentin is heavily phosphorylated by several different kinases, it has been hypothesized that vimentin may function as a "phosphorylation sink," competing with other substrates like GEF-H1, which is involved in actin reorganization via the RhoA pathway. Vimentin also acts upon GTPase Rac1, which is a factor that works in actin polymerization. Hence, vimentin can also indirectly weaken the actin cytoskeleton by interfering with its phosphorylation status [62, 63].

Tumor progression and metastasis are also significantly influenced by the tumor microenvironment. During tumor development, the integrity of the neighboring basement membrane is an important factor. Hence, the ECM plays a vital role in regulating the process of EMT. An essential matrix metalloproteinase is the enzyme MMP3, which has the ability to break down the matrix that stromal fibroblasts secrete. It can cause mammary epithelial cells to undergo EMT in both in vivo and

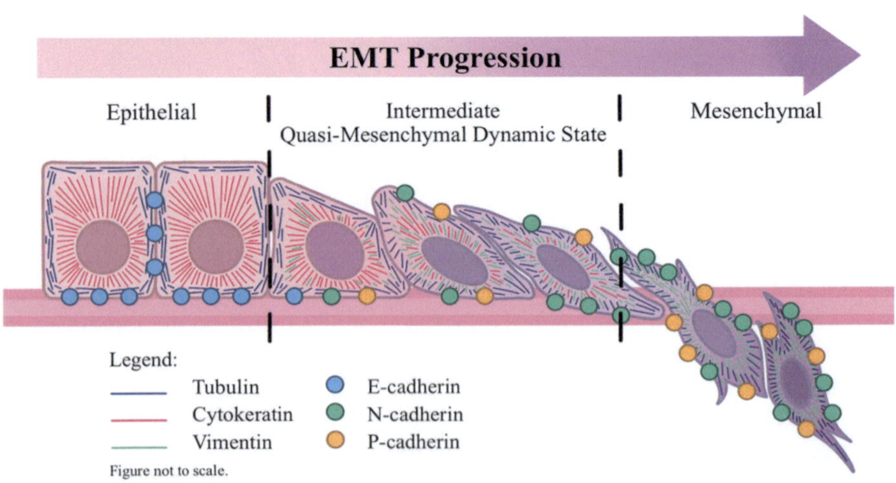

Fig. 1 A cartoon showing the dynamic process of epithelial–mesenchymal transition

in vitro conditions [64]. Human carcinoma cell lines that have activated EGFR exhibit increased MMP-9 activity, which is linked to an increase in in vitro cell invasion [65].

Collagen is a key molecule that is known to affect ECM composition. Hepatic inflammation, fibrosis, and chronic liver injury are the main causes of hepatocellular carcinoma, and liver fibrosis is a precondition for cancerization. As fibrillation progresses, the configuration of different types of collagens is altered significantly. They undergo changes and become fibrillar types I and III instead of the normal non-fibrillar types IV and V [66]. It was found out that Discoidin Domain Receptor 2 (DDR2) is a widely expressed collagen receptor tyrosine kinase that can be upregulated due to matrix stiffness and hence drive invasion [67].

Other than the ECM, the nature of the microenvironment itself is important for the development of a tumor. The tumor microenvironment favors rapid growth, which can quickly lead to the development of hypoxic conditions. This is the reason why sustained growing tumors require their own local vasculature systems for oxygen and metabolite delivery. Due to the hypoxic conditions of the microenvironment, tumor cells are quick to regulate genes such as HIF-1α which are used to combat the downsides of hypoxia. This, in turn, can drive cancer progression further [68, 69]. Blocking TGF-β signaling, or knocking down HIF-1α, can reverse hypoxia-induced EMT and reduce the stemness of cancer cells [30, 70]. It was also reported that conditional suppression of TGF-β signaling in a bladder cancer model led to a reduction in cancer stem cell populations, thereby inhibiting tumor progression and invasion [71]. Since this kind of EMT is also mediated by transcription factors such as the Snail family, ZEB1/2, etc., it results in the loss of E-cadherin as well [72–74].

Moreover, studies have also shown that EMT is associated with modifications that alter the chromatin structure, thereby regulating the accessibility of transcription factors to their target genes. After induction of EMT via TGF-β in mouse hepatocytes (AML12 cell line), changes were observed in the transcriptional mark H3 Lys36 trimethylation (H3K36Me3), the euchromatin mark H3 Lys4 trimethylation (H3K4Me3), and the heterochromatin mark H3 Lys9 dimethylation (H3K9Me2). H3K9Me2 and H3K4Me3 were globally reduced, while H3K36Me3 was globally increased [75]. It has also been suggested that an "EMT acetylome" regulates epithelial maintenance and EMT [76].

Other than DNA methylation and histone modifications, other epigenetic regulators include microRNAs, among various other competing endogenous RNAs (ceRNAs). MicroRNAs (miRNAs) are small noncoding RNA molecules that are approximately 18–22 nt long. Both transcriptional and posttranscriptional gene expression regulation are maintained by these small RNAs. For example, the miR-200 family exhibits abnormal expression in cancer and plays a significant role in the development and progression of malignant transformation [77]. Histone demethylase KDM5B controls the expression of members of the miR-200 family [78]. A 2013 study found that the miR-200a-200b-429 cluster was found to be silenced through polycomb group-mediated histone modifications in stem-like populations in breast cancer cells. The study also observed that the miR-200c-141 cluster was controlled by DNA methylation [79]. Meanwhile, miR-200 repressor, transcription factor GRHL2, acts as an EMT suppressor by maintaining the epithelial status of ovarian cancer by modulation of histone modifications [80]. Hence, the entire regulatory landscape of EMT is further complicated by the epigenetic control of these EMT miRNA-TF networks. Comprehending the intricate relationship between these small microRNAs and EMT processes is also essential for creating viable cancer treatment approaches since dysregulation of EMT-associated microRNAs can also contribute to chemoresistance [81].

Long noncoding RNAs (lncRNAs) are another type of ceRNA of length more than 200 nt that are known to change their expression during EMT. The induction of EMT causes global reprogramming of the lncRNA transcriptome, with thousands of lncRNAs exhibiting changed expression [82]. Important EMT transcription factors, such as Twist, ZEB, TGF-β, and Snail, are known to be regulated by them. Because of their high expression specificity and increased abundance, these molecules have the potential to be a significant source of biomarkers and possible therapeutic targets [83].

In a 2021 study, Deshmukh et al. found that mRNAs of many stemness-related factors such as *sox4*, *myc*, *fos*, and *dnmt1* are upregulated after EMT induction via TGF-β1 in immortalized MCF10A cells [84]. This is an important observation, as the steps involved in EMT, along with the cross talk of various molecular pathways, collectively enable cells to detach from the primary tumor, invade surrounding tissues, and potentially enter the bloodstream, leading to metastasis in the context of cancer. The cells that gain this ability to migrate to distal sites have been termed circulating tumor cells (CTCs). Twist and Snail are proteins that are important for inducing EMT, and they have been found to be increased in CTCs. This is an interesting observation as this upregulation enables the cancer cell phenotype to return to a quiescent state that can be like the one observed in stem cells [85]. A 2011 study gathered CTCs based on the expression of an epithelial marker called EpCAM from advanced breast and prostate cancer. It found out that 80% of these EpCAM-expressing CTCs also expressed a combination of mesenchymal and epithelial markers, indicating that these CTCs were transitioning along the EMT axis, further corroborating the connection between EMT and CTCs [86].

4 Targeting EMT Using Phytochemicals as Therapeutic Strategy

Many naturally occurring, plant-derived compounds called phytochemicals have emerged as promising agents for targeting EMT. Various phytochemicals have been observed to act as toxins on competing species and hence are able to indirectly promote the growth of their origin species. This quality has garnered interest in them as possible therapeutic agents. Because of their anti-EMT characteristics and comparatively low toxicities, phytochemicals are particularly interesting as additional treatments and chemo-preventative options, despite their low bioavailability. As many as 4000 compounds have been termed as phytochemicals [87, 88].

Numerous phytochemicals, like phenylpropanoids, can interfere with different EMT-associated signaling pathways (Table 1). These compounds suppress EMT by modulating transcription factors like Snail, Slug, Twist, and Zeb by interfering with important signaling pathways such as TGF-β, NF-κB, Wnt, and Notch [89, 90]. Dietary phytochemicals can also act upon histone modifications via these TFs. In particular, they can be used in chemopreventive strategies to target NF-κB signaling [91, 92]. Dietary phytoestrogens, such as genistein, resveratrol, kaempferol, and 3,3′-diindolylmethane (DIM), found in commonly consumed vegetables and tea, have shown potential in inhibiting EMT by upregulating epithelial markers and downregulating mesenchymal markers. These phytoestrogens may be used as adjuvants or alternative therapies alongside traditional HRT [93].

A lot of emerging research has shown that not only can phytochemicals influence EMT, but that they may also be able to reverse the process, and even reverse chemoresistance. They have all these abilities due to their ability to modulate coding and noncoding genes related to cancer hallmarks like apoptosis, angiogenesis, and metastasis [94]. Furthermore, phytochemicals have the ability to alter tumor suppressors and oncogenes by modulating miRNA expression. For example, in a 2017 study, sulforaphane exerted an anti-metastatic activity by decreasing miR-616-5p levels in both in vitro and in vivo models of lung cancer [95]. Similarly, genistein displayed anti-EMT action in hepatocellular carcinoma by upregulating miR-1275. Upregulating this microRNA led to the inhibition of the EIF5A2/PI3K/Akt pathway as it was found to target the *EIF5A2* gene [96].

Many dietary phytochemicals such as Epigallocatechin gallate (EGCG), sulforaphane, and genistein have been noted to act upon CSCs as well, making them appropriate potential supplements to traditional cancer therapies. This is vital, as CSCs are currently not targeted completely by presently available therapies and often evade treatment [97].

5 Conclusion

More and more data indicate that signaling pathways are important for controlling EMT. Cross talk between these pathways is essential for driving the process, and EMT is fundamental for metastasis. Hence, it is essential to deeply study these pathways, especially TGF-β1 signaling. Developing mainstream molecular therapies, as well as phytochemical-based adjuvants, can help target the drivers of this process. Dietary phytochemicals derived from traditional remedies and natural sources can also supplement traditional therapies for better therapeutic outcomes. Their usage might even lessen the side effects that commonly occur in patients due to high dose radio- and chemotherapy [118].

Suppressing EMT is a potentially effective way to stop metastases; however, it is important to remember that this process is dynamic and that malignant cells can undergo MET once they penetrate a tissue. Hence, suppressing EMT should be done in a well-regulated and cautious way. Suppressing a particular effector could affect multiple pathways and might even exacerbate the process of MET induction

Table 1 Phytochemicals which target EMT in various cancers

Active compound	Plant source	Cancer type	EMT targets	Model used		Reference
				In vitro	In vivo	
Morin (3, 2, 4, 5, 7-pentahydroxyflavone)	*Moraceae* genus (Figs, etc.)	Breast cancer	Suppression of Akt pathway and MMP9	MDA-MB-231 cell line	Nude mice	[98]
			Suppression of VCAM and N-cadherin; promotion of antimetastatic activity	MDA-MB-231 cell line	Nude mice	[99]
		Ovarian cancer	Anti-metastatic, anti-adhesive, and anti-proliferative effect by targeting genes that regulate cell migration and motility, and cell adhesion	A2780 and SKOV3 cell lines		[100]
Kaempferol	*Kaempferia galanga*, and other edible plants	Breast cancer	Inhibition of migration and invasion via regulation of EMT-associated genes and proteins	MCF7 cell line		[101]
		Lung cancer	Suppression of EMT via TGF-β; inhibition of Akt1-mediated phosphorylation of Smad3	A549 cell line		[102]
		Osteosarcoma	Suppression of metastasis by inhibition of ERK-p38-JNK and AP-1 signaling pathways	U2-OS cell line		[103]
		Renal cancer	Inhibition of migration and invasion via downregulation of Akt and FAK pathways	786-O cell line	SCID mice	[104]
		Pancreatic cancer	Suppression of migration by blocking EGFR signaling	Miapaca-2, Panc-1, and SNU-213 cell lines		[105]

Active compound	Plant source	Cancer type	EMT targets	Model used		Reference
				In vitro	In vivo	
Genistein	Soy	Colorectal cancer	Inhibition of migration and distant metastasis by suppression of MMP2 and FLT4	HCT116, SW620, and HT 29 cell lines	Nude mice	[106]
			Causation of apoptosis via reversal of EMT via a Notch1/NF-κB/slug/E-cadherin pathway	HT29 cell line		[107]
		Thyroid cancer	Reversal of EMT by preventing nucleus translocation of β-catenin	BHP10-3, BCPAP, and IHH4 cell lines		[108]
		Hepatocellular carcinoma	Restriction of EMT and stemness by upregulation of miR-1275 to block Akt pathway	HCC-LM3, MHCC-97H, Hep-3B, Hep-G2, and Huh-7 cell lines Patient samples ($N = 70$)	Nude mice	[96]
Sulforaphane	Cruciferous plants	Lung cancer	Suppression of EMT and metastasis through miR-616-5p-mediated GSK3β/β-catenin signaling pathway	H1299, 95C, and 95D cell lines Patient samples ($N = 179$)	Nude mice	[95]
			Inhibition of EMT via activation of ERK5	A549 and H1299 cell lines	Nude mice	[109]
		Breast cancer	Inhibition of stemness and metastasis via downregulation of sirtuins-mediated EMT signaling axis	MDA-MB-231 and MDA-MB-468 cell lines		[110]
		Hepatocellular carcinoma	Inhibition of TGF-β-induced EMT via a ROS-dependent pathway	HepG2 cell line	Nude mice	[111]

(continued)

Table 1 (continued)

Active compound	Plant source	Cancer type	EMT targets	Model used		Reference
				In vitro	In vivo	
Resveratrol	Grapes, apples, blueberries, peanuts, etc.	Breast cancer	Inhibition of migration and metastasis by reversal of TGF-β1-induced EMT	MDA-MB-231 cell line		[112]
		Gastric cancer	Reversal of Doxorubicin resistance by inhibition of EMT via Akt signaling	SGC7901 cell line	Nude mice	[113]
		Glioblastoma	Suppression of EMT via regulation of Smad-dependent signaling	LN18 and U87 cell lines	Nude mice	[114]
Epigallocatechin gallate (EGCG)	Tea plant (*Camellia sinensis*)	Thyroid cancer	Suppression of EMT via blockade of TGF-β1/Smad signaling	8505C cell line		[115]
		Lung cancer	Suppression of TGF-β1 mediated EMT	A549 and NCI-H1299 cell lines		[116]
Withaferin A (WFA)	*Withania somnifera*	Lung cancer	Inhibition of TGF-β1 TNFα-induced EMT; mechanistic suppression of phosphorylation and nuclear translocation of Smad2/3 and NF-κB	A549 and H1299 cell lines		[117]

in already metastasized tissues. For this reason, it is pharmacologically challenging to directly target EMT effector molecules. Therefore, further in-depth research is required in this direction.

Acknowledgments Grants received by ABT from the Jawaharlal Nehru University as LRE and DBT builder are gratefully acknowledged. ML is thankful to the University Grants Commission for the fellowship.

Conflict of Interest The authors report no conflict of interest.

References

1. Théry M, et al. Anisotropy of cell adhesive microenvironment governs cell internal organization and orientation of polarity. Proc Natl Acad Sci. 2006;103:19771–6. https://doi.org/10.1073/pnas.0609267103.
2. Galic M, Matis M. Polarized trafficking provides spatial cues for planar cell polarization within a tissue. BioEssays. 2015;37:678–86. https://doi.org/10.1002/bies.201400196.
3. Kalluri R, Neilson EG. Epithelial-mesenchymal transition and its implications for fibrosis. J Clin Invest. 2003;112:1776–84. https://doi.org/10.1172/JCI20530.
4. Hay ED. An overview of epithelio-mesenchymal transformation. Acta Anat (Basel). 1995;154:8–20. https://doi.org/10.1159/000147748.
5. Lamouille S, et al. Molecular mechanisms of epithelial-mesenchymal transition. Nat Rev Mol Cell Biol. 2014;15:178–96. https://doi.org/10.1038/nrm3758.
6. Thiery JP, Sleeman JP. Complex networks orchestrate epithelial-mesenchymal transitions. Nat Rev Mol Cell Biol. 2006;7:131–42. https://doi.org/10.1038/nrm1835.
7. Thiery JP, et al. Epithelial-mesenchymal transitions in development and disease. Cell. 2009;139:871–90. https://doi.org/10.1016/j.cell.2009.11.007.
8. Miyazono K. Transforming growth factor-β signaling in epithelial-mesenchymal transition and progression of cancer. Proc Jpn Acad Ser B Phys Biol Sci. 2009;85:314–23. https://doi.org/10.2183/pjab.85.314.
9. Chaffer CL, Weinberg RA. A perspective on cancer cell metastasis. Science. 2011;331:1559–64. https://doi.org/10.1126/science.1203543.
10. Huber MA, et al. Molecular requirements for epithelial-mesenchymal transition during tumor progression. Curr Opin Cell Biol. 2005;17:548–58. https://doi.org/10.1016/j.ceb.2005.08.001.
11. Lavin DP, Tiwari VK. Unresolved complexity in the gene regulatory network underlying EMT. Front Oncol. 2020;10:1–19. https://doi.org/10.3389/fonc.2020.00554.
12. Ribatti D, et al. Epithelial-mesenchymal transition in cancer: a historical overview. Transl Oncol. 2020;13:100773. https://doi.org/10.1016/j.tranon.2020.100773.
13. Birchmeier W, Behrens J. Cadherin expression in carcinomas: role in the formation of cell junctions and the prevention of invasiveness. Biochim Biophys Acta. 1994;1198:11–26. https://doi.org/10.1016/0304-419x(94)90003-5.
14. Zhu W, et al. Cytoplasmic O-glycosylation prevents cell surface transport of E-cadherin during apoptosis. EMBO J. 2001;20:5999–6007. https://doi.org/10.1093/emboj/20.21.5999.
15. Shamir ER, et al. Twist1-induced dissemination preserves epithelial identity and requires E-cadherin. J Cell Biol. 2014;204:839–56. https://doi.org/10.1083/jcb.201306088.
16. Shamir ER, Ewald AJ. Chapter eleven – Adhesion in mammary development: novel roles for E-cadherin in individual and collective cell migration. In: Yap AS, editor. Cellular adhesion in development and disease. Academic Press; 2015. p. 353–82. https://doi.org/10.1016/bs.ctdb.2014.12.001.

17. Na T-Y, et al. The functional activity of E-cadherin controls tumor cell metastasis at multiple steps. Proc Natl Acad Sci. 2020;117:5931–7. https://doi.org/10.1073/pnas.1918167117.
18. Serrano-Gomez SJ, et al. Regulation of epithelial-mesenchymal transition through epigenetic and post-translational modifications. Mol Cancer. 2016;15:1–14. https://doi.org/10.1186/s12943-016-0502-x.
19. Mandal M, et al. Modeling continuum of epithelial mesenchymal transition plasticity. Integr Biol. 2016;8:167–76. https://doi.org/10.1039/c5ib00219b.
20. Willipinski-Stapelfeldt B, et al. Changes in cytoskeletal protein composition indicative of an epithelial-mesenchymal transition in human micrometastatic and primary breast carcinoma cells. Clin Cancer Res. 2005;11:8006–14. https://doi.org/10.1158/1078-0432.CCR-05-0632.
21. Kokkinos MI, et al. Vimentin and epithelial-mesenchymal transition in human breast cancer-observations *in vitro* and *in vivo*. Cell Tissue Organ. 2007;185:191–203. https://doi.org/10.1159/000101320.
22. Joyce JA, Pollard JW. Microenvironmental regulation of metastasis. Nat Rev Cancer. 2009;9:239–52. https://doi.org/10.1038/nrc2618.
23. Guo S, Deng C. Effect of stromal cells in tumor microenvironment on metastasis initiation. Int J Biol Sci. 2018;14:2083–93. https://doi.org/10.7150/ijbs.25720.
24. Lindsey S, Langhans SA. Crosstalk of oncogenic signaling pathways during epithelial-mesenchymal transition. Front Oncol. 2014;4:1–10. https://doi.org/10.3389/fonc.2014.00358.
25. Scott LE, et al. Mechanochemical signaling of the extracellular matrix in epithelial-mesenchymal transition. Front Cell Dev Biol. 2019;7:135. https://doi.org/10.3389/fcell.2019.00135.
26. Bonnans C, et al. Remodelling the extracellular matrix in development and disease to cite this version: HAL Id: hal-01952416 NIH Public Access. Nat Rev Mol Cell Biol. 2018;15:786–801. https://doi.org/10.1038/nrm3904.Remodelling.
27. Stamenkovic I. Extracellular matrix remodelling: the role of matrix metalloproteinases. J Pathol. 2003;200:448–64. https://doi.org/10.1002/path.1400.
28. Brassart-Pasco S, et al. Tetrastatin, the NC1 domain of the α4(IV) collagen chain: a novel potent anti-tumor matrikine. PLoS One. 2012;7:e29587. https://doi.org/10.1371/journal.pone.0029587.
29. Colorado PC, et al. Anti-angiogenic cues from vascular basement membrane collagen. Cancer Res. 2000;60:2520–6.
30. Hamano Y, et al. Physiological levels of tumstatin, a fragment of collagen IV α3 chain, are generated by MMP-9 proteolysis and suppress angiogenesis *via* αVβ3 integrin. Cancer Cell. 2003;3:589–601. https://doi.org/10.1016/S1535-6108(03)00133-8.
31. Kamphaus GD, et al. Canstatin, a novel matrix-derived inhibitor of angiogenesis and tumor growth. J Biol Chem. 2000;275:1209–15. https://doi.org/10.1074/jbc.275.2.1209.
32. Dong B, et al. Tackle epithelial-mesenchymal transition with epigenetic drugs in cancer. Front Pharmacol. 2020;11:596239. https://doi.org/10.3389/fphar.2020.596239.
33. Yamazaki T, et al. Preparation of mitochondria- and epigenetics-targeting nanoparticles for suppression of cancer metastasis. Part Part Syst Charact. 2021;38:2100003. https://doi.org/10.1002/ppsc.202100003.
34. Li CH, et al. Stationed or relocating: the seesawing EMT/MET determinants from embryonic development to cancer metastasis. Biomedicines. 2021;9:1265. https://doi.org/10.3390/biomedicines9091265.
35. Chan S-H, Wang L-H. Regulation of cancer metastasis by microRNAs. J Biomed Sci. 2015;22:9. https://doi.org/10.1186/s12929-015-0113-7.
36. Yue J, Mulder KM. Requirement of Ras/MAPK pathway activation by transforming growth factor beta for transforming growth factor beta 1 production in a Smad-dependent pathway. J Biol Chem. 2000;275:30765–73. https://doi.org/10.1074/jbc.M000039200.
37. Gui T, et al. The roles of mitogen-activated protein kinase pathways in TGF-β-induced epithelial-mesenchymal transition. J Signal Transduct. 2012;2012:289243. https://doi.org/10.1155/2012/289243.
38. Platanias LC. Map kinase signaling pathways and hematologic malignancies. Blood. 2003;101:4667–79. https://doi.org/10.1182/blood-2002-12-3647.

39. Lee RJ, et al. Induction of cyclin D1 requires collaborative interactions between the extracellular signal-regulated kinase, p38, and Jun kinase pathways: a role for cAMP response element-binding protein and activating transcript. J Biol Chem. 1999;274:7341–50. https://doi.org/10.1074/jbc.274.11.7341.
40. Ricote M, et al. The p38 transduction pathway in prostatic neoplasia. J Pathol. 2006;208:401–7. https://doi.org/10.1002/path.1910.
41. Recio JA, Merlino G. Hepatocyte growth factor/scatter factor activates proliferation in melanoma cells through p38 MAPK, ATF-2 and cyclin D1. Oncogene. 2002;21:1000–8. https://doi.org/10.1038/sj.onc.1205150.
42. Kage H, Borok Z. EMT and interstitial lung disease: a mysterious relationship. Curr Opin Pulm Med. 2012;18:517–23. https://doi.org/10.1097/MCP.0b013e3283566721.
43. Moustakas A, Heldin C-H. Signaling networks guiding epithelial-mesenchymal transitions during embryogenesis and cancer progression. Cancer Sci. 2007;98:1512–20. https://doi.org/10.1111/j.1349-7006.2007.00550.x.
44. Kudo-Saito C, et al. Cancer metastasis is accelerated through immunosuppression during Snail-induced EMT of cancer cells. Cancer Cell. 2009;15:195–206. https://doi.org/10.1016/j.ccr.2009.01.023.
45. Azuma H, et al. Effect of Smad7 expression on metastasis of mouse mammary carcinoma JygMC(A) cells. J Natl Cancer Inst. 2005;97:1734–46. https://doi.org/10.1093/jnci/dji399.
46. Deckers M, et al. The tumor suppressor Smad4 is required for transforming growth factor β-induced epithelial to mesenchymal transition and bone metastasis of breast cancer cells. Cancer Res. 2006;66:2202–9. https://doi.org/10.1158/0008-5472.CAN-05-3560.
47. Takano S, et al. Smad4 is essential for Down-regulation of E-cadherin induced by TGF-β in pancreatic cancer cell line PANC-1. J Biochem. 2007;141:345–51. https://doi.org/10.1093/jb/mvm039.
48. Kaimori A, et al. Transforming growth factor-β1 induces an epithelial-to-mesenchymal transition state in mouse hepatocytes *in vitro*. J Biol Chem. 2007;282:22089–101. https://doi.org/10.1074/jbc.M700998200.
49. Gu J, et al. Gastric cancer exosomes trigger differentiation of umbilical cord derived mesenchymal stem cells to carcinoma-associated fibroblasts through TGF-β/Smad pathway. PLoS One. 2012;7:e52465. https://doi.org/10.1371/journal.pone.0052465.
50. Zhao Y, et al. 2-hydroxypropyl-B-cyclodextrin regulates the epithelial to mesenchymal transition in breast cancer cells by modulating cholesterol homeostasis and endoplasmic reticulum stress. Metabolites. 2021;11:562. https://doi.org/10.3390/metabo11080562.
51. Nakajima S, et al. N-cadherin expression and epithelial-mesenchymal transition in pancreatic carcinoma. Clin Cancer Res. 2004;10:4125–33. https://doi.org/10.1158/1078-0432.CCR-0578-03.
52. Huber MA, et al. Wirth, NF-kappaB is essential for epithelial-mesenchymal transition and metastasis in a model of breast cancer progression. J Clin Invest. 2004;114:569–81. https://doi.org/10.1172/JCI21358.
53. Arsura M, et al. Transient activation of NF-kappaB through a TAK1/IKK kinase pathway by TGF-beta1 inhibits AP-1/SMAD signaling and apoptosis: implications in liver tumor formation. Oncogene. 2003;22:412–25. https://doi.org/10.1038/sj.onc.1206132.
54. Mirzaei S, et al. NF-κB as a regulator of cancer metastasis and therapy response: a focus on epithelial-mesenchymal transition. J Cell Physiol. 2022;237:2770–95. https://doi.org/10.1002/jcp.30759.
55. Peinado H, et al. Snail, ZEB and bHLH factors in tumour progression: an alliance against the epithelial phenotype? Nat Rev Cancer. 2007;7:415–28. https://doi.org/10.1038/nrc2131.
56. Zhang J, et al. Magnesium Isoglycyrrhizinate induces an inhibitory effect on progression and epithelial-mesenchymal transition of laryngeal cancer *via* the NF-κB/Twist signaling. Drug Des Devel Ther. 2020;14:5633–44. https://doi.org/10.2147/dddt.s272323.
57. Spiering D, Hodgson L. Dynamics of the Rho-family small GTPases in actin regulation and motility. Cell Adh Migr. 2011;5:170–80. https://doi.org/10.4161/cam.5.2.14403.

58. Meyer G, Feldman EL. Signaling mechanisms that regulate actin-based motility processes in the nervous system. J Neurochem. 2002;83:490–503. https://doi.org/10.1046/j.1471-4159.2002.01185.x.
59. Nurmagambetova A, et al. Morphological and cytoskeleton changes in cells after EMT. Sci Rep. 2023;13:22164. https://doi.org/10.1038/s41598-023-48279-y.
60. Shi R, et al. Downregulation of cytokeratin 18 induces cellular partial EMT and stemness through increasing EpCAM expression in breast cancer. Cell Signal. 2020;76:109810. https://doi.org/10.1016/j.cellsig.2020.109810.
61. Kuburich NA, et al. Vimentin and cytokeratin: good alone, bad together. Semin Cancer Biol. 2022;86:816–26. https://doi.org/10.1016/j.semcancer.2021.12.006.
62. Jiu Y, et al. Vimentin intermediate filaments control actin stress fiber assembly through GEF-H1 and RhoA. J Cell Sci. 2017;130:892–902. https://doi.org/10.1242/jcs.196881.
63. Colburn ZT, Jones JCR. Complexes of α6β4 integrin and vimentin act as signaling hubs to regulate epithelial cell migration. J Cell Sci. 2018;131:jcs214593. https://doi.org/10.1242/jcs.214593.
64. Sternlicht MD, et al. The stromal proteinase MMP3/stromelysin-1 promotes mammary carcinogenesis. Cell. 1999;98:137–46. https://doi.org/10.1016/s0092-8674(00)81009-0.
65. Zuo J-H, et al. Activation of EGFR promotes squamous carcinoma SCC10A cell migration and invasion *via* inducing EMT-like phenotype change and MMP-9-mediated degradation of E-cadherin. J Cell Biochem. 2011;112:2508–17. https://doi.org/10.1002/jcb.23175.
66. Baglieri J, et al. The role of fibrosis and liver-associated fibroblasts in the pathogenesis of hepatocellular carcinoma. Int J Mol Sci. 2019;20:1723. https://doi.org/10.3390/ijms20071723.
67. Kim D, et al. DDR2 controls the epithelial-mesenchymal-transition-related gene expression *via* c-Myb acetylation upon matrix stiffening. Sci Rep. 2017;7:6847. https://doi.org/10.1038/s41598-017-07126-7.
68. Vaupel P. The role of hypoxia-induced factors in tumor progression. Oncologist. 2004;9(S5):10–7. https://doi.org/10.1634/theoncologist.9-90005-10.
69. Axelson H, et al. Hypoxia-induced dedifferentiation of tumor cells–a mechanism behind heterogeneity and aggressiveness of solid tumors. Semin Cell Dev Biol. 2005;16:554–63. https://doi.org/10.1016/j.semcdb.2005.03.007.
70. Moen I, et al. Hyperoxic treatment induces mesenchymal-to-epithelial transition in a rat adenocarcinoma model. PLoS One. 2009;4:e6381. https://doi.org/10.1371/journal.pone.0006381.
71. Yu L, et al. Conditional ablation of TGF-β signaling inhibits tumor progression and invasion in an induced mouse bladder cancer model. Sci Rep. 2016;6:29479. https://doi.org/10.1038/srep29479.
72. Lu X, Kang Y. Hypoxia and hypoxia-inducible factors: master regulators of metastasis. Clin Cancer Res. 2010;16:5928–35. https://doi.org/10.1158/1078-0432.CCR-10-1360.
73. Martin A, Cano A. Tumorigenesis: Twist1 links EMT to self-renewal. Nat Cell Biol. 2010;12:924–5. https://doi.org/10.1038/ncb1010-924.
74. Yang M-H, et al. Direct regulation of TWIST by HIF-1alpha promotes metastasis. Nat Cell Biol. 2008;10:295–305. https://doi.org/10.1038/ncb1691.
75. McDonald OG, et al. Genome-scale epigenetic reprogramming during epithelial-to-mesenchymal transition. Nat Struct Mol Biol. 2011;18:867–74. https://doi.org/10.1038/nsmb.2084.
76. Jordan NV, et al. Tracking the intermediate stages of epithelial-mesenchymal transition in epithelial stem cells and cancer. Cell Cycle. 2011;10:2865–73. https://doi.org/10.4161/cc.10.17.17188.
77. Zaravinos A. The regulatory role of MicroRNAs in EMT and cancer. J Oncol. 2015;2015:865816. https://doi.org/10.1155/2015/865816.
78. Enkhbaatar Z, et al. KDM5B histone demethylase controls epithelial-mesenchymal transition of cancer cells by regulating the expression of the microRNA-200 family. Cell Cycle. 2013;12:2100–12. https://doi.org/10.4161/cc.25142.

79. Lim Y-Y, et al. Epigenetic modulation of the miR-200 family is associated with transition to a breast cancer stem cell-like state. J Cell Sci. 2013;126:2256–66. https://doi.org/10.1242/jcs.122275.
80. Chung VY, et al. GRHL2-miR-200-ZEB1 maintains the epithelial status of ovarian cancer through transcriptional regulation and histone modification. Sci Rep. 2016;6:19943. https://doi.org/10.1038/srep19943.
81. Pan G, et al. EMT-associated microRNAs and their roles in cancer stemness and drug resistance. Cancer Commun. 2021;4:199–217. https://doi.org/10.1002/cac2.12138.
82. Liao J-Y, et al. Deep sequencing reveals a global reprogramming of lncRNA transcriptome during EMT. Biochim Biophys Acta, Mol Cell Res. 2017;1864:1703–13. https://doi.org/10.1016/j.bbamcr.2017.06.003.
83. Gugnoni M, Ciarrocchi A. Long noncoding RNA and epithelial mesenchymal transition in cancer. Int J Mol Sci. 2019;20:1–25. https://doi.org/10.3390/ijms20081924.
84. Deshmukh AP, et al. Identification of EMT signaling cross-talk and gene regulatory networks by single-cell RNA sequencing. Proc Natl Acad Sci. 2021;118:e2102050118. https://doi.org/10.1073/pnas.2102050118.
85. Li Y-M, et al. Epithelial-mesenchymal transition markers expressed in circulating tumor cells in hepatocellular carcinoma patients with different stages of disease. Cell Death Dis. 2013;4:e831. https://doi.org/10.1038/cddis.2013.347.
86. Armstrong AJ, et al. Circulating tumor cells from patients with advanced prostate and breast cancer display both epithelial and mesenchymal markers. Mol Cancer Res. 2011;9:997–1007. https://doi.org/10.1158/1541-7786.MCR-10-0490.
87. Cragg GM, et al. Impact of natural products on developing new anti-cancer agents. Chem Rev. 2009;109:3012–43. https://doi.org/10.1021/cr900019j.
88. Fitter A. Making allelopathy respectable. Science. 2003;301:1337–8. https://doi.org/10.1126/science.1089291.
89. Illam SP, et al. Epithelial mesenchymal transition in cancer progression: prev entive phytochemicals. Recent Pat Anti-Cancer Drug Discov. 2017;12:234–46. https://doi.org/10.2174/1574892812666170424150407.
90. Shanmugam MK, et al. Cancer prevention and therapy through the modulation of transcription factors by bioactive natural compounds. Semin Cancer Biol. 2016;40–41:35–47. https://doi.org/10.1016/j.semcancer.2016.03.005.
91. Aliyu M, et al. Dietary phytochemicals targeting Nf-KB signalling pathways: potential cancer chemoprevention strategy. Dutse J Pure Appl Sci. 2023;9:179–90. https://doi.org/10.4314/dujopas.v9i2a.18.
92. Xu M, Zhang Y. Morin inhibits ovarian cancer growth through the inhibition of NF-κB signaling pathway. Anticancer Agents Med Chem. 2019;19:2243–50. https://doi.org/10.2174/1871521409666191014164742.
93. Lee G-A, et al. Roles of dietary phytoestrogens on the regulation of epithelial-mesenchymal transition in diverse cancer metastasis. Toxins (Basel). 2016;8:162. https://doi.org/10.3390/toxins8060162.
94. Budişan L, et al. Dietary intervention by phytochemicals and their role in modulating coding and non-coding genes in cancer. Int J Mol Sci. 2017;18:1178. https://doi.org/10.3390/ijms18061178.
95. Wang DX, et al. Sulforaphane suppresses EMT and metastasis in human lung cancer through miR-616-5p-mediated GSK3β/β-catenin signaling pathways. Acta Pharmacol Sin. 2017;38:241–51. https://doi.org/10.1038/aps.2016.122.
96. Yang X, et al. Genistein restricts the epithelial mesenchymal transformation (EMT) and stemness of hepatocellular carcinoma *via* upregulating miR-1275 to inhibit the EIF5A2/PI3K/Akt pathway. Biology (Basel). 2022;11:1383. https://doi.org/10.3390/biology11101383.
97. Naujokat C, McKee DL. The "Big Five" phytochemicals targeting cancer stem cells: curcumin, EGCG, sulforaphane, resveratrol and genistein. Curr Med Chem. 2021;28:4321–42. https://doi.org/10.2174/0929867327666200228110738.

98. Jin H, et al. Morin, a flavonoid from moraceae, suppresses growth and invasion of the highly metastatic breast cancer cell line MDA-MB-231 partly through suppression of the Akt pathway. Int J Oncol. 2014;45:1629–37. https://doi.org/10.3892/ijo.2014.2535.
99. Lee J, et al. Morin, a flavonoid from moraceae, inhibits cancer cell adhesion to endothelial cells and EMT by downregulating VCAM1 and ncadherin. Asian Pac J Cancer Prev. 2016;17:3071–5.
100. Nowak E, et al. Morin exerts anti-metastatic, anti-proliferative and anti-adhesive effect in ovarian cancer cells: an *in vitro* studies. Mol Biol Rep. 2020;47:1965–78. https://doi.org/10.1007/s11033-020-05293-x.
101. Lee G-A, et al. Kaempferol, a phytoestrogen, suppressed triclosan-induced epithelial-mesenchymal transition and metastatic-related behaviors of MCF-7 breast cancer cells. Environ Toxicol Pharmacol. 2017;49:48–57. https://doi.org/10.1016/j.etap.2016.11.016.
102. Jo E, et al. Kaempferol suppresses transforming growth factor-β1-induced epithelial-to-mesenchymal transition and migration of A549 lung cancer cells by inhibiting Akt1-mediated phosphorylation of Smad3 at threonine-179. Neoplasia. 2015;17:525–37. https://doi.org/10.1016/j.neo.2015.06.004.
103. Chen H-J, et al. Kaempferol suppresses cell metastasis *via* inhibition of the ERK-p38-JNK and AP-1 signaling pathways in U-2 OS human osteosarcoma cells. Oncol Rep. 2013;30:925–32. https://doi.org/10.3892/or.2013.2490.
104. Hung T-W, et al. Kaempferol inhibits the invasion and migration of renal cancer cells through the downregulation of AKT and FAK pathways. Int J Med Sci. 2017;14:984–93. https://doi.org/10.7150/ijms.20336.
105. Lee J, Kim JH. Kaempferol inhibits pancreatic cancer cell growth and migration through the blockade of EGFR-related pathway *in vitro*. PLoS One. 2016;11:e0155264. https://doi.org/10.1371/journal.pone.0155264.
106. Xiao X, et al. Genistein suppresses FLT4 and inhibits human colorectal cancer metastasis. Oncotarget. 2015;6:3225–39. https://doi.org/10.18632/oncotarget.3064.
107. Zhou P, et al. Genistein induces apoptosis of colon cancer cells by reversal of epithelial-to-mesenchymal *via* a Notch1/NF-κB/slug/E-cadherin pathway. BMC Cancer. 2017;17:813. https://doi.org/10.1186/s12885-017-3829-9.
108. Zhang C, et al. Genistein inhibits human papillary thyroid cancer cell detachment, invasion and metastasis. J Cancer. 2019;10:737–48. https://doi.org/10.7150/jca.28111.
109. Chen Y, et al. Sulforaphane inhibits epithelial-mesenchymal transition by activating extracellular signal-regulated kinase 5 in lung cancer cells. J Nutr Biochem. 2019;72:108219. https://doi.org/10.1016/j.jnutbio.2019.108219.
110. Sinha S, et al. Sulforaphane-cisplatin combination inhibits the stemness and metastatic potential of TNBCs *via* down regulation of sirtuins-mediated EMT signaling axis. Phytomedicine. 2021;84:153492. https://doi.org/10.1016/j.phymed.2021.153492.
111. Wu J, et al. Sulforaphane inhibits TGF-β-induced epithelial-mesenchymal transition of hepatocellular carcinoma cells *via* the reactive oxygen species-dependent pathway. Oncol Rep. 2016;35:2977–83. https://doi.org/10.3892/or.2016.4638.
112. Sun Y, et al. Resveratrol inhibits the migration and metastasis of MDA-MB-231 human breast cancer by reversing TGF-β1-induced epithelial-mesenchymal transition. Molecules. 2019;24:1131. https://doi.org/10.3390/molecules24061131.
113. Xu J, et al. Resveratrol reverses doxorubicin resistance by inhibiting epithelial-mesenchymal transition (EMT) through modulating PTEN/Akt signaling pathway in gastric cancer. J Exp Clin Cancer Res. 2017;36:19. https://doi.org/10.1186/s13046-016-0487-8.
114. Song Y, et al. Resveratrol suppresses epithelial-mesenchymal transition in GBM by regulating Smad-dependent signaling. Biomed Res Int. 2019;2019:1321973. https://doi.org/10.1155/2019/1321973.
115. Li Z, et al. Exosomal lncRNA ZFAS1 regulates esophageal squamous cell carcinoma cell proliferation, invasion, migration and apoptosis *via* microRNA-124/STAT3 axis. J Exp Clin Cancer Res. 2019;38:477. https://doi.org/10.1186/s13046-019-1473-8.

116. Liu LC, et al. EGCG inhibits transforming growth factor-β-mediated epithelial-to-mesenchymal transition *via* the inhibition of smad2 and Erk1/2 signaling pathways in nonsmall cell lung cancer cells. J Agric Food Chem. 2012;60:9863–73. https://doi.org/10.1021/jf303690x.
117. Kyakulaga AH, et al. Withaferin A inhibits epithelial to mesenchymal transition in non-small cell lung cancer cells. Sci Rep. 2018;8:1–14. https://doi.org/10.1038/s41598-018-34018-1.
118. Ahmed MB, et al. Phytochemicals as chemo-preventive agents and signaling molecule modulators: current role in cancer therapeutics and inflammation. Int J Mol Sci. 2022;23:15765. https://doi.org/10.3390/ijms232415765.

Targeting Metabolic Reprogramming: A New Frontier in Cancer Therapy

Ravi Thakur and Niti Kumari

1 Introduction

Cancer cells exhibit a range of metabolic adaptations, collectively termed metabolic reprogramming, a key hallmark of cancer. This reprogramming encompasses a multitude of adaptations, including increased glucose uptake and the Warburg effect, altered mitochondrial and lipid metabolism, enhanced glutamine catabolism, heightened hexosamine biosynthesis pathway (HBP) and pentose phosphate pathway (PPP), reprogrammed amino acid and one-carbon metabolism, and opportunistic nutrient uptake [1]. These coordinated attributes achieve critical objectives for cancer progression, including bolstering biomass production, sustaining metabolic fluxes via nicotinamide adenine dinucleotide (NAD^+) homeostasis, managing oxidative stress, ensuring efficient nutrient and reaction intermediate allocation, and enabling survival and proliferation within nutrient-poor and hypoxic tumor microenvironment (TME) [2]. Growing evidence indicates that this intricate metabolic reprogramming is not merely a consequence of malignant transformation but also actively contributes to the initiation and maintenance of core cancer hallmarks, including sustained proliferation, evasion of growth suppressors, resistance to cell death, enabling replicative immortality, inducing angiogenesis, activating invasion and metastasis, evading immune destruction, and promoting genome instability and mutations [3]. These observations have rekindled the metabolic theory of cancer in recent years [4]. Although different

R. Thakur (✉)
Department of Oncology Science, University of Oklahoma Health Sciences Center, Oklahoma City, OK, USA

Department of Life Sciences, Christ University, Bengaluru, India
e-mail: ravi-thakur@ouhsc.edu; iamravithakur@live.com

N. Kumari
Biomedical Sciences Department, Creighton University School of Medicine, Omaha, NE, USA

A. S. Ramteke et al. (eds.), *Molecular Targets in Cancer Therapy*,
https://doi.org/10.1007/978-981-96-7336-0_9

cancers frequently share some metabolic similarities, there is significant diversity in how they reprogram their metabolism. Different subtypes within a single cancer type can rely on distinct metabolic pathways to fuel their growth and survival. Factors like the cancer's origin, stage, and size can dramatically influence its metabolic dependencies [1, 5, 6]. Therefore, effectively targeting cancer metabolism requires a detailed understanding of the specific metabolic changes that occur during cancer development. Accurate modeling, possibly using individual patient tumor data, is essential to design personalized treatment strategies. This chapter begins with a concise overview of the diverse changes associated with cancer cells and TME metabolism, followed by an overview of therapeutic strategies targeting these metabolic vulnerabilities, involving synergistic combinations with radiotherapy, chemotherapy, and immunotherapy, and finally highlights the challenges and future research directions in this dynamic field.

2 Hallmarks of Metabolic Reprogramming in Cancer

Cancer metabolic reprogramming, a cornerstone of tumorigenesis, represents a deeply orchestrated rewiring of cellular metabolism, instigated by oncogenic molecular events and selective pressures within the tumor microenvironment [7]. This section will discuss how reprogramming supports rapid proliferation by altering pathways of energy (ATP) and biomass (nucleotides, lipids, amino acids) production for uncontrolled cell division. We will also discuss how it augments cancer cell survival in stressful environments, such as hypoxia, by promoting adaptation and preventing cell death. Simultaneously, we will explore how metabolic reprogramming facilitates invasion and metastasis by providing the necessary energy and resources. Additionally, we will discuss the influence of altered metabolism on the complex interplay between cancer cells and other cells within the TME, followed by its contribution to the development of resistance to cancer treatments, which can ultimately lead to treatment failure (Fig. 1).

The Warburg effect, a cardinal feature, epitomizes this reprogramming, wherein cancer cells exhibit elevated glycolysis and lactate production, even under normoxic conditions [8, 9]. This metabolic phenotype, driven by upregulated expression of glucose transporters (GLUTs) and enzymes like hexokinase (HK), phosphofructokinase (PFK), pyruvate kinase M2 (PKM2), and lactate dehydrogenase A (LDHA), ensures rapid ATP generation and, more critically, the buildup of glycolytic intermediates that are strategically shunted into anabolic pathways [2]. Specifically, glucose-6-phosphate is channeled into the pentose phosphate pathway (PPP), generating NADPH, a critical reductant for fatty acid synthesis and reactive oxygen species (ROS) detoxification, and ribose-5-phosphate, a precursor for nucleotide biosynthesis [10]. Fructose-6-phosphate enters the hexosamine biosynthesis pathway (HBP), producing uridine diphosphate N-acetylglucosamine (UDP-GlcNAc) for protein glycosylation entailed in cell signaling and adhesion [11]. Furthermore, the glycolytic intermediate 3-phosphoglycerate serves as the precursor for serine and glycine biosynthesis, feeding into one-carbon metabolism, vital for nucleotide and amino acid synthesis, as well as substrate synthesis for epigenetic modifications [12]. Lactate, the terminal product of Warburg glycolysis, acidifies the TME,

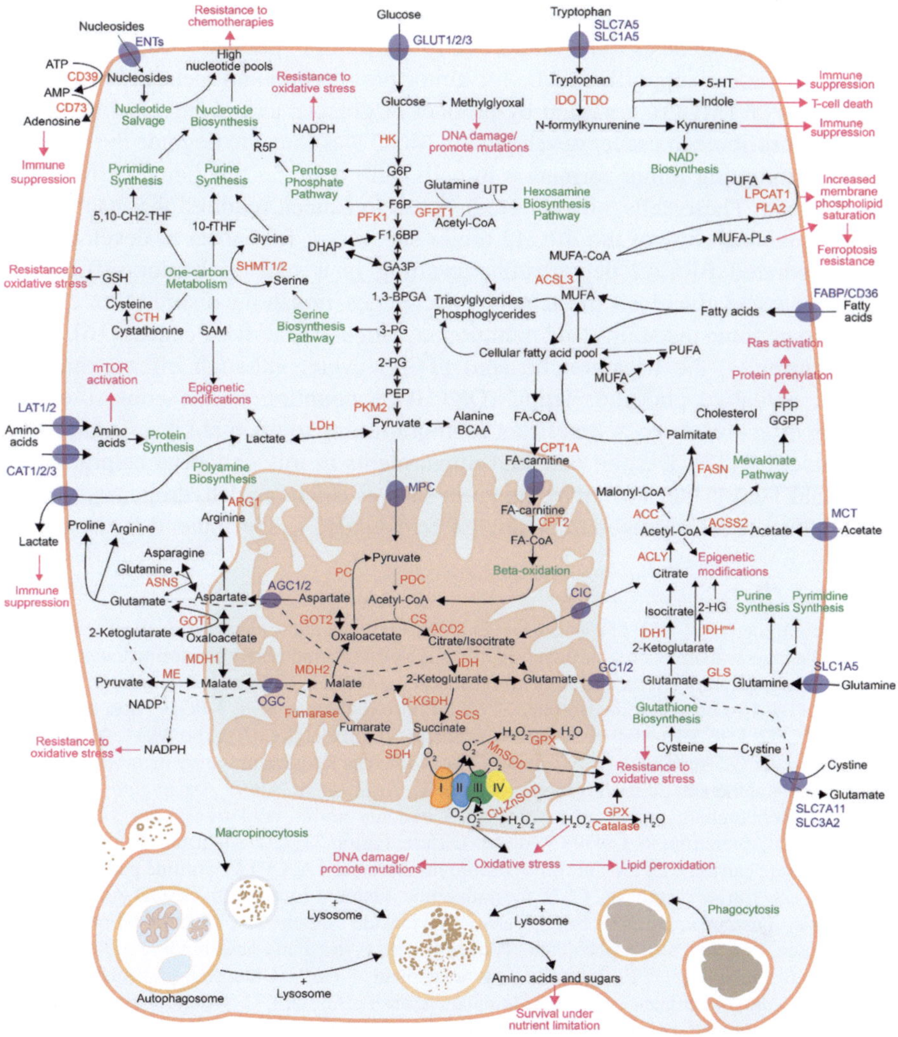

Fig. 1 Metabolic reprogramming and its consequences: Cancer cells reprogram their metabolism by altering key enzymes (red) and utilizing metabolic intermediates in alternative ways (black) to impact multiple pathways (green), ultimately achieving numerous benefits (pink) that support their growth and progression. This metabolic reprogramming encompasses alterations in glucose metabolism (entailing higher glucose uptake via GLUT1/2/3, the Warburg effect with increased glycolysis, lactate production via LDH, and modulation of key glycolytic enzymes like HK, PFK1, PKM2), lipid metabolism (involving FA uptake via CD36, FA synthesis via FASN, FA oxidation via CPT1A/CPT2, and regulation of lipid-related enzymes like ACLY, ACSL3, LPCAT1, PLA2), amino acid metabolism (encompassing glutamine uptake and catabolism via GLS, asparagine synthesis via ASNS, BCAA metabolism, and modulation of amino acid transporters like CAT1/2/3, LAT1/2), nucleotide metabolism (involving the pentose phosphate pathway for R5P production and one-carbon metabolism with SHMT1/2 and 10-fTHF), redox homeostasis (with increased NADPH production via the PPP and malic enzyme (ME), and antioxidant defense via GPX, Cu,ZnSOD, MnSOD), and other crucial processes like the TCA cycle (with enzymes like CS, ACO_2, IDH, α-KGDH, SCS, SDH, MDH1), and the hexosamine biosynthesis pathway

facilitating extracellular matrix degradation via matrix metalloproteinases (MMPs), promoting angiogenesis, and suppressing antitumor immune responses, thus actively shaping a supportive niche for tumor progression and metastasis [13–15]. Methylglyoxal (MGO), a natural by-product of glucose and glycolysis intermediates, can contribute to cancer development. MGO was found to degrade the BRCA2 protein, promoting tumor formation in cells with a single functional copy of the *BRCA2* gene. These cells, already predisposed to cancer, required the inactivation of the remaining copy of the *BRCA2* tumor suppressor for cancer to develop. This MGO-mediated BRCA2 degradation, resulting in a state of haploinsufficiency, linked enhanced glycolysis (hastened by oncogenes, metabolic disorders, or diet) to increased genome instability and mutation, a critical hallmark of cancer [16].

Concurrently, the tricarboxylic acid (TCA) cycle, although often exhibiting reduced oxidative phosphorylation (OXPHOS) coupling in the context of the Warburg effect, undergoes significant reprogramming to maintain its essential anabolic function [17]. Despite potential impairments in mitochondrial respiration, a functional TCA cycle, steered by enzymes such as isocitrate dehydrogenase (IDH), α-ketoglutarate dehydrogenase complex (α-KGDHC), succinate dehydrogenase

Fig. 1 (continued) (via GFPT1). These metabolic adaptations allow cancer cells to sustain proliferation, manage oxidative stress, efficiently allocate nutrients and reaction intermediates, survive in the challenging tumor microenvironment, and modulate TME. 1,3-BPGA 1,3-Bisphosphoglyceric acid, 10-fTHF 10-Formyltetrahydrofolate, 2PG 2-Phosphoglyceric acid, 3-PG 3-Phosphoglyceric acid, 5,10-CH2-THF 5,10-Methylenetetrahydrofolate, ACC Acetyl-CoA carboxylase, ACLY ATP-citrate lyase, ACO2 Aconitase, ACSL3 Acyl-CoA synthetase long chain family member 3, ACSS2 Acetyl-CoA synthetase 2,Fig. 1 (continued) ARG1 Arginase 1, ASNS Asparagine synthetase, BCAA Branched-chain amino acids, CAT1/2/3 cationic amino acid transporter 1/2/3, CD36 Cluster of differentiation 36, CD39 Cluster of Differentiation 39, CD73 Cluster of differentiation 73, CIC Citrate carrier, CPT1A Carnitine palmitoyltransferase 1A, CPT2 Carnitine palmitoyltransferase 2, CS Citrate synthase, CTH Cystathionine Gamma-Lyase, Cu,ZnSOD Copper Zinc Superoxide Dismutase, ENT Equilibrative nucleoside transporter, F1,6BP Fructose-1,6-bisphosphate, F6P Fructose 6-phosphate, FA Fatty acid, FABP Fatty-acid-binding protein, FASN Fatty acid synthase, FPP Farnesyl pyrophosphate, G6P Glucose-6-phosphate, GA3P Glyceraldehyde 3-phosphate, GC1/2 Glutamate carrier 1/2, GFPT1 Glucosamine-fructose-6-phosphate aminotransferase isomerizing 1, GGPP Geranylgeranyl diphosphate, GLS Glutaminase, GLUT1/2/3 Glucose transporter 1/2/3, GOT1 Glutamic-oxaloacetic transaminase 1, GOT2 Glutamic-oxaloacetic transaminase 2, GPX Glutathione peroxidase, GSH Glutathione, HK Hexokinase, IDH1 Isocitrate dehydrogenase 1, IDH2 Isocitrate dehydrogenase 2, IDO Indoleamine 2,3-dioxygenase, LAT1/2 L-Type Amino Acid Transporter 1/2, LDH Lactate dehydrogenase, LPCAT1 Lysophosphatidylcholine Acyltransferase 1, MCT Monocarboxylate transporter, MDH1/2 Malate Dehydrogenase 1/2, ME Malic enzyme, MnSOD Manganese superoxide dismutase, MPC Mitochondrial pyruvate carrier, MUFA Monounsaturated fatty acids, NAD^+ Nicotinamide adenine dinucleotide, NADP Nicotinamide adenine dinucleotide phosphate, NADPH Nicotinamide adenine dinucleotide phosphate hydrogen, PC Pyruvate carboxylase, PDC Pyruvate dehydrogenase complex, PEP Phosphoenolpyruvic acid, PFK1 Phosphofructokinase 1, PKM2 Pyruvate kinase muscle isozyme M2, PLA2 Phospholipase A2, PUFA Polyunsaturated fatty acid, R5P Ribose-5-phosphate, SAM S-adenosylmethionine, SCS Succinyl coenzyme A synthetase, SDH Succinate dehydrogenase, SHMT1/2 Serine hydroxymethyltransferase 1/2, TDO Tryptophan 2,3-dioxygenase, UTP Uridine triphosphate, α-KGDH Alpha-ketoglutarate dehydrogenase

(SDH), and fumarate hydratase (FH), remains indispensable for providing biosynthetic precursors [18]. Anaplerotic reactions, notably glutaminolysis and pyruvate carboxylation, are upregulated to replenish TCA cycle intermediates consumed for biomass production [19]. Glutamine, catabolized via glutaminase (GLS), produces α-ketoglutarate, a key TCA cycle intermediate and nitrogen donor [20]. Pyruvate carboxylation by pyruvate carboxylase (PC) generates oxaloacetate, another crucial TCA cycle intermediate, and a precursor for aspartate biosynthesis, which itself is vital for nucleotide synthesis [21]. Acetyl-CoA from fatty acid oxidation (FAO) is utilized to fuel the TCA cycle under nutrient stress conditions. Reciprocally, under nutrient abundance this pathway generates precursors for fatty acid synthesis [17]. Moreover, dysfunctional TCA cycle metabolism usually resulting from mutations in enzymes like SDH, FH, and IDH, can lead to the accumulation of oncometabolites such as succinate, fumarate, and 2-hydroxyglutarate (2-HG), respectively [14]. These metabolites competitively inhibit α-ketoglutarate-dependent dioxygenases (α-KGDDs), a family of enzymes involved in epigenetic regulation, including histone and DNA demethylation, resulting in widespread epigenetic changes that promote oncogenesis and impair cellular differentiation [22]. Under hypoxic conditions, reductive glutamine metabolism, directed by hypoxia-inducible factor 2 alpha (HIF-2α)-induced upregulation of glutamine transporters and metabolic enzymes, becomes vital for replenishing citrate in the TCA cycle, which is then utilized for de novo lipogenesis, ensuring membrane biosynthesis and survival in oxygen-poor regions of the tumor [23]. Moreover, the TCA cycle and the electron transport chain (ETC) in mitochondria represent major endogenous sources of reactive oxygen species (ROS) [24]. While excessive ROS is cytotoxic, sublethal levels of mitochondrial ROS often activate pro-tumorigenic pathways like nuclear factor kappa-light-chain-enhancer of activated B cells (NF-κB), mitogen-activated protein kinase (MAPK), and phosphatidylinositol 3-kinase/protein kinase B (PI3K-AKT), which promote cell proliferation, invasion, angiogenesis, and metastasis, paradoxically harnessing a normally damaging by-product to fuel malignant progression [25–28].

Lipid metabolism is also profoundly reprogrammed with cancer cells exhibiting enhanced de novo lipogenesis, prompted by enzymes such as acetyl-CoA carboxylase (ACC), fatty acid synthase (FASN), and ATP-citrate lyase (ACLY), alongside increased uptake of exogenous fatty acids [29, 30]. This metabolic shift addresses the high demand for membrane lipid synthesis required for rapid cell proliferation and replenishing lipid stores [31]. Acetyl-CoA, the building block for fatty acid synthesis, is generated from glucose and glutamine metabolism, as well as acetate assimilation via acetyl-CoA synthetase 2 (ACSS2), particularly relevant under hypoxic conditions [32, 33]. NADPH, abundantly produced by the PPP, serves as an essential cofactor for de novo lipogenesis [23]. Cancer cells often use the stored lipids to meet their metabolic demands when nutrients are limited [31]. Additionally, various cancer cells efficiently remodel their lipidome, increasing the proportion of saturated and monounsaturated fatty acids (MUFAs), which are less susceptible to lipid peroxidation than polyunsaturated fatty acids, thus boosting resistance to ROS-induced damage and contributing to therapy resistance [34]. It is facilitated by long-chain acyl-coenzyme A (CoA) synthetase 3 (ACSL3)-mediated synthesis of saturated and monounsaturated phospholipids [35]. The mevalonate pathway, responsible for cholesterol and isoprenoid synthesis, is frequently hyperactivated, providing isoprenoids like farnesyl pyrophosphate (FPP) and geranylgeranyl pyrophosphate (GGPP), vital for protein prenylation and activation of oncogenic RAS signaling, directly driving proliferation and survival [36].

Furthermore, amino acid metabolism is extensively reconfigured, with cancer cells exhibiting a pronounced "addiction" to conditionally essential amino acids (CEAAs) like glutamine, arginine, serine, and glycine [37]. Glutamine, beyond its role in anaplerosis, serves as a nitrogen source for nucleotide and nonessential amino acid biosynthesis and is critical for glutathione synthesis, a major cellular antioxidant linking amino acid metabolism directly to redox homeostasis [5, 38, 39]. Serine and glycine, synthesized via the upregulated serine biosynthesis pathway and interconverted by serine hydroxymethyltransferases (SHMTs), fuel one-carbon metabolism, providing crucial one-carbon units for nucleotide biosynthesis, methylation reactions, and NADPH production [40]. Arginine metabolism supports polyamine biosynthesis, essential for cell proliferation and ribosome biogenesis [41]. Branched-chain amino acid (BCAA) metabolism activates the mammalian target of rapamycin complex 1 (mTORC1) signaling, a central regulator of cell growth and anabolism [42, 43]. Tryptophan catabolism, particularly through the kynurenine pathway, contributes to immune evasion by generating immunosuppressive metabolites like kynurenine and by providing precursors for NAD^+ biosynthesis, enhancing stress resistance and potentially contributing to therapy resistance by fueling DNA repair mechanisms [44–47]. Amino acids also act as signaling molecules, directly modulating mTORC1 signaling, further integrating nutrient availability with cellular growth and metabolism [48].

Nucleotide metabolism is stringently enhanced to meet the immense demand for DNA and RNA precursors for rapid cell division [49–51]. This heightened nucleotide biosynthesis is critical for supporting the growth of cancer cells and plays a central role in therapy resistance [52]. Pyrimidine and purine nucleotides are generated via distinct de novo and salvage pathways [50]. The de novo route relies on ATP, glutamine-derived nitrogen, aspartate, and ribose-5-phosphate [50, 53]. Crucially, de novo synthesis is intimately linked to one-carbon (1C) metabolism and the folate pathway [54]. The 1C-units are utilized by thymidylate synthase (TYMS) to catalyze the rate-limiting conversion of dUMP to dTMP, a key target for antifolate chemotherapeutics and the 5-fluorouracil metabolite, fluorodeoxyuridine monophosphate (FdUMP) [55–59]. Complementing de novo synthesis, salvage pathways support cellular nucleotide pools [60]. Pyrimidine salvage involves uridine–cytidine kinases 1/2 (UCK1/2), deoxycytidine kinase (DCK), and thymidine kinases 1/2 (TK1/2), which phosphorylate (deoxy)nucleosides to (deoxy)ribonucleoside monophosphates (d)NMPs. It is noteworthy that mammalian cells exhibit limited salvage of free pyrimidine nucleobases (uracil, cytosine, thymine) due to the absence of PRPP-dependent phosphoribosyl transferase activity for these bases and the inability to reconvert them to orotate for de novo synthesis [61, 62]. Purine ribonucleoside salvage, however, utilizes purine nucleoside phosphorylase (PNP) to convert ribonucleosides to nucleobases (adenine, guanine, hypoxanthine), followed by PRPP-dependent phosphoribosyl transferases, adenine phosphoribosyl transferase (APRT) and hypoxanthine-guanine phosphoribosyl transferase (HGPRT), which generate NMPs [60, 63]. Purine deoxynucleosides can be directly salvaged by deoxycytidine kinase (DCK) to dAMP and dGMP [60]. Finally, nucleoside transporters, including equilibrative nucleoside transporters (ENTs) and concentrative nucleoside transporters (CNTs), mediate the uptake of nucleobases and nucleosides for salvage pathways [63].

In addition to these core metabolic pathway alterations, cancer cells employ sophisticated opportunistic nutrient acquisition strategies in the nutrient-scarce TME, comprising macropinocytosis, entosis, phagocytosis, and autophagy, to scavenge and recycle nutrients, ensuring survival under nutrient stress conditions [64–66]. Complementing these opportunistic scavenging mechanisms, cancer cells enhance nutrient uptake via upregulated transporters for glucose (GLUT1/3/SGLT1), amino acids (ASCT2/LAT1/xCT/SLC6A14/SLC1A3), fatty acids (FATPs/CD36), and nucleotides (ENT/CNTs) [67–70]. This creates a nutrient-deprived microenvironment. For instance, higher glucose and glutamine consumption by cancer cells leads to glucose and glutamine-limited TMEs that starve the neighboring cells of these indispensable fuel sources [67–70]. Immune cell metabolism is intrinsically linked to their function, with distinct metabolic profiles characterizing different subsets. Naïve T cells primarily rely on OXPHOS for basal energy needs. In contrast, effector T cells undergo a metabolic switch toward glycolysis upon activation to meet the bioenergetic and biosynthetic demands of rapid proliferation and cytokine production [71]. Memory T cells favor FAO for long-term survival and metabolic efficiency, whereas regulatory T cells (Tregs) utilize both OXPHOS and FAO to support their suppressive functions [72]. Similarly, M1-polarized macrophages, crucial for antitumor immunity, depend on glycolysis for their pro-inflammatory and phagocytic activities, while M2-polarized macrophages, commonly associated with tumor promotion, favor OXPHOS and FAO [73]. Consequently, glucose deprivation within the TME directly impairs the growth, activation, and effector functions of both antitumor effector T cells and M1 macrophages, which are metabolically reliant on glycolysis. Moreover, the substantial lactate production characteristic of the Warburg effect in cancer cells leads to lactate accumulation in the TME [74]. Elevated lactate levels have been demonstrated to exert multiple immunosuppressive effects, such as, inhibiting T-cell proliferation and cytokine production, promoting differentiation and polarization of macrophages toward the M2 phenotype, stimulating angiogenesis to support tumor expansion and nutrient supply, and contributing to the acidification of the TME, which further compromise immune cell activity and facilitate cancer immune evasion [75]. Interestingly, cancer-associated fibroblasts (CAFs) can metabolically adapt to utilize lactate secreted by cancer cells as a carbon source [15]. This metabolic symbiosis, termed "reverse Warburg effect," allows CAFs to generate energy and biomass through the oxidative metabolism of lactate, effectively preventing the TME from becoming excessively acidic [15, 76]. Similarly, tryptophan depletion and the concomitant accumulation of its catabolites, kynurenine mediated by enzymes like indoleamine-2,3-dioxygenase (IDO) and tryptophan-2,3-dioxygenase (TDO) expressed by cancer cells and certain stromal cells, also contribute to T-cell suppression within the TME [77]. Adenosine produced by the ectonucleotidases, CD39 and CD73, has a strong immunosuppressive property [78]. The complex metabolic milieu of the TME is further shaped by a diverse pool of released metabolites, encompassing molecules such as formate, acetate, 2-hydroxyglutarate, methylthioadenosine, and succinate [14]. These metabolites can exert varied effects in TME—providing substrates for biosynthesis, activating intracellular signaling pathways in both cancer and stromal cells and modulating cell growth and survival dynamics within the complex niche of the tumor microenvironment [14].

Furthermore, metabolic reprogramming contributes significantly to therapy resistance. The Warburg effect, with reduced reliance on mitochondrial respiration, can diminish the efficacy of drugs targeting mitochondrial function [79]. Enhanced antioxidant systems, induced by altered glucose and amino acid metabolism and increased NADPH production, can neutralize ROS-mediated damage induced by radiotherapy and certain chemotherapies [80, 81]. Metabolic flexibility allows cancer cells to adapt to therapy-induced metabolic stress, potentially switching metabolic dependencies to bypass drug targets and ensuring continued survival [81]. The complex interplay of these reprogrammed metabolic pathways, tightly orchestrated and interconnected, is not merely a secondary phenomenon in cancer but a fundamental driver of malignancy, underpinning core cancer hallmarks and presenting both vulnerabilities and challenges for therapeutic intervention [82].

3 Targeting Metabolic Vulnerabilities

The metabolic pathways and enzymes that are hyperactive in cancer or are indispensable for rapid cell proliferation could be attractive drug targets. This section will discuss various efforts that have been made to target cancer metabolism. It will explore different strategies and highlights both the potential benefits and challenges associated with each approach. We will also discuss how targeting specific metabolic pathways in the tumor cells and other cells in TME improves the efficacy of the existing and experimental therapies.

3.1 Targeting Glycolysis and the Warburg Effect

Targeting the Warburg effect and glycolysis offers a promising therapeutic strategy in cancer. Current efforts focus on disrupting glucose uptake and inhibiting key glycolytic enzymes. Various inhibitors targeting GLUTs (WZB117 and BAY-876) [83], HK (2-DG, lonidamine, and 3BrPA), LDH (gossypol, FX11, oxamate, and stiripentol), ENO-1 (ENOblock, POMHEX, and SF2312), PFKFB3 (PFK15 and PFK-158), and PKM2 (shikonin) are currently under preclinical and clinical investigations [84]. Besides, monocarboxylate transporters (MCT1 and MCT4), essential for lactate export, are being targeted with inhibitors like AZD3965 (for MCT1) and syrosingopine, VB124, and AZD0095 (for MCT4) to trap lactate within tumor cells and hinder their growth [85]. Inhibiting enzymes of the PPP, glucose-6-phosphate dehydrogenase (G6PD) with 6-AN, DHEA, its analog (8354 and Polydatin), and transketolase (TKT) with oxythiamine, is also being explored [86]. Finally, targeting HBP with inhibitors against glucosamine-fructose-6-phosphate aminotransferase isomerizing 1 (GFPT1) (azaserine), O-GlcNAc transferase (OGT) (OSMI-4), and O-GlcNAcase (OGA) (Thiamet-G) has shown promise in reducing tumor cell proliferation [87]. Targeting glycolysis appears promising in some preclinical settings. However, monotherapy recurrently fails due to metabolic plasticity and phenotypic heterogeneity in cancer cells [85]. Therefore, approaches combining glycolysis inhibitors with other metabolic or conventional therapies are increasingly being explored [85, 88]. Specifically, modulating glycolysis to overcome resistance to chemotherapy and radiotherapy has shown promise. For example, combining 3-bromopyruvate (3BrPA) with daunorubicin increased its efficacy in breast cancer [89], while 2-deoxyglucose (2-DG) increased the potency of radiotherapy in pancreatic cancer models [90]. In trastuzumab-resistant-HER2-positive breast cancer, targeting elevated LDHA with oxamate improves trastuzumab efficacy [90]. Inhibition of ENO-1 enhances cisplatin sensitivity, and PFK158 overcomes carboplatin resistance in ovarian cancer [91]. Thiamet-G improves the efficacy of paclitaxel against leukemia cells [92]. Similarly, metabolic reprogramming-induced immunosuppressive tumor microenvironments, characterized by glucose deprivation and T-cell dysfunction, can be potentially reversed by glycolysis inhibition [93]. Combining the hexokinase inhibitor 2-DG or GLUT1 inhibitor BAY-876 with immunotherapy boosts T-cell function and infiltration [93, 94]. Preclinical evidence suggests that the inhibition of GAPD, PFKFB3, PKM2, and ENO-1 can enhance immunotherapy efficacy, potentially by disrupting cancer cell metabolism and/or improving immune cell function [95]. In conclusion, strategic targeting of glycolysis and its modulation within the tumor microenvironment represents a compelling approach to improve the efficacy of chemotherapy, radiotherapy, and immunotherapy, necessitating further clinical evaluation.

3.2 Targeting Mitochondrial OXPHOS and Biogenesis

Targeting mitochondrial metabolism, central to cancer cell bioenergetics and biosynthesis, represents a promising therapeutic approach [96]. Inhibiting mitochondrial Complex I, a key electron transport chain component, is a prime approach. Biguanides like metformin and phenformin (antidiabetic agents and Complex I inhibitors), showed promising anticancer effects in preclinical settings [97]. Newer biguanides (e.g., IM156) [98] and non-biguanide inhibitors (e.g., IACS-010759) [99] have demonstrated improved potency. Targeting mitochondrial pyruvate carrier (MPC) to limit pyruvate entry into mitochondria, using inhibitors like UK5099 is also under exploration [100]. Modulation of mitochondrial biogenesis via PGC-1α is another area of investigation. However, the lack of specific inhibitors poses a challenge [101]. Importantly, mitochondrial targeting, particularly OXPHOS inhibition, can synergize with conventional therapies [101]. However, various mitochondria-targeting regimens showed toxicities in clinical settings [102]. IACS-010759, for instance, disrupts respiration and ATP synthesis, but despite showing preclinical promise, raised concerns due to underlying neurological and hyperlactatemia toxicities in early trials [99]. Thus, while OXPHOS and mitochondrial biogenesis modulation could offer a multifaceted strategy to potentiate chemo-, immuno-, and radiotherapy, careful clinical development is crucial to ensure minimal toxicity and maximal therapeutic benefits [101].

3.3 Targeting the TCA Cycle

The TCA cycle, a central metabolic hub in cancer, is an important target for therapeutic intervention. Key approaches include enzyme inhibition and exploiting mutation-specific vulnerabilities. For example, CPI-613 (devimistat) inhibits pyruvate dehydrogenase (PDH) and α-ketoglutarate dehydrogenase (α-KGDH), disrupting TCA cycle flux and ATP production. CPI-613 has shown clinical efficacy in acute myeloid leukemia (AML) and hematological malignancies [103, 104]. Similarly, glutaminase inhibition, using CB-839 (telaglenastat), restricts α-ketoglutarate supply, limiting TCA cycle anaplerosis, and is under clinical trial for solid tumors like triple-negative breast cancer [105]. Ivosidenib and enasidenib target mutant IDH1 and IDH2, respectively, in IDH-mutant cancers such as AML. By reducing the oncometabolite 2-HG, these drugs restore cellular differentiation [106]. Targeting SDH/FH mutant tumors could also be prudent and is an active area of research. Importantly, TCA cycle modulation has shown synergistic effects with other therapies. For example, CPI-613 potentiates chemotherapy (e.g., FOLFIRINOX in pancreatic cancer) [107], and glutaminase inhibition enhances antitumor immunity [108]. Additionally, TCA cycle disruption can also enhance the effects of radiation therapy [109].

3.4 Targeting Amino Acid Metabolism in Cancer

Targeting glutamine metabolism, crucial for cancer cell growth, is a therapeutic strategy focused on inhibiting glutamine transport and utilization. Glutamine uptake blockade, utilizing amino acid transporter antagonists like V-9302, demonstrated synergistic effects with CB-839 [110, 111] and JPH203, a LAT1-selective inhibitor [112] in preclinical cancer models. Blocking glutamine utilization using 6-diazo-5-oxo-L-norleucine (DON), a glutamine analog, and its prodrugs JHU-083 [113] and DRP-104 [114] is another promising avenue. JHU-083 showed preclinical promise in enhancing antitumor immunity by inhibiting checkpoint inhibitors, and DRP-104 is in clinical trials with atezolizumab (targets and blocks the PD-L1 on cancer cells). Glutaminase (GLS1/2) inhibitor, CB-839, enhanced antitumor immunity in clinical settings [115], and IPN60090, which synergizes with mTORC1/2 inhibitors, displayed improved pharmacokinetics in Phase I trials [116]. Proline synthesis inhibitors have shown limited preclinical efficacy and need further development [117]. Serine, glycine, and tryptophan metabolism inhibitors are discussed in the subsequent sections. Importantly, glutamine metabolism inhibitors can potentiate conventional cancer therapies, for instance, inhibiting glutamine uptake potentiated paclitaxel efficacy [118], JHU-083 boosted T-cell response and checkpoint blockade [119], and GLS1 inhibition increased radiosensitivity by impairing DNA repair [120].

3.5 Targeting Lipid Metabolism in Cancer

Targeting de novo fatty acid synthesis, often upregulated in cancer, offers a promising therapeutic avenues. ACLY, crucial for acetyl-CoA production, can be targeted by bempedoic acid (FDA-approved for hyperlipidemia), NDI-091143, and cucurbitacin B [121]. These inhibitors showed promising results in in vitro studies and require further in vivo validation [121]. Genetic ablation or pharmacological inhibition of ACSS2 using VY-3-135 shows tumor-suppressive effects [32]. ACC1/ACC2 inhibitors, such as TOFA [122], ND-646, and PF-05221304 [33] could be explored for anticancer therapies. FASN targeting using omeprazole [123], and various newer agents like AZ22, AZ65 [124], TVB-3664 [125], and TVB-2640 (in clinical trials) [126] is currently under investigation. The SCD1 inhibitor, GSK1940029, is in Phase I dermatologic trials and may have potential use in cancer [127]. Targeting fatty acid synthesis also shows promise for improving existing cancer treatments. FASN inhibitors (e.g., TVB-3664) can improve the efficacy of paclitaxel [128]. SCD1 inhibition may modulate immunotherapy responses [129], and FASN inhibition can radiosensitize cancer cells [130]. Hence, targeting fatty acid synthesis could emerge as a potential therapeutic avenue.

3.6 Targeting Nucleotide Metabolism for Cancer Therapy

Targeting nucleotide metabolism remains a cornerstone of cancer therapeutics since the discovery of antifolate drugs by Sidney Farber to inhibit DNA synthesis in cancer cells [131]. Folate antagonists, including methotrexate and pemetrexed, exemplify clinically established inhibitors of 1C metabolism. While methotrexate competitively inhibits dihydrofolate reductase (DHFR), impairing the regeneration of tetrahydrofolate (THF), pemetrexed exhibits multi-target inhibition against TYMS and 5-aminoimidazole-4-carboxamide-1-β-D-ribofuranoside (AICAR) transformylase, collectively disrupting 1C unit transfer reactions essential for nucleotide synthesis [132]. The delivery of 1C units is critically dependent on serine and glycine metabolism. Serine, synthesized de novo from 3-phosphoglycerate via phosphoglycerate dehydrogenase (PHGDH) and phosphoserine aminotransferase 1 (PSAT1), undergoes enzymatic conversion to glycine by SHMT1/2, generating 5,10-methylene-tetrahydrofolate, which is a crucial 1C donor for thymidylate and purine biosynthesis [59]. Pharmacological PHGDH inhibitors, encompassing ketothioamides [133], BI-4924 [134], indole amides [135], piperazine-1-thiourea derivatives [136], NCT-503 [137], and PH-755 [138] are promising preclinical candidates. Specific SHMT inhibitors, such as AGF347, with dual SHMT1/2 inhibitory activity, exhibit in vivo antitumor effects [139], SHIN1 and SHIN2 were synergistic with methotrexate [140], and repurposed agents like sertraline (an antidepressant) in combination with artemether are promising antineoplastic candidates [141]. Dietary manipulation of serine and glycine intake is an additional investigational strategy, with serine starvation shown to enhance oxidative phosphorylation and prolong survival in lymphoma murine models when combined with phenformin [142], and to synergize with PHGDH inhibition in curbing tumor growth across various xenograft models [142]. Targeting purine and pyrimidine de novo biosynthesis pathways presents another viable option. The inosine 5′-monophosphate dehydrogenase 2 (IMPDH2) inhibitor, mycophenolic acid, shows antiproliferative activity. However, its limited clinical efficacy in early-phase studies prompted the development of more potent IMPDH2 inhibitors [143]. Leflunomide, a rheumatoid arthritis therapeutic agent with dihydroorotate dehydrogenase (DHODH) inhibitory properties, has shown preclinical antitumor activity in murine models and manageable toxicity in early clinical trials for refractory myeloma [144]. Various next-generation DHODH inhibitors, including AG-636, are being developed [145]. Despite the therapeutic efficacy of nucleotide metabolism targeting, inherent toxicities, encompassing myelosuppression and gastrointestinal mucositis, necessitate continued refinement of therapeutic strategies through the development of more selective inhibitors, exploration of combination regimens consisting of immunotherapy, dietary interventions, and a deeper understanding of tumor microenvironment metabolic interplay to optimize efficacy and mitigate off-target effects.

3.7 Targeting Redox Homeostasis in Cancer

Targeting redox homeostasis aims at increasing ROS to cytotoxic levels via two main therapeutic strategies, ROS induction and antioxidant inhibition. ROS-inducing agents like piperlongumine [146], motexafin gadolinium (MGd) [147], β-phenylethyl isothiocyanate (PEITC) [147], and eleutheroside B [148] aim to elevate ROS to cytotoxic levels through various mechanisms. Conversely, antioxidant inhibitors, such as buthionine sulfoximine (BSO) [149], PX-12 and TRX-26 (thioredoxin reductase inhibitors) [150], 2-methoxy estradiol and ATN-224 (SOD inhibitors) [151], compromise redox buffering in cancer cells. A distinct approach involves ferroptosis induction, leveraging the iron-dependent cell death incited by lipid peroxidation. Inhibiting glutathione peroxidase 4 (GPX4), a key ferroptosis suppressor, with agents like (1S,3R)-RSL3, ML162/ML210, and FIN56/FINO2 triggers ferroptosis [152]. Glutathione depletion, via BSO or xCT inhibitors (erastin and sulfasalazine), which limit cysteine import, also sensitizes cancer cells to ferroptosis [153]. Ferroptosis-inducing therapies (FITs) are in early clinical development, with siramesine and sorafenib showing ferroptosis-inducing activity [153]. Redox and ferroptosis targeting can enhance conventional cancer therapies. It is to be noted that various chemotherapies induce ROS to kill cancer cells. Therefore, targeting the antioxidant redox mechanisms becomes a promising strategy [154]. Initial evidence suggests that glutathione synthesis inhibitors (BSO) and GPX4 inhibitors (RSL3) potentiated chemotherapy [155, 156]. Similarly, ROS enhancers (MGd) and ferroptosis inducers (RSL3) augmented radiosensitivity [157], and ferroptosis induction could enhance immunotherapy [158].

3.8 Targeting Metabolic Crosstalk in TME

A crucial strategy for improving cancer immunotherapy involves targeting the production of immunosuppressive metabolites within the TME. A prime example is the adenosine pathway, where neoplastic cells frequently overexpress ectonucleotidases, CD39 and CD73, catalyzing the conversion of ATP to adenosine. Adenosine acts as a potent immunosuppressant via the engagement of A2A and A2B receptors on immune cells, thereby dampening their antitumor function [78]. Counteracting this immunosuppressive axis is pursued through the development of anti-CD39 antibodies, such as BMS-986179, anti-CD73 antibodies like MEDI9447, and small-molecule inhibitors like AB680 [159]. Preclinical investigations demonstrated that CD73 inhibition potentiates T-cell activity and improves response to immune checkpoint blockade [160, 161]. Complementarily, A2A/A2B receptor antagonists-CPI-444, AZD4635, and PBF-509, are in clinical development to impede adenosine signaling on immune cells [160, 162]. IDH represents another important target; FDA-approved IDH inhibitors, such as enasidenib (for mutant IDH2) and ivosidenib (for mutant IDH1), are employed to reduce 2-HG levels, potentially restoring immune cell competence, thus enhancing immunotherapy efficacy in glioma models [163]. Moreover, IDO and TDO enzymes catabolizing tryptophan to immunosuppressive metabolites are being targeted. While initial clinical trials of IDO inhibitors like epacadostat and navoximod in combination with checkpoint inhibitors were unsuccessful, the development of alternative IDO and TDO inhibitors continues, with preclinical studies demonstrating that TDO inhibition can augment immunotherapy efficacy [164]. Lactate modulation is also being studied with strategies to mitigate lactate-mediated immunosuppression by inhibiting lactate dehydrogenase A (LDHA) to suppress lactate production and use or employ monocarboxylate transporter (MCT) inhibitors to block lactate transport [165]. Additionally, MCT4 inhibition in CAFs can suppress tumor growth [166]. Lastly, targeting methylthioadenosine phosphorylase (MTAP) insufficiency is under investigation. MTAP deficiency, frequent in various malignancies, results in the accumulation of methylthioadenosine (MTA), which can exert immunosuppressive effects. Potent inhibitors capable of selectively killing MTAP-deficient cells are in preclinical development [167, 168]. Additionally, enzyme-mediated depletion of MTA is being explored [167, 168]. These interventions aim to diminish the concentration of immunosuppressive metabolites within the TME, thereby alleviating constraints on immune-cell activity and enhancing the effectiveness of antitumor immunotherapy.

Beyond the direct targeting of cancer cell metabolism, modulating the metabolic landscape of stromal cells represents another promising approach to potentiate cancer therapy. Metabolic reprogramming of T cells is a key area of focus, with strategies such as mTOR modulation using low-dose rapamycin shown to promote the generation of memory $CD8^+$ T cells exhibiting enhanced antitumor activity in preclinical models [169]. Similarly, the antidiabetic agent metformin can boost T-cell metabolic fitness and enhance immunotherapy efficacy [170]. In the context of chimeric antigen receptor (CAR) T-cell therapy, metabolic engineering approaches to enhance CAR T cell glycolysis or mitochondrial function are under investigation [171]. Targeting glutamine metabolism in immune cells also presents a potential therapeutic avenue, with glutaminase inhibitors like CB-839 or JHU-083 under evaluation for their capacity to enhance immune cell function through the modulation of glutamine availability [108]. Beyond T cells, other stromal components are being targeted. Targeting glycolysis, glutaminase, or fatty acid oxidation in tumor-associated macrophages (TAMs) may facilitate a phenotypic shift from immunosuppressive M2-like macrophages to pro-inflammatory, antitumor M1-like macrophages [172]. Enhancing glycolysis or modulating glutamine metabolism in natural killer (NK) cells also holds the potential to boost their antitumor cytotoxicity [173]. Furthermore, metabolic reprogramming of dendritic cells (DCs), critical for antigen presentation, by enhancing glycolysis or modulating fatty acid metabolism may increase their capacity to stimulate T-cell responses [174]. In conclusion, modulating metabolic interactions within the TME, encompassing both inhibition of immunosuppressive pathways and enhancement of immune cell metabolism, holds substantial promise for potentiating cancer therapy.

4 Dietary Interventions

Dietary interventions represent a complementary approach to target cancer metabolism, aiming to modulate systemic and tumor-intrinsic metabolism through nutritional modifications. These interventions exploit cancer cells' metabolic dependencies by restricting the availability of specific nutrients or altering macronutrient ratios to create a metabolically unfavorable environment for tumor growth. Various dietary strategies, encompassing short-term fasting (STF), ketogenic diets (KDs), and fasting-mimicking diets (FMDs) are being tested. STF (3–5 days) induces systemic metabolic shifts such as reduced glucose and elevated ketone bodies [175]. In vivo, STF inhibited tumor growth, reduced chemotherapy toxicity, and synergized with chemotherapy/radiotherapy. It also potentiated tyrosine kinase inhibitor-targeted therapy [175]. KDs (very low carbohydrate, high fat) aim to shift energy substrate utilization away from glucose and toward ketone bodies. While studies suggest a potential for KDs in cancer treatment, variability in diet composition hinders definitive conclusions. Ongoing trials are evaluating KDs with chemotherapy or radiotherapy in glioblastoma [176, 177]. Caloric restriction and intermittent fasting are also being explored for their effects on systemic metabolism and the potential to reduce growth factor signaling and induce metabolic stress in tumors [175, 178]. The FMD, a plant-based, low-calorie, low-protein KD, mimics

STF's metabolic effects; it reduces glucose levels and increases ketone bodies [175, 179, 180]. Specific nutrient restrictions, such as methionine or glutamine, aim to limit the availability of essential nutrients that cancer cells may be highly dependent on for growth and proliferation [181]. However, clinical evidence for dietary interventions as standalone cancer therapies remains limited. Clinical trials are ongoing to rigorously assess the efficacy of these dietary strategies, often in combination with conventional therapies, and to identify predictive biomarkers for patient selection and optimal implementation [175].

5 Challenges, Opportunities, and Future Directions

Cancers display diverse metabolic profiles influenced by factors such as disease stage, genetic background, and tissue of origin [6]. Therefore, successfully understanding and targeting cancer metabolism requires a deep understanding of how metabolic characteristics and dependencies change as cancer progresses from early, premalignant stages to metastasis. Tissue-specific metabolic traits are retained in early premalignant lesions because the tissue of origin and developmental context influence tumor initiation and the effects of metabolic enzyme mutations [6]. Some tumors exhibit stereotyped metabolic properties even without canonical mutations, while others display distinct metabolic shifts in late premalignant stages [183–185]. Primary invasive cancers often converge on shared metabolic pathways, including those governing energy production, macromolecule synthesis, and redox control, frequently instigated by common oncogenes [1, 186]. However, metabolic heterogeneity within these cancers arises from both intrinsic factors, such as co-occurring mutations (e.g., *KRAS*, *STK11*, *KEAP1*), and extrinsic factors like the dynamically changing tumor microenvironment [1]. Metastasis introduces unique metabolic demands involving extracellular acidification, and heightened antioxidant responses [187, 188]. Dormancy and subsequent reactivation of micrometastatic growth can involve shifts in metabolic dependencies, potentially an increased reliance on oxidative phosphorylation (OXPHOS) [189]. Consequently, effective metabolism-targeted therapies must account for these diverse and evolving factors. Subtype-specific and personalized therapies offer the most promising path forward. Despite the complexities associated with the therapeutic targeting of metabolic vulnerabilities, opportunities for therapeutic intervention exist, for instance, the IDH mutation-associated metabolic dependency [106] or GPX4 dependence in persistent cancer cells, and exploiting various metabolic interactions between cancer cells and the microenvironment [152]. The increasing reliance on OXPHOS in advanced cancers also presents a potential vulnerability [190]. On the therapeutic end, the advancement of numerous metabolic inhibitors through preclinical and clinical stages, with some gaining FDA approval, is encouraging (Table 1). However, overcoming challenges, including metabolic plasticity and resistance, mitigating off-target toxicities, and establishing predictive biomarkers remain important goals.

Future cancer metabolism research should prioritize: (1) Deeper exploration of tumor metabolic heterogeneity and plasticity using multi-omics and single-cell analyses to define context-specific dependencies and resistance mechanisms. (2)

Table 1 Metabolic inhibitor-associated clinical trials in cancer

Metabolic pathway	Target/inhibitor or drug(s) (clinical trial stage)	Cancer type/key observations, outcomes/clinical trial ID
Glycolysis	HK/2-DG (I/II)	Breast, gastric, head and neck, lung, prostate, and pancreatic cancer/mixed results, some toxicity reported, often combined with chemotherapy or radiation/NCT00096707, NCT00633087
	HK/Lonidamine (I/II/III)	Breast, lung, ovarian, head and neck and several advanced cancers/improved response rates in combination with chemotherapy, limited survival impact [182]
	PFKFB3/PFK-158 (Phase I/II)	Advanced solid tumors/acceptable safety profile, potential clinical activity, promising preliminary data/NCT02044861
	LDH/Gossypol (I/II)	Various cancers/ limited efficacy, some toxicity. NCT00848016, NCT00286793, NCT00544596, NCT00275431, NCT00286806, NCT00540722, NCT00666666, NCT00286780, NCT00397293, NCT00891072, NCT00773955, NCT00440388, NCT00390403, NCT00390403, NCT02697344, NCT00571675
Glutamine Metabolism	GLS1/ CB-839 (I/II/III)	Solid tumors, hematologic malignancies, renal and breast cancer/some clinical activity, ineffective with carbozantinib in renal cell carcinoma/NCT03163667, NCT03057600, NCT03428217, NCT03965845
	Glutamine utilization/Drp-104 (I)	Advanced solid tumors/early-stage trials, combined with atezolizumab/NCT04471415
Mitochondrial Metabolism and TCA cycle	Complex I/ metformin (I/II/III)	Various cancers/mixed results, often combined with other therapies/NCT03109873, NCT02325401, NCT02949700, NCT00930579, NCT01312467, NCT02028221, NCT02176161, NCT01579812, NCT02035787, NCT01620593, NCT01447927, NCT03800602, NCT01310231, NCT01433913, NCT02949700, NCT01666730, NCT01968317, NCT01101438
	Complex I/IM156 (I)	Solid tumors, AML/early signs of clinical response, some neurological side effects and hyperlactatemia reported/NCT05497778, NCT03272256
	PDK/DCA (I/II)	Various cancers/mixed results with neurotoxicity concerns/ NCT00566410, NCT01111097, NCT01163487, NCT00540176, NCT01386632
	PDH, α-KGDH/ CPI-613 (I/II/III)	Pancreatic cancer, AML, others/poor responses in pancreatic cancer (with FOLFIRINOX), signs of response in AML/NCT04593758, NCT04203160

(continued)

Table 1 (continued)

Metabolic pathway	Target/inhibitor or drug(s) (clinical trial stage)	Cancer type/key observations, outcomes/clinical trial ID
	IDH1/2 (mutant)/ enasidenib, ivosidenib (approved)	AML, IDH-mutated cancers/FDA-approved for AML with IDH mutations, trials in other IDH-mutated cancers/ NCT02677922, NCT03839771, NCT02632708
	Complex I/ metformin (I/II/III)	Various cancers/mixed results, often combined with other therapies/NCT03109873, NCT02325401, NCT02949700, NCT00930579, NCT01312467, NCT02028221, NCT02176161, NCT01579812, NCT02035787, NCT01620593, NCT01447927, NCT03800602, NCT01310231, NCT01433913, NCT02949700, NCT01666730, NCT01968317, NCT01101438
FA metabolism	FASN/TVB-2640 (II)	Solid tumors/early results show some tumor shrinkage/NCT02223247, NCT03179904
Nucleotide metabolism	DHODH/ leflunomide	Refractory myeloma, others/antitumor activity in mouse models, manageable side effects in myeloma trial/NCT00003775, NCT01611675, NCT04370483, NCT02509052, NCT00004071
	DHODH/AG-636 (early phase)	Diffuse large B-cell lymphoma, mantle cell lymphoma, others/completed early phase trials/ NCT03834584
	IMPDH/ mycophenolate mofetil	Liver cancer/showed some efficacy in a small clinical study for liver cancer/NCT05049863
	TYMS/pemetrexed (approved)	Various solid tumors, mesothelioma, NSCLC/ approved antifolate drug, widely used in chemotherapy regimens, often combined with other agents/NCT00410904, NCT00454194, NCT00691301, NCT00234052, NCT00503997, NCT00798603, NCT00117962, NCT01107626, NCT00660816, NCT00226577, NCT01344824, NCT00407550, NCT00301808, NCT00702299, NCT02426658, NCT01085630, NCT01001910, NCT00101283, NCT00950365, NCT01338792, NCT00698815
	DHFR/methotrexate (approved)	Various hematological and solid malignancies/ classic antifolate drug, used for decades, especially in leukemia, lymphoma, and some solid tumors, often in combination regimens/ NCT00992602, NCT00003702, NCT01345682, NCT02347332, NCT01176981

(continued)

Table 1 (continued)

Metabolic pathway	Target/inhibitor or drug(s) (clinical trial stage)	Cancer type/key observations, outcomes/clinical trial ID
Tumor microenvironment	CD73/MEDI9447, AB680 (I/II)	Solid tumors/CD73 inhibitors targeting adenosine production in TME, in clinical trials to enhance immunotherapy/NCT03736473, NCT03773666, NCT02503774, NCT03267589, NCT03611556
	Adenosine A2A/A2B receptors/CPI-444, AZD4635, PBF-509 (I/II)	Solid tumors/adenosine receptor antagonists block immunosuppressive adenosine signaling in the TME, being tested in clinical trials/NCT02655822, NCT04280328, NCT03454451, NCT02740985, NCT04495179, NCT04089553, NCT02403193
	IDO/epacadostat, navoximod, others (clinical trials—Mostly failed/re-evaluating)	Various cancers/initial trials of IDO inhibitors and checkpoint inhibitors showed limited benefit for epacadostat, ongoing trials with other IDO/TDO inhibitors and combination strategies/NCT01604889, NCT01195311, NCT03602586, NCT03196232, NCT02318277, NCT03322540, NCT03361865, NCT03374488, NCT03322384, NCT02178722, NCT02327078, NCT02862457, NCT01961115, NCT02042430, NCT03085914, NCT03006302, NCT03322566, NCT02752074, NCT02118285, NCT02166905
	MCT1/AZD3965 (I)	Solid tumors/MCT1 inhibitor in Phase I trials, targeting lactate transport in TME/NCT01791595

Development of more selective and potent inhibitors for emerging targets. (3) Further investigation of rational combination therapies, particularly integrating metabolic inhibitors with immunotherapy and targeted agents to identify synergistic combinations that enhance antitumor immunity and overcome resistance. (4) Biomarker development to personalize metabolic therapies and monitor response. (5) Exploration of novel drug delivery to improve tumor-specific accumulation and reduce systemic toxicity. (6) Study of dietary interventions and lifestyle modifications as adjuncts to metabolic therapies to modulate systemic metabolism and enhance therapeutic windows.

Acknowledgments The figure was drawn in Adobe Illustrator.

References

1. Pavlova NN, et al. The hallmarks of cancer metabolism: still emerging. Cell Metab. 2022;34(3):355–77.
2. DeBerardinis RJ, Chandel NS. Fundamentals of cancer metabolism. Sci Adv. 2016;2(5):e1600200.
3. Zhang S, et al. Tumor initiation and early tumorigenesis: molecular mechanisms and interventional targets. Signal Transduct Target Ther. 2024;9(1):149.

4. Seyfried TN, Shelton LM. Cancer as a metabolic disease. Nutr Metab (Lond). 2010;7:7.
5. Pavlova NN, Thompson CB. The emerging hallmarks of cancer metabolism. Cell Metab. 2016;23(1):27–47.
6. Faubert BA, et al. Metabolic reprogramming and cancer progression. Science. 2020;368(6487):eaaw5473.
7. Huang K, et al. Tumor metabolic regulators: key drivers of metabolic reprogramming and the promising targets in cancer therapy. Mol Cancer. 2025;24(1):7.
8. Warburg O. On respiratory impairment in cancer cells. Science. 1956;124(3215):269–70.
9. Warburg O, et al. The metabolism of tumors in the body. J Gen Physiol. 1927;8(6):519–30.
10. Kaushik N, et al. Blockade of cellular energy metabolism through 6-aminonicotinamide reduces proliferation of non-small lung cancer cells by inducing endoplasmic reticulum stress. Biology (Basel). 2021;10(11):1088.
11. Ying H, et al. Oncogenic Kras maintains pancreatic tumors through regulation of anabolic glucose metabolism. Cell. 2012;149(3):656–70.
12. Li AM, Ye J. The PHGDH enigma: do cancer cells only need serine or also a redox modulator? Cancer Lett. 2020;476:97–105.
13. Apostolova P, Pearce EL. Lactic acid and lactate: revisiting the physiological roles in the tumor microenvironment. Trends Immunol. 2022;43(12):969–77.
14. Baryla M, et al. Oncometabolites-A link between cancer cells and tumor microenvironment. Biology (Basel). 2022;11(2):270.
15. Fiaschi T, et al. Reciprocal metabolic reprogramming through lactate shuttle coordinately influences tumor-stroma interplay. Cancer Res. 2012;72(19):5130–40.
16. Kong LR, et al. A glycolytic metabolite bypasses "two-hit" tumor suppression by BRCA2. Cell. 2024;187(9):2269–2287 e16.
17. Anderson NM, et al. The emerging role and targetability of the TCA cycle in cancer metabolism. Protein Cell. 2018;9(2):216–37.
18. Martinez-Reyes I, Chandel NS. Mitochondrial TCA cycle metabolites control physiology and disease. Nat Commun. 2020;11(1):102.
19. Cheng T, et al. Pyruvate carboxylase is required for glutamine-independent growth of tumor cells. Proc Natl Acad Sci USA. 2011;108(21):8674–9.
20. Arnold PK, Finley LWS. Regulation and function of the mammalian tricarboxylic acid cycle. J Biol Chem. 2023;299(2):102838.
21. Sellers K, et al. Pyruvate carboxylase is critical for non-small-cell lung cancer proliferation. J Clin Invest. 2015;125(2):687–98.
22. Lsman JA, et al. 2-Oxoglutarate-dependent dioxygenases in cancer. Nat Rev Cancer. 2020;20(12):710–26.
23. Metallo CM, et al. Reductive glutamine metabolism by IDH1 mediates lipogenesis under hypoxia. Nature. 2011;481(7381):380–4.
24. Cui Q, et al. Modulating ROS to overcome multidrug resistance in cancer. Drug Resist Updat. 2018;41:1–25.
25. Benhar MD, et al. ROS, stress-activated kinases and stress signaling in cancer. EMBO Rep. 2002;3(5):420–5.
26. Cheung EC, et al. Dynamic ROS control by TIGAR regulates the initiation and progression of pancreatic cancer. Cancer Cell. 2020;37(2):168–182 e4.
27. Cheung EC, Vousden KH. The role of ROS in tumour development and progression. Nat Rev Cancer. 2022;22(5):280–97.
28. Cichon MA, Radisky DC. ROS-induced epithelial-mesenchymal transition in mammary epithelial cells is mediated by NF-kB-dependent activation of snail. Oncotarget. 2014;5(9):2827–38.
29. Rohrig F, Schulze A. The multifaceted roles of fatty acid synthesis in cancer. Nat Rev Cancer. 2016;16(11):732–49.
30. Mashek DG, Coleman RA. Cellular fatty acid uptake: the contribution of metabolism. Curr Opin Lipidol. 2006;17(3):274–8.

31. Beloribi-Djefaflia S, et al. Lipid metabolic reprogramming in cancer cells. Oncogenesis. 2016;5(1):e189.
32. Miller KD, et al. Targeting ACSS2 with a transition-state mimetic inhibits triple-negative breast cancer growth. Cancer Res. 2021;81(5):1252–64.
33. Svensson RU, et al. Inhibition of acetyl-CoA carboxylase suppresses fatty acid synthesis and tumor growth of non-small-cell lung cancer in preclinical models. Nat Med. 2016;22(10):1108–19.
34. Wang Y, Yu G, Chen X. Mechanism of ferroptosis resistance in cancer cells. Cancer Drug Resist. 2024;7:47.
35. Klasson TD, et al. ACSL3 regulates lipid droplet biogenesis and ferroptosis sensitivity in clear cell renal cell carcinoma. Cancer Metab. 2022;10(1):14.
36. Mullen PJ, et al. The interplay between cell signalling and the mevalonate pathway in cancer. Nat Rev Cancer. 2016;16(11):718–31.
37. Chen J, et al. Amino acid metabolism in tumor biology and therapy. Cell Death Dis. 2024;15(1):42.
38. Bhutia YD, et al. Amino acid transporters in cancer and their relevance to "glutamine addiction": novel targets for the design of a new class of anticancer drugs. Cancer Res. 2015;75(9):1782–8.
39. Yoo HC, et al. Glutamine reliance in cell metabolism. Exp Mol Med. 2020;52(9):1496–516.
40. Yang M, Vousden KH. Serine and one-carbon metabolism in cancer. Nat Rev Cancer. 2016;16(10):650–62.
41. Albaugh VL, et al. Arginine-dual roles as an onconutrient and immunonutrient. J Surg Oncol. 2017;115(3):273–80.
42. Ananieva EA, Wilkinson AC. Branched-chain amino acid metabolism in cancer. Curr Opin Clin Nutr Metab Care. 2018;21(1):64–70.
43. Ericksen RE, et al. Loss of BCAA catabolism during carcinogenesis enhances mTORC1 activity and promotes tumor development and progression. Cell Metab. 2019;29(5):1151–1165 e6.
44. Abd El-Fattah EE. IDO/kynurenine pathway in cancer: possible therapeutic approaches. J Transl Med. 2022;20(1):347.
45. Della Chiesa M, et al. The tryptophan catabolite L-kynurenine inhibits the surface expression of NKp46- and NKG2D-activating receptors and regulates NK-cell function. Blood. 2006;108(13):4118–25.
46. Liu XH, Zhai XY. Role of tryptophan metabolism in cancers and therapeutic implications. Biochimie. 2021;182:131–9.
47. Tummala KS, et al. Inhibition of de novo NAD(+) synthesis by oncogenic URI causes liver tumorigenesis through DNA damage. Cancer Cell. 2014;26(6):826–39.
48. Jewell JL, et al. Amino acid signalling upstream of mTOR. Nat Rev Mol Cell Biol. 2013;14(3):133–9.
49. Mullen NJ, Singh PK. Nucleotide metabolism: a pan-cancer metabolic dependency. Nat Rev Cancer. 2023;23(5):275–94.
50. Ali ES, Ben-Sahra I. Regulation of nucleotide metabolism in cancers and immune disorders. Trends Cell Biol. 2023;33(11):950–66.
51. Villa E, et al. Cancer cells tune the signaling pathways to empower de novo synthesis of nucleotides. Cancers (Basel). 2019;11(5):688.
52. Shukla SK, et al. MUC1 and HIF-1alpha signaling crosstalk induces anabolic glucose metabolism to impart gemcitabine resistance to pancreatic cancer. Cancer Cell. 2017;32(1):71–87 e7.
53. Tong X, et al. The molecular determinants of de novo nucleotide biosynthesis in cancer cells. Curr Opin Genet Dev. 2009;19(1):32–7.
54. MacFarlane AJ, et al. Nuclear localization of de novo thymidylate biosynthesis pathway is required to prevent uracil accumulation in DNA. J Biol Chem. 2011;286(51):44015–22.
55. Newman AC, Maddocks ODK. One-carbon metabolism in cancer. Br J Cancer. 2017;116(12):1499–504.
56. Tibbetts AS, Appling DR. Compartmentalization of mammalian folate-mediated one-carbon metabolism. Annu Rev Nutr. 2010;30:57–81.

57. Goulian M, et al. Methotrexate-induced misincorporation of uracil into DNA. Proc Natl Acad Sci USA. 1980;77(4):1956–60.
58. Woeller CF, et al. Evidence for small ubiquitin-like modifier-dependent nuclear import of the thymidylate biosynthesis pathway. J Biol Chem. 2007;282(24):17623–31.
59. Ser Z, et al. Targeting one carbon metabolism with an antimetabolite disrupts pyrimidine homeostasis and induces nucleotide overflow. Cell Rep. 2016;15(11):2367–76.
60. Wang L. Mitochondrial purine and pyrimidine metabolism and beyond. Nucleosides Nucleotides Nucleic Acids. 2016;35(10–12):578–94.
61. Li J, et al. Identification and characterization of human uracil phosphoribosyltransferase (UPRTase). J Hum Genet. 2007;52(5):415–22.
62. Perignon JL, et al. Metabolism of pyrimidine bases and nucleosides by pyrimidine-nucleoside phosphorylases in cultured human lymphoid cells. Biochim Biophys Acta. 1987;928(2):130–6.
63. Tran DH, et al. De novo and salvage purine synthesis pathways across tissues and tumors. Cell. 2024;187(14):3602–18. e20.
64. Commisso C, et al. Macropinocytosis of protein is an amino acid supply route in Ras-transformed cells. Nature. 2013;497(7451):633–7.
65. Florey O, et al. Autophagy machinery mediates macroendocytic processing and entotic cell death by targeting single membranes. Nat Cell Biol. 2011;13(11):1335–43.
66. Mukhopadhyay S, et al. The metabolic landscape of RAS-driven cancers from biology to therapy. Nat Cancer. 2021;2(3):271–83.
67. Peppicelli S, et al. FDG uptake in cancer: a continuing debate. Theranostics. 2020;10(7):2944–8.
68. Cognet G, Muir A. Identifying metabolic limitations in the tumor microenvironment. Sci Adv. 2024;10(40):eadq7305.
69. Reinfeld BI, et al. Cell-programmed nutrient partitioning in the tumour microenvironment. Nature. 2021;593(7858):282–8.
70. Deng S, et al. Glucose partitioning in the bone marrow micro-environment in acute myeloid leukaemia. Leukemia. 2023;37(7):1407–12.
71. van der Windt GJ, Pearce EL. Metabolic switching and fuel choice during T-cell differentiation and memory development. Immunol Rev. 2012;249(1):27–42.
72. Fox CJ, et al. Fuel feeds function: energy metabolism and the T-cell response. Nat Rev Immunol. 2005;5(11):844–52.
73. Kelly B, O'Neill LA. Metabolic reprogramming in macrophages and dendritic cells in innate immunity. Cell Res. 2015;25(7):771–84.
74. Feng Q, et al. Severely polarized extracellular acidity around tumour cells. Nat Biomed Eng. 2024;8(6):787–99.
75. Chen S, et al. The emerging role of lactate in tumor microenvironment and its clinical relevance. Cancer Lett. 2024;590:216837.
76. Liang L, et al. 'Reverse Warburg effect' of cancer-associated fibroblasts (review). Int J Oncol. 2022;60(6):67.
77. Venkateswaran N, et al. MYC promotes tryptophan uptake and metabolism by the kynurenine pathway in colon cancer. Genes Dev. 2019;33(17–18):1236–51.
78. Allard B, et al. The ectonucleotidases CD39 and CD73: novel checkpoint inhibitor targets. Immunol Rev. 2017;276(1):121–44.
79. Porporato PE, et al. Mitochondrial metabolism and cancer. Cell Res. 2018;28(3):265–80.
80. Matsuoka Y, et al. The antioxidative stress regulator Nrf2 potentiates radioresistance of oral squamous cell carcinoma accompanied with metabolic modulation. Lab Investig. 2022;102(8):896–907.
81. Cort A, et al. Redox control of multidrug resistance and its possible modulation by antioxidants. Oxidative Med Cell Longev. 2016;2016:4251912.
82. Brunner JS, Finley LWS. Metabolic determinants of tumour initiation. Nat Rev Endocrinol. 2023;19(3):134–50.

83. Pliszka M, Szablewski L. Glucose transporters as a target for anticancer therapy. Cancers (Basel). 2021;13(16):4184.
84. Akins NS, et al. Inhibition of glycolysis and glutaminolysis: an emerging drug discovery approach to combat cancer. Curr Top Med Chem. 2018;18(6):494–504.
85. Chelakkot C, et al. Modulating glycolysis to improve cancer therapy. Int J Mol Sci. 2023;24(3):2606.
86. Ghanem N, et al. The pentose phosphate pathway in cancer: regulation and therapeutic opportunities. Chemotherapy. 2021;66(5–6):179–91.
87. Lam C, et al. The hexosamine biosynthetic pathway and cancer: current knowledge and future therapeutic strategies. Cancer Lett. 2021;503:11–8.
88. Huang Y, et al. The potential of lonidamine in combination with chemotherapy and physical therapy in cancer treatment. Cancers (Basel). 2020;12(11):3332.
89. Nakano A, et al. Glycolysis inhibition inactivates ABC transporters to restore drug sensitivity in malignant cells. PLoS One. 2011;6(11):e27222.
90. Coleman MC, et al. 2-deoxy-D-glucose causes cytotoxicity, oxidative stress, and radiosensitization in pancreatic cancer. Free Radic Biol Med. 2008;44(3):322–31.
91. Mondal S, et al. Therapeutic targeting of PFKFB3 with a novel glycolytic inhibitor PFK158 promotes lipophagy and chemosensitivity in gynecologic cancers. Int J Cancer. 2019;144(1):178–89.
92. Ding N, et al. Thiamet-G-mediated inhibition of O-GlcNAcase sensitizes human leukemia cells to microtubule-stabilizing agent paclitaxel. Biochem Biophys Res Commun. 2014;453(3):392–7.
93. Tang X, et al. Glycolysis inhibition induces anti-tumor central memory CD8(+)T cell differentiation upon combination with microwave ablation therapy. Nat Commun. 2024;15(1):4665.
94. Ren X, et al. Inhibition of glycolysis-driven immunosuppression with a nano-assembly enhances response to immune checkpoint blockade therapy in triple negative breast cancer. Nat Commun. 2023;14(1):7021.
95. Gill KS, et al. Glycolysis inhibition as a cancer treatment and its role in an anti-tumour immune response. Biochim Biophys Acta. 2016;1866(1):87–105.
96. Kalyanaraman B, et al. OXPHOS-targeting drugs in oncology: new perspectives. Expert Opin Ther Targets. 2023;27(10):939–52.
97. Ashton TM, et al. Oxidative phosphorylation as an emerging target in cancer therapy. Clin Cancer Res. 2018;24(11):2482–90.
98. Janku F, et al. First-in-human study of IM156, a novel potent biguanide oxidative phosphorylation (OXPHOS) inhibitor, in patients with advanced solid tumors. Investig New Drugs. 2022;40(5):1001–10.
99. Molina JR, et al. An inhibitor of oxidative phosphorylation exploits cancer vulnerability. Nat Med. 2018;24(7):1036–46.
100. Bader DA, et al. Mitochondrial pyruvate import is a metabolic vulnerability in androgen receptor-driven prostate cancer. Nat Metab. 2019;1(1):70–85.
101. LaGory EL, et al. Suppression of PGC-1alpha is critical for reprogramming oxidative metabolism in renal cell carcinoma. Cell Rep. 2015;12(1):116–27.
102. Yap TA, et al. Complex I inhibitor of oxidative phosphorylation in advanced solid tumors and acute myeloid leukemia: phase I trials. Nat Med. 2023;29(1):115–26.
103. Pardee TS, et al. A phase I study of CPI-613 in combination with high-dose cytarabine and mitoxantrone for relapsed or refractory acute myeloid Leukemia. Clin Cancer Res. 2018;24(9):2060–73.
104. Anderson R, et al. Phase II trial of cytarabine and mitoxantrone with devimistat in acute myeloid leukemia. Nat Commun. 2022;13(1):1673.
105. Gross MI, et al. Antitumor activity of the glutaminase inhibitor CB-839 in triple-negative breast cancer. Mol Cancer Ther. 2014;13(4):890–901.
106. Stein EM, et al. Ivosidenib or enasidenib combined with intensive chemotherapy in patients with newly diagnosed AML: a phase 1 study. Blood. 2021;137(13):1792–803.

107. Philip PA, et al. Devimistat (CPI-613) with modified fluorouracil, oxaliplatin, irinotecan, and leucovorin (FFX) versus FFX for patients with metastatic adenocarcinoma of the pancreas: the phase III AVENGER 500 study. J Clin Oncol. 2024;42(31):3692–701.
108. Varghese S, et al. The glutaminase inhibitor CB-839 (telaglenastat) enhances the antimelanoma activity of T-cell-mediated immunotherapies. Mol Cancer Ther. 2021;20(3):500–11.
109. Khan HY, et al. Targeting cellular metabolism with CPI-613 sensitizes pancreatic cancer cells to radiation therapy. Adv Radiat Oncol. 2023;8(1):101122.
110. Schulte ML, et al. Pharmacological blockade of ASCT2-dependent glutamine transport leads to antitumor efficacy in preclinical models. Nat Med. 2018;24(2):194–202.
111. Jin H, et al. A powerful drug combination strategy targeting glutamine addiction for the treatment of human liver cancer. elife. 2020;9:e56749.
112. Okano N, et al. First-in-human phase I study of JPH203, an L-type amino acid transporter 1 inhibitor, in patients with advanced solid tumors. Investig New Drugs. 2020;38(5):1495–506.
113. Hanaford AR, et al. Orally bioavailable glutamine antagonist prodrug JHU-083 penetrates mouse brain and suppresses the growth of MYC-driven medulloblastoma. Transl Oncol. 2019;12(10):1314–22.
114. Rais R, et al. Discovery of DRP-104, a tumor-targeted metabolic inhibitor prodrug. Sci Adv. 2022;8(46):eabq5925.
115. Cederkvist H, et al. Identification and characterization of a novel glutaminase inhibitor. FEBS Open Bio. 2022;12(1):163–74.
116. Soth MJ, et al. Discovery of IPN60090, a clinical stage selective glutaminase-1 (GLS-1) inhibitor with excellent pharmacokinetic and physicochemical properties. J Med Chem. 2020;63(21):12957–77.
117. D'Aniello C, et al. Proline metabolism in tumor growth and metastatic progression. Front Oncol. 2020;10:776.
118. Kim G, et al. Inhibition of glutamine uptake resensitizes paclitaxel resistance in SKOV3-TR ovarian cancer cell *via* mTORC1/S6K signaling pathway. Int J Mol Sci. 2022;23(15):8761.
119. Praharaj M, et al. Metabolic reprogramming of tumor-associated macrophages using glutamine antagonist JHU083 drives tumor immunity in myeloid-rich prostate and bladder cancers. Cancer Immunol Res. 2024;12(7):854–75.
120. Patel R, et al. Targeting glutamine metabolism improves sarcoma response to radiation therapy *in vivo*. Commun Biol. 2024;7(1):608.
121. Montesdeoca N, et al. Inhibitors of lipogenic enzymes as a potential therapy against cancer. FASEB J. 2020;34(9):11355–81.
122. Li S, et al. TOFA suppresses ovarian cancer cell growth *in vitro* and *in vivo*. Mol Med Rep. 2013;8(2):373–8.
123. Li Y, et al. Omeprazole suppresses aggressive cancer growth and metastasis in mice through promoting Snail degradation. Acta Pharmacol Sin. 2022;43(7):1816–28.
124. Syamprasad NP, et al. Aldose reductase and cancer metabolism: the master regulator in the limelight. Biochem Pharmacol. 2023;211:115528.
125. Zaytseva YY, et al. Preclinical evaluation of novel fatty acid synthase inhibitors in primary colorectal cancer cells and a patient-derived xenograft model of colorectal cancer. Oncotarget. 2018;9(37):24787–800.
126. Kelly W, et al. Phase II investigation of TVB-2640 (Denifanstat) with bevacizumab in patients with first relapse high-grade astrocytoma. Clin Cancer Res. 2023;29(13):2419–25.
127. Brigandi RA, et al. A phase 1 randomized, placebo-controlled trial with a topical inhibitor of Stearoyl-coenzyme A desaturase 1 under occluded and nonoccluded conditions. Clin Pharmacol Drug Dev. 2019;8(3):270–80.
128. Heuer TS, et al. FASN inhibition and taxane treatment combine to enhance anti-tumor efficacy in diverse xenograft tumor models through disruption of tubulin palmitoylation and microtubule organization and FASN inhibition-mediated effects on oncogenic signaling and gene expression. EBioMedicine. 2017;16:51–62.

129. Katoh Y, et al. Inhibition of stearoyl-CoA desaturase 1 (SCD1) enhances the antitumor T cell response through regulating beta-catenin signaling in cancer cells and ER stress in T cells and synergizes with anti-PD-1 antibody. J Immunother Cancer. 2022;10(7):e004616.
130. Zhan N, et al. Inhibition of FASN expression enhances radiosensitivity in human non-small cell lung cancer. Oncol Lett. 2018;15(4):4578–84.
131. Farber S, Diamond LK. Temporary remissions in acute leukemia in children produced by folic acid antagonist, 4-aminopteroyl-glutamic acid. N Engl J Med. 1948;238(23):787–93.
132. Gonen N, Assaraf YG. Antifolates in cancer therapy: structure, activity and mechanisms of drug resistance. Drug Resist Updat. 2012;15(4):183–210.
133. Ravez S, et al. Alpha-Ketothioamide derivatives: a promising tool to interrogate phosphoglycerate dehydrogenase (PHGDH). J Med Chem. 2017;60(4):1591–7.
134. Weinstabl H, et al. Intracellular trapping of the selective phosphoglycerate dehydrogenase (PHGDH) inhibitor BI-4924 disrupts serine biosynthesis. J Med Chem. 2019;62(17):7976–97.
135. Mullarky E, et al. Inhibition of 3-phosphoglycerate dehydrogenase (PHGDH) by indole amides abrogates de novo serine synthesis in cancer cells. Bioorg Med Chem Lett. 2019;29(17):2503–10.
136. Rohde JM, et al. Discovery and optimization of piperazine-1-thiourea-based human phosphoglycerate dehydrogenase inhibitors. Bioorg Med Chem. 2018;26(8):1727–39.
137. Pacold ME, et al. A PHGDH inhibitor reveals coordination of serine synthesis and one-carbon unit fate. Nat Chem Biol. 2016;12(6):452–8.
138. Ngo B, et al. Limited environmental serine and glycine confer brain metastasis sensitivity to PHGDH inhibition. Cancer Discov. 2020;10(9):1352–73.
139. O'Connor C, et al. Role of mitochondrial and cytosolic Folylpolyglutamate Synthetase in one-carbon metabolism and antitumor efficacy of mitochondrial-targeted antifolates. Mol Pharmacol. 2024;106(4):173–87.
140. Garcia-Canaveras JC, et al. SHMT inhibition is effective and synergizes with methotrexate in T-cell acute lymphoblastic leukemia. Leukemia. 2021;35(2):377–88.
141. Duarte D, Vale N. Antidepressant drug sertraline against human cancer cells. Biomol Ther. 2022;12(10):1513.
142. Gravel SP, et al. Serine deprivation enhances antineoplastic activity of biguanides. Cancer Res. 2014;74(24):7521–33.
143. Hirunsatitpron P, et al. Mycophenolic acid and cancer risk in solid organ transplant recipients: systematic review and meta-analysis. Br J Clin Pharmacol. 2022;88(2):476–89.
144. Rosenzweig M, et al. Repurposing leflunomide for relapsed/refractory multiple myeloma: a phase 1 study. Leuk Lymphoma. 2020;61(7):1669–77.
145. McDonald G, et al. Selective vulnerability to pyrimidine starvation in hematologic malignancies revealed by AG-636, a novel clinical-stage inhibitor of dihydroorotate dehydrogenase. Mol Cancer Ther. 2020;19(12):2502–15.
146. Kung FP, et al. Piperlongumine, a potent anticancer phytotherapeutic, induces cell cycle arrest and apoptosis *in vitro* and *in vivo* through the ROS/Akt pathway in human thyroid cancer cells. Cancers (Basel). 2021;13(17):4266.
147. Magda D, Miller RA. Motexafin gadolinium: a novel redox active drug for cancer therapy. Semin Cancer Biol. 2006;16(6):466–76.
148. Lee CH, et al. Reactive oxygen species mediate the chemopreventive effects of syringin in breast cancer cells. Phytomedicine. 2019;61:152844.
149. Delbart W, et al. Disturbing the redox balance using buthionine sulfoximine radiosensitized somatostatin receptor-2 expressing pre-clinical models to peptide receptor radionuclide therapy with (177)Lu-DOTATATE. Cancers (Basel). 2023;15(8):2332.
150. Ramanathan RK, et al. A phase I pharmacokinetic and pharmacodynamic study of PX-12, a novel inhibitor of thioredoxin-1, in patients with advanced solid tumors. Clin Cancer Res. 2007;13(7):2109–14.
151. Lin J, et al. A non-comparative randomized phase II study of 2 doses of ATN-224, a copper/zinc superoxide dismutase inhibitor, in patients with biochemically recurrent hormone-naive prostate cancer. Urol Oncol. 2013;31(5):581–8.

152. Zhou Q, et al. Ferroptosis in cancer: from molecular mechanisms to therapeutic strategies. Signal Transduct Target Ther. 2024;9(1):55.
153. Nie Q, et al. Induction and application of ferroptosis in cancer therapy. Cancer Cell Int. 2022;22(1):12.
154. Yang H, et al. The role of cellular reactive oxygen species in cancer chemotherapy. J Exp Clin Cancer Res. 2018;37(1):266.
155. Vanhoefer U, et al. d,l-buthionine-(S,R)-sulfoximine potentiates *in vivo* the therapeutic efficacy of doxorubicin against multidrug resistance protein-expressing tumors. Clin Cancer Res. 1996;2(12):1961–8.
156. Zhang J, et al. Class II ferroptosis inducers are a novel therapeutic approach for t(4;14)-positive multiple myeloma. Blood Adv. 2024;8(19):5022–38.
157. Ye LF, et al. Radiation-induced lipid peroxidation triggers ferroptosis and synergizes with ferroptosis inducers. ACS Chem Biol. 2020;15(2):469–84.
158. Lu S, et al. RAS-selective lethal 3-induced ferroptosis promotes the antitumor efficiency of anti-programmed cell death protein 1 treatment in colorectal cancer. Turk J Gastroenterol. 2024;35(4):288–98.
159. Thompson EA, Powell JD. Inhibition of the adenosine pathway to potentiate cancer immunotherapy: potential for combinatorial approaches. Annu Rev Med. 2021;72:331–48.
160. Beavis PA, et al. CD73: a potential biomarker for anti-PD-1 therapy. Onco Targets Ther. 2015;4(11):e1046675.
161. Young A, et al. Co-inhibition of CD73 and A2AR adenosine signaling improves anti-tumor immune responses. Cancer Cell. 2016;30(3):391–403.
162. Beavis PA, et al. Targeting the adenosine 2A receptor enhances chimeric antigen receptor T cell efficacy. J Clin Invest. 2017;127(3):929–41.
163. Kohanbash G, et al. Isocitrate dehydrogenase mutations suppress STAT1 and CD8+ T cell accumulation in gliomas. J Clin Invest. 2017;127(4):1425–37.
164. Schramme F, et al. Inhibition of tryptophan-dioxygenase activity increases the antitumor efficacy of immune checkpoint inhibitors. Cancer Immunol Res. 2020;8(1):32–45.
165. Brand A, et al. LDHA-associated lactic acid production blunts tumor immunosurveillance by T and NK cells. Cell Metab. 2016;24(5):657–71.
166. Pertega-Gomes N, et al. A glycolytic phenotype is associated with prostate cancer progression and aggressiveness: a role for monocarboxylate transporters as metabolic targets for therapy. J Pathol. 2015;236(4):517–30.
167. Marjon K, et al. MTAP deletions in cancer create vulnerability to targeting of the MAT2A/PRMT5/RIOK1 axis. Cell Rep. 2016;15(3):574–87.
168. Gjuka D, et al. Enzyme-mediated depletion of methylthioadenosine restores T cell function in MTAP-deficient tumors and reverses immunotherapy resistance. Cancer Cell. 2023;41(10):1774–1787 e9.
169. Gammon JM, et al. Low-dose controlled release of mTOR inhibitors maintains T cell plasticity and promotes central memory T cells. J Control Release. 2017;263:151–61.
170. Wabitsch S, et al. Metformin treatment rescues CD8(+) T-cell response to immune checkpoint inhibitor therapy in mice with NAFLD. J Hepatol. 2022;77(3):748–60.
171. Kawalekar OU, et al. Distinct signaling of coreceptors regulates specific metabolism pathways and impacts memory development in CAR T cells. Immunity. 2016;44(2):380–90.
172. Xiao L, et al. Tumor-associated macrophages: new insights on their metabolic regulation and their influence in cancer immunotherapy. Front Immunol. 2023;14:1157291.
173. He C, et al. Vitamin B6 competition in the tumor microenvironment hampers antitumor functions of NK cells. Cancer Discov. 2024;14(1):176–93.
174. Moller SH, et al. Metabolic programming in dendritic cells tailors immune responses and homeostasis. Cell Mol Immunol. 2022;19(3):370–83.
175. Vernieri C, et al. Targeting cancer metabolism: dietary and pharmacologic interventions. Cancer Discov. 2016;6(12):1315–33.
176. Poff AM, et al. Ketone supplementation decreases tumor cell viability and prolongs survival of mice with metastatic cancer. Int J Cancer. 2014;135(7):1711–20.

177. Tiwari S, et al. Effect of fasting on cancer: a narrative review of scientific evidence. Cancer Sci. 2022;113(10):3291–302.
178. Pomatto-Watson LCD, et al. Daily caloric restriction limits tumor growth more effectively than caloric cycling regardless of dietary composition. Nat Commun. 2021;12(1):6201.
179. Brandhorst S. Fasting and fasting-mimicking diets for chemotherapy augmentation. Geroscience. 2021;43(3):1201–16.
180. Vernieri C, et al. Fasting-mimicking diet is safe and reshapes metabolism and antitumor immunity in patients with cancer. Cancer Discov. 2022;12(1):90–107.
181. Wanders D, et al. Methionine restriction and cancer biology. Nutrients. 2020;12(3):684.
182. Cervantes-Madrid D, et al. Reviving lonidamine and 6-Diazo-5-oxo-L-norleucine to be used in combination for metabolic cancer therapy. Biomed Res Int. 2015;2015:690492.
183. Xu J, et al. Abnormal oxidative metabolism in a quiet genomic background underlies clear cell papillary renal cell carcinoma. elife. 2019;8:e38986.
184. Selak MA, et al. Succinate links TCA cycle dysfunction to oncogenesis by inhibiting HIF-alpha prolyl hydroxylase. Cancer Cell. 2005;7(1):77–85.
185. Hu CM, et al. High glucose triggers nucleotide imbalance through O-GlcNAcylation of key enzymes and induces KRAS mutation in pancreatic cells. Cell Metab. 2019;29(6):1334–49. e10.
186. Kim J, DeBerardinis RJ. Mechanisms and implications of metabolic heterogeneity in cancer. Cell Metab. 2019;30(3):434–46.
187. Wang X, et al. UDP-glucose accelerates SNAI1 mRNA decay and impairs lung cancer metastasis. Nature. 2019;571(7763):127–31.
188. Schafer ZT, et al. Antioxidant and oncogene rescue of metabolic defects caused by loss of matrix attachment. Nature. 2009;461(7260):109–13.
189. Fischer GM, et al. Molecular profiling reveals unique immune and metabolic features of melanoma brain metastases. Cancer Discov. 2019;9(5):628–45.
190. Viale A, et al. Oncogene ablation-resistant pancreatic cancer cells depend on mitochondrial function. Nature. 2014;514(7524):628–32.

Genomic Instability as a Therapeutic Target: Advancements from Experimental Research to Clinical Application

Anowar Hussain, Seydur Rahman, and Anand Shankar Ramteke

1 Introduction

Genomic instability (GI), the hallmark of cancer, can be defined as an elevated frequency of mutations in the genetic repository. This instability leads to substantial alterations in chromosome number and structure, enabling genetic modifications that propel tumorigenesis and promotes disease aggressiveness, including metastasis to distant organs, immune evasion, and emergence of drug resistance [1, 2]. In normal cells, active DNA repair mechanisms like homologous recombination (HR), mismatch repair (MMR) ensure genomic integrity. However, malfunctioning of these pathways leads to sustained DNA damage, rendering GI, a key contributor to the development of various cancers such as ovarian, breast, and colorectal [3].

An in-depth understanding of GI is significant, as it can be used as a target for therapeutic intervention in a clinical setting. While GI is a major driver of cancer growth, it also carries susceptibilities that may be selectively exploited for therapeutic gains [4]. Several strategies have been developed by exploiting GI as therapeutic targets, and numbers of molecules have been tested in various preclinical and clinical settings. Induction of synthetic lethality is one such strategy that kills cancer cells by inhibiting the DNA repair pathways exploited by genomically unstable cells for their survival. Targeting deficits in HR mechanisms in *BRCA1*- and *BRCA2*-mutated cancers is an example of such selective exploitation of GI. Poly

A. Hussain (✉)
Department of Biophysics, Pub Kamrup College, Baihata Chariali, Kamrup, Assam, India

S. Rahman
Department of Zoology, Faculty of Science, Assam Down Town University, Panikhaiti, Guwahati, Kamrup (M), Assam, India

A. S. Ramteke
Department of Molecular Biology and Biotechnology, Tezpur University, Tezpur, Sonitpur, Assam, India

A. S. Ramteke et al. (eds.), *Molecular Targets in Cancer Therapy*,
https://doi.org/10.1007/978-981-96-7336-0_10

(ADP-ribose) polymerase (PARP) inhibitors such as Olaparib, Rucaparib, etc., showed promising results in preclinical and clinical settings. The Food and Drug Administration (FDA), USA, has approved them for use in clinics to treat *BRCA1/2*-, *HER-2*-mutated breast, and HRD-mutated ovarian cancers [4, 5].

Though the preclinical and clinical research has demonstrated the promise of many therapeutic strategies developed based on GI targets, their use in clinics is restricted because of limitations associated with them. The major obstacles associated with GI-targeted therapeutics include off-target toxicity, differential treatment response due to a heterogeneous intra- and inter-tumor genetic landscape, the development of resistance, etc. [3, 4]. Moreover, the onset of drug resistance can reduce the effectiveness of therapeutic strategies intended to address GI [6]. Also, precise identification of patient populations that may benefit depends on their distinct genomic profiles, making this strategy even more complicated. Continuous investigations into the mechanisms underlying GI hold promise for newer therapeutic options with enhanced patient outcomes, leading to more personalized and effective approaches to cancer treatment [7]. In this review, we explored the recent advancements in understanding GI as a therapeutic target, focusing on the transition from experimental research to clinical applications for treatment of cancers. The review will provide insights into the pivotal role of GI in cancer, the latest research and development for clinical applications that may pave the foundation of more effective strategies for better patient outcomes.

2 Mechanisms and Source of GI in Cancer

GI, characterized by chromosomal abnormalities and an increased frequency of mutations, arises through multiple mechanisms and plays a key role in tumor evolution, emerging as a driving force in tumor progression by accumulating more and more abnormalities and mutations [8]. A few sources and mechanisms of GI are listed below.

2.1 Mutations

GI often emerges as mutations at the nucleotide level, starting from single nucleotide mutations such as base modifications to small-scale DNA fragment alterations such as insertions, deletions, etc. These mutations typically arise from DNA replication errors, environmental factors (e.g., UV radiation, carcinogens), or inefficiency in DNA repair mechanisms [9]. Oncogenes and tumor suppressor genes are more susceptible and known to play a role in various types of cancers. Activation of oncogenes such as *K-Ras*, *myc*, etc., and loss of function of tumor suppressor genes (TSG) such as *p53*, *Rb*, etc., are well-established examples of gene mutations that not only play a role in cancer initiation, but also promote adaptation of tumor cells to adverse conditions including therapeutic insults [9, 10].

2.2 Chromosomal Instability (CIN)

The phenomenon of CIN is characterized by alterations in the chromosomal landscape and is found in more than 90% of solid tumors and blood cancers. It can interrupt normal gene expression, through gaining extra genes or losing normal genes, and is often found in 60–80% of human cancers such as triple-negative breast cancer, high-grade serous ovarian carcinoma, etc. [3, 4, 8]). CIN can arise in two different forms, viz., numerical (changes in ploidy level) and structural CIN (rearrangements). The numerical changes arise because of defective mechanisms involved in chromosome segregation such as failure of mitotic checkpoint, defective cohesion, centromere amplifications, etc., forming daughter cells with unequal chromosomal configurations, whereas structural changes arise due to deletion or extra-amplification of chromosomal fragments. In many types of cancers, both kinds of CIN coexist and interplay [3].

2.3 Replication Stress

Replication stress arises when obstacles hinder the DNA replication mechanism, causing the replication forks to stall. As the cancer cells replicate rapidly, they encounter strain on replication machinery more frequently because of oxidative stress, depletion of nucleotide pools, difficulty in replicating repetitive sequences which disrupt the progression of the replication fork. This replication stress leads to incomplete replication and ultimately causes DNA damage such as double-strand breaks (DSB), intensified by oncogene activation such as *Myc*, *Ras*, etc., which is prevalent in several cancers [11, 12]. Cellular mechanism evolved to overcome these types of stress includes molecules such as Ataxia Telangiectasia and Rad3-related Kinase (ATR), CHK1, *RAD51*, etc.; however, abnormalities in these molecules could lead to DNA breaks, fueling tumor heterogeneity, and ultimately increasing resistance to therapies [11].

2.4 Role of DNA Repair Pathways and Deficiencies

Normal cells maintain genomic integrity in which DNA repair pathways play a crucial role. Whenever this balance is disrupted, cells accumulate DNA damages, resulting in instabilities. Major repair mechanisms include mismatched repair (MMR), homologous recombination (HR), nucleotide excision repair (NER), base excision repair (BER), etc., each of which addresses specific types of DNA damage. For example, small-scale DNA damages involving single nucleotides are repaired by NER and BER. MMR amends errors arising (mispaired bases) during replication. Malfunctioning of this pathway, particularly due to mutations in *MLH1* or *MSH2*, results in microsatellite instability (MSI), as seen in Lynch syndrome-associated endometrial and colorectal cancers [13, 14]. DSBs are repaired through the HR mechanism, using the undamaged sister chromatid as a template. Flaws in

HR, especially in *BRCA1* and *BRCA2*, cause accumulation of DSB and CIN. Tumors with HR deficiencies, such as *BRCA*-mutated ovarian and breast cancers, show notable sensitivity to therapies like PARP inhibitors, which target DNA damage [15]. Repair of DSBs by nonhomologous end joining (NHEJ) is less accurate as compared to HR which often leads to deletions or insertions. Such inaccuracies are known to lead to chromosomal translocations, additionally contributing to instability, and ultimately lead to the development of cancer [13].

2.5 GI and Cancer Progression

Beside initiating tumors, GI also accelerate progression of cancer through enhancement of intra-tumor genetic diversity, enabling the cells to better adapt to the selective pressures such as immune surveillance and escaping from therapeutics [6, 7]. Tumors having such high GI, as seen in cancer exhibiting CIN, show more aggressive behavior and emerge as therapeutically resistant, impacting the effectiveness of conventional therapeutics. Accumulation of mutations over time promotes resistance against radiation and chemotherapy [2]. Though genetic instability creates considerable challenges for treatment, it also provides opportunities for newer therapeutics. By targeting the genetically unstable susceptible cancer cells, particularly those with defective DNA repair pathways, novel therapies can be developed, laying the foundation for innovative cancer treatments [8].

3 Preclinical Research and Experimental Approaches

3.1 Laboratory Models to Study GI

Several experimental models have been developed to explore the mechanistic insights and therapeutic potential of GI in cancer. Cell lines derived from tumor patients, such as the *HeLa* cell line (cervical cancer), which exhibits high chromosomal instability; the *HCT116* cell line (colon cancer), which bears *MLH1* gene mutations; and the *MDA-MB-231* cell line (breast cancer), which carries *BRCA1* mutations, serve as useful models to explore the role of GIs in different types of cancers [16–18]. Transgenic mouse models are crucial for in vivo exploration of GI as mutation can also be introduced into their genes. A mouse model with engineered (knockout) DNA repair genes, namely in *BRCA1* or *P53*, resembles the GI similar to human cancer, enabling researchers to evaluate PARP inhibitors [4]. Patient-derived organoids (PDOs) serve as a promising and better representative experimental model for exploring GI as they retain both genetic and phenotypic characteristics of human tumors. These types of models are best as they mimic the original 3D behavior tumor and their microenvironments [19]. Ovarian cancer-derived organoids with mutations in *BRCA1*, *PIK3CA*, *ARID1A*, *TP53* facilitate the screening of drugs targeting GI, bolstering their contribution in personalized medicine [20]. By using CRISPR/Cas9 or RNAi technology, the patient-derived cell lines, organoid models, etc., can be further modified in order to study the effects of disrupted genes and/or screen and identify drugs that address GI, thus facilitating therapeutic discovery.

3.2 Techniques for Detecting GI

Recent developments in diagnostic techniques enhanced the applicability of targeting GI in cancers.

3.2.1 Classical Techniques

Classical techniques such as karyotyping, fluorescence in situ hybridization (FISH), comparative genomic hybridization (CGH), micronuclei assay, comet assay, etc., were employed to study the change in ploidy level, copy number variations, chromosomal fragments, etc. These techniques, though having limited applicability, reveal fundamental insights of GI, particularly in the development and progression of cancer.

3.2.2 Sequencing Technologies

The advent of high-throughput sequencing methods like next-generation sequencing (NGS) and targeted gene panel sequencing have changed the landscape of GI detection. These techniques can identify multiple mutations, variations in copy number, and structural alterations throughout the genome, providing rapid and multidimensional platforms to elucidate the impact of GI in tumor evolution pinpointing therapeutic targets [21].

3.2.3 CRISPR/Cas9-Based Approaches

Isogenic cell lines having knocked-out genes such as in DNA repair pathways can be precisely created using CRISPR/Cas9 technology in order to evaluate their effects on genomic stability. Moreover, using CRISPR/Cas9 technology, specific DNA damages can be induced which allow to study their response on basic cellular processes and DNA repair mechanisms in unstable genomic states [22].

3.3 Key Preclinical Findings

Recent preclinical researches involving multiple models and techniques have changed the existing knowledge on the role of GIs in cancer. Studies have demonstrated enhanced therapeutic efficacy of several compounds targeting GI. Cancer cells with *BRCA* deficiency are highly sensitive to PARP inhibitors and induce selective lethality while sparing normal cells, indicating potential for clinical use in the treatment of ovarian and breast cancer [4, 5]. Multiple microenvironmental factors like nutrients, oxygen availability, etc., aggravate GI and thus the tumor cell's responses toward therapeutics, paving the way for discovering more and more innovative strategies targeting GI in order to elevate the treatment outcomes (Fig. 1).

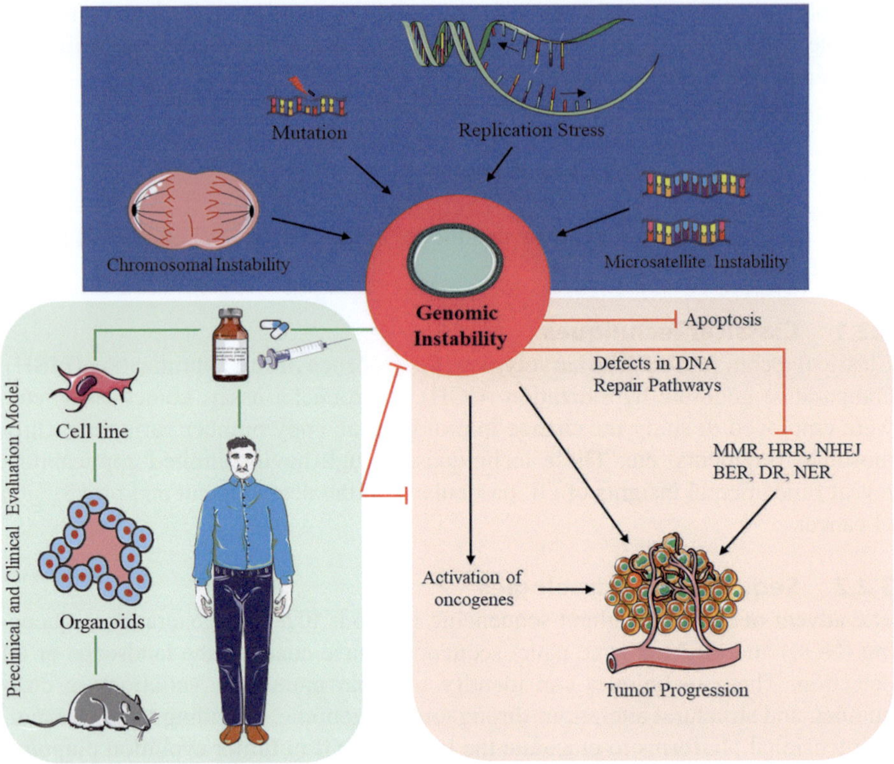

Fig. 1 Diagram illustrating the relationship between genomic instability and its impact on cancer progression

4 Targeting GI in Cancer Therapy

GI plays a significant role in initiation of tumors, accelerates the progression of disease, and also affects treatment responses. Therefore, a precise understanding of this instability and targeting them provide unique and better opportunities for effective management of cancers. By exploiting GI, several strategies have been developed such as synthetic lethality, chromosomal instability, DNA damage response (DDR) pathways, etc., and some of them are discussed below.

4.1 Synthetic Lethality and PARP Inhibitors

Synthetic lethality, particularly through the use of poly (ADP-ribose) polymerase inhibitors, gained significant importance. In cancer cells with deficiency in HR due to mutations in genes like *BRCA1*, *BRCA2*, PARP proteins are responsible for repairing single-strand breaks in DNA. The use of PARP inhibitors leads to the accumulation of irreparable DNA damage, ultimately resulting in cell death. PARP inhibitor, such as Olaparib and Niraparib, have demonstrated clinical efficacy against ovarian and breast cancers with *BRCA* mutations [5]. Recent studies have demonstrated that PARP inhibitors also showed efficacy against wide numbers of HR gene mutations such as *RAD51C*, *RAD51D*, and *PALB2* [23]. Studies have also shown that combinatorial use of PARP inhibitors with immune- and chemotherapy enhances therapeutic outcomes by inducing synthetic lethality in multiple types of cancer. For example, use of PARP inhibitors with immune checkpoint inhibitors (e.g., anti-PD-L1) has demonstrated promising results in improving immune-resistant ovarian tumor response rates [24]. Combination of PARP inhibitors with ATR and DNA-PK inhibitors; HDAC and BET inhibitors; CDK12 inhibitors, radiotherapy have shown promising efficacy in various types of cancers, including resistance types [25]. Combinatorial approaches such as use of tankyrase inhibitors along with PARP inhibitors re-sensitize the PARP resistance cancers [26]. These advancements highlight the potentials of PARP inhibitors in inducing synthetic lethality for effective management of multiple cancer types.

4.2 Exploiting DNA Damage Response (DDR) Pathways

DNA damage response (DDR) pathways are known to play a critical role in maintaining genomic integrity. To cope with and survive in GI states, cancer cells depend on DDR pathways, and therefore inhibiting them represents a promising approach in the treatment of cancer. For instance, targeting DDR through the use of checkpoint kinase 1 (CHK1) and ataxia telangiectasia and Rad3-related protein (ATR) inhibitors has emerged as a promising therapeutic option, as they sensitize ovarian and breast cancer cells to DNA-damaging treatment, such as radiation and chemotherapy [27]. Studies have demonstrated that cancer cells treated with CHK1 and WEE1 kinase synergistically increase the sensitivity of radiation therapies [28]. These findings emphasized the importance of personalized treatment approaches to enhance therapeutic impacts and limit resistance.

4.3 Targeting CIN and Mitotic Checkpoints

In many cancers, CIN induces aneuploidy, ultimately leading to origin of tumor heterogeneity. The consequences of CIN, such as poor prognosis and treatment resistance, can be effectively counteracted by targeting mitotic checkpoints, particularly the spindle assembly checkpoint (SAC). Such a strategy includes inhibiting mitotic kinases, such as Aurora kinases and PLK1, which are required for mitotic progression causing death of cancer cells through the induction of mitotic catastrophe [29]. This approach demonstrates significant potentials in both preclinical and clinical settings. Recent research has also demonstrated elevated therapeutic effects of chemotherapeutic agents when combined with Aurora kinase inhibitors (PLK1) and checkpoint kinase 1 (CHK1) inhibitors, particularly in breast and lung cancers with a high degree of CIN [29, 30]. This combinatorial treatment has potential applications in treating resistant cancers whose survival depends on abnormal cell division.

4.4 Emerging Therapeutic Targets

Besides conventional GI targets, numerous emerging targets like replication stress, telomere maintenance have gained significant impetus in recent years. Replication stress occurs as a result of disruption in DNA replication process, leading to GI, and targeting proteins responsible for replication stress, such as the MCM helicase complex, has demonstrated significance results in preclinical studies [11]. Inhibiting proteins involved in telomere maintenance mechanisms, like those involved in the alternative lengthening of telomeres (ALT), has been shown to inhibit proliferation of ALT-positive tumors [31]. Recent research has identified new targets such as the MRN complex; reactivation of TSGs such as *P53*, *RB1*, inhibition of *CDK4/6*; disruption of tumor–microenvironmental interaction through targeting cancer-associated fibroblast as newer strategies that enhance therapeutic outcomes [3, 32]. These inhibitors served as an innovative therapeutic strategy and are currently being explored in clinical trials.

Exploiting targets of GI through diverse therapeutic strategies opens up promising and transformative avenues in cancer therapeutics. By harnessing these vulnerable targets, clinicians have the potential to significantly improve patient outcomes in various types of cancers. However, sustained research and clinical trials are essential to optimize these approaches and determine their effectiveness in various types of cancers, ultimately leading to more effective and personalized cancer therapies.

5 Translational Research: From Bench to Bedside

Translational research in the GI area has bridged the gap between laboratory findings and clinical applications. Several experimental works on GI have been successfully translated into cancer treatments, with some achieving success, while others have faced challenges hindering their clinical implementation. For example, PARP inhibitors such as Olaparib, Rucaparib, Talazoparib, etc., effectively target *BRCA1/2*-, *HER-2*-mutated breast, and homologous recombination deficiency (HRD) mutated ovarian cancers [5, 33]. An early phase clinical study showed the efficacy of berzosertib, an ATR inhibitor, in inducing synthetic lethality in ATR- and ATM-mutated cancers [34]. In recent years, genome instability signatures also find use in precision medicine to identify patients upon whom immune checkpoint inhibitors, such as pembrolizumab, effectively work, further underscoring the potential of GI as a therapeutic target [4].

6 Clinical Applications of Targeting GI

Advancements in research have exploited the genomic vulnerability of cancer cells and established GI as an effective and promising therapeutic target. Based on GI, the Food and Drug Administration (FDA) has approved several personalized and combinatorial therapies and many more are in the pipeline.

6.1 Current FDA-Approved Therapies

Development of inhibitors for PARP represents a significant advancement in exploiting GI of cancer cells and the FDA has approved them for clinical applications (Table 1). Agents such as Olaparib, Niraparib, and Rucaparib are highly effective in cancers, in which impaired homologous recombination mechanisms due to *BRCA1* or *BRCA2* mutations cause cancer cells more susceptible to DNA damage [36]. The use of PARP inhibitors in clinical trials exhibit better survival outcome compared to conventional chemotherapies for ovarian and breast cancer patients

Table 1 List of FDA approved drugs developed based on GI target (FDA [35])

Drug name	Targeted genomic instability	Mechanism of action	Types of cancer	Clinical trial phase
Olaparib (Lynparza)	*BRCA*1/2 mutations, DNA repair	PARP inhibitor targeting homologous recombination	Ovarian, breast cancer	Phase 3
Niraparib (Zejula)	*BRCA*1/2 mutations, DNA repair	PARP inhibitor	Ovarian cancer	Phase 3
Rucaparib (Rubraca)	*BRCA*1/2 mutations, DNA repair	PARP inhibitor	Ovarian cancer	Phase 2/3
Talazoparib (Talzenna)	*BRCA*1/2 mutations, DNA repair	PARP inhibitor	Breast cancer	Phase 3
Pembrolizumab (Keytruda)	MSI, DNA mismatch repair	PD-1 checkpoint inhibitor	Melanoma, lung cancer	Phase 3
Atezolizumab (Tecentriq)	MSI, DNA mismatch repair	PD-L1 checkpoint inhibitor	Bladder, lung cancer	Phase 3
Alpelisib (Piqray)	PIK3CA mutations, DNA replication	PI3K inhibitor	Breast cancer	Phase 3
Dostarlimab (Jemperli)	MSI, DNA mismatch repair	PD-1 checkpoint inhibitor	Endometrial cancer	Phase 3
Trabectedin (Yondelis)	DNA damage, transcription block	DNA alkylating agent	Soft tissue sarcoma	Phase 2
Durvalumab (Imfinzi)	MSI, DNA mismatch repair	PD-L1 checkpoint inhibitor	Non-small cell lung cancer	Phase 3

having *BRCA* mutation. Recently, the FDA has broadened the scope of these drugs, approving their use in cancer patients and thereby enhancing treatment accessibility for these patients [37].

6.2 Ongoing Clinical Trials Investigating GI

Several ongoing clinical trials are currently examining the efficacy of various agents that target GI. For example, berzosertib, an ATR inhibitor, exhibits promising results in treating solid tumors with impaired DDR pathways [38]. Clinical trials are also being conducted to examine the efficacy of checkpoint kinase inhibitors, such as Chk1 inhibitors [39]. These trials are essential to evaluate and establish therapeutic efficacy of new agents, as well as to understand and develop therapeutic protocols, alone and/or in combination with conventional therapeutics.

6.3 Tumor Genotype-Based Personalized Medicine

Cancers exhibit significant variation in genetic and phenotypic landscape across and within tumors, necessitating personalized therapeutic options that specifically target individual GI. Clinicians can make decision on the use of particular therapeutic options based on the altered genetic profile of patients. Pembrolizumab, an immune checkpoint inhibitor, exemplifies such personalized treatment. It has been approved to treat cancer patients having high MSI or MMR genes mutation [40]. Personalized treatment modalities improve overall patient outcomes by increasing therapeutic efficacy with minimum and/or no exposure to ineffective treatments. Integration of genetic profile of patients refines the available therapeutic options and delivers the most suitable one to patients based on the unique genetic landscape [7].

6.4 Combination Therapies

Targeting GI when combined with existing therapeutics shows better efficacy compared to individual treatment options, thereby improving patient outcome. Several preclinical and clinical studies have demonstrated that PARP inhibitors, when used in combination with chemotherapy or immune checkpoint inhibitors, synergistically enhance effectiveness [41]. For instance, PARP inhibitors when used with DNA damaging agents increase cancer cell death, thereby improving treatment responses [38]. DDR inhibitors disrupt DNA repair pathways and increase the effectiveness of immune-based therapies by enhancing tumor immunogenicity. Besides overcoming treatment resistance, these combinatorial approaches also improve therapeutic outcomes of cancer patients [42].

7 Challenges and Limitations in Clinical Applications

Therapeutic strategies developed based on GI targets faced several challenges such as drug resistance, unwanted toxicity, and variable patient responses which must be addressed to improve patient outcomes.

7.1 Development of Drug Resistance

Cancer cells often exhibit adaptive responses to therapeutic stress through secondary mutations in target genes and/or activation of alternative pathways tending survival under therapeutic conditions. This adaptive behavior leads to the development of resistance, ultimately reducing treatment efficacy and outcomes [36]. Like other therapeutics, strategies developed based on GI, such as PARP inhibitors, also face drug resistance challenges through additional mutations in *BRCA1/2* and other homologous recombination genes, making the subsequent therapies ineffective [43]. Therefore, understanding the mechanism involved in drug resistance is crucial and combinatorial/additive approaches must be considered to overcome drug resistance challenges.

7.2 Toxicity and Off-Target Effects

Like other cancer therapies, treatments targeting GI also exhibit potential toxicity and unwanted side effects. These therapies, beside targeting the vulnerable cancer cells, also adversely affect nearby and distantly located normal cells. For instance, PARP inhibitors have been found to cause gastrointestinal symptoms, hematologic toxicities, fatigue, and other side effects, impacting the patient's quality of life and treatment adherence [44]. Therefore, the identification and mitigation of potential adverse effect is crucial for enhancing the therapeutic index.

7.3 Challenges in Identification of the Ideal Target Patients

In a clinical setting, heterogenetic behaviors of cancer cells complicate the response of GI-based therapies. Therefore, the ideal patient population must be identified along with a personalized treatment protocol for which diagnostic and prognosis biomarkers having a high degree of reproducibility are required. However, such accurate biomarkers are still under investigation, making this personalized therapeutic approach remain in its infant stage [45]. To enhance the effectiveness of personalized GI therapeutics, combinatorial approaches must be developed. In brief, for effective clinical applications of GI-based therapeutic strategies, these obstacles must be addressed for which numbers of preclinical and clinical studies are in progress.

8 Future Directions of GI-Targeted Therapeutics

Precise targeting of GI holds the key potential to improve the therapeutic outcomes for cancer patients. However, the current understanding of the mechanism of GI and their exploitation for therapeutic purposes mostly revolves around PARP proteins. Therefore, identification of novel GI targets and their validation in preclinical and clinical setting have gained traction in recent years. Additionally, the use of circulating tumor DNA through liquid biopsies will enhance real-time monitoring of treatment responses [46].

Combinatorial approaches targeting GI alongside conventional therapeutic strategies hold promise for clinical use, as targeting additional pathways leaves little room for survival of cancer cells under therapeutic shocks and in drug resistance states. Therefore, more synergistic combinations of treatments must be explored in preclinical and clinical settings to maximize efficacy. Integration of genomic data with artificial intelligence and modeling using algorithms will accelerate the development of increasingly precise and robust therapeutic options [47]. These integrated and innovative approaches hold promise for expediting personalized and combinatorial treatment plans based on individual and tumor-specific genetic profiles, thereby enhancing clinical outcomes.

9 Conclusions

Genomic instabilities (GI), besides playing a crucial role in the initiation of cancer, also promote aggressiveness and evade immune surveillance and therapeutic insults. Targeting GI disrupts impaired and vulnerable genomic pathways leading to the death of heterogenetic cancer. Several therapeutic strategies, like the induction of synthetic lethality, mitotic catastrophe, etc., have been developed along with the evaluation of various prodrugs/drugs (e.g., inhibitors of PARP, checkpoints, etc.) in preclinical settings. The encouraging results of some of the strategies and drugs, like Olaparib, Dostarlimab, etc., hold promise in clinical avenues. This transformative approach also has the potential to underscore the limitations of conventional therapeutic modalities. The impact of targeting GI can further be aggravated through combinatorial approaches that integrate conventional therapies with GI-targeted strategies. This approach maximizes the effectiveness of therapy, increases patient survival rates, and enhances their quality of life by integrating precision medicine techniques.

Acknowledgment The author (AH) would like to extend his sincere gratitude to the authority of Pub Kamrup College for its unwavering support. The support from SERB-DST, DBT, and ICMR, New Delhi, is also acknowledged.

Conflict of Interest The authors declare that there are no conflicts of interest.

References

1. Negrini S, et al. Genomic instability-an evolving hallmark of cancer. Nat Rev Mol Cell Biol. 2010;11:220–8.
2. Chen X, et al. Chromosomal instability as a driver of cancer progression. Nat Rev Genet. 2024;26:31.
3. Hosea R, et al. The two sides of chromosomal instability: drivers and brakes in cancer. Signal Transduct Target Ther. 2024;9:75.
4. Al-Rawi DH, et al. Targeting chromosomal instability in patients with cancer. Nat Rev Clin Oncol. 2024;21:645–59.
5. Patel PS, et al. Exploiting synthetic lethality to target *BRCA*1/2-deficient tumors: where we stand. Oncogene. 2021;40:3001–14.
6. Dagogo-Jack I, Shaw A. Tumour heterogeneity and resistance to cancer therapies. Nat Rev Clin Oncol. 2018;15:81–94.
7. Riedl JM, et al. Molecular diagnostics tailoring personalized cancer therapy-an oncologist's view. Virchows Arch. 2024;484(2):169–79.
8. Ferguson LR, et al. Genomic instability in human cancer: molecular insights and opportunities for therapeutic attack and prevention through diet and nutrition. Semin Cancer Biol. 2015;35(Suppl):S40–8.
9. Sinkala M. Mutational landscape of cancer-driver genes across human cancers. Sci Rep. 2023;13:12742.
10. Kontomanolis EN, et al. Role of oncogenes and tumor-suppressor genes in carcinogenesis: a review. Anticancer Res. 2020;40(11):6009–15.
11. Gaillard H, et al. Replication stress and cancer. Nat Rev Cancer. 2015;15:276–89.
12. da Costa AAB, et al. Targeting replication stress in cancer therapy. Nat Rev Drug Discov. 2023;22:38–58.
13. Li LY, et al. DNA repair pathways in cancer therapy and resistance. Front Pharmacol. 2021;11:629266.
14. Abildgaard AB, et al. Lynch syndrome, molecular mechanisms and variant classification. Br J Cancer. 2023;128:726–34.
15. Huang R, Zhou PK. DNA damage repair: historical perspectives, mechanistic pathways and clinical translation for targeted cancer therapy. Signal Transduct Target Ther. 2021;6:254.
16. Frattini A, et al. High variability of genomic instability and gene expression profiling in different HeLa clones. Sci Rep. 2015;5:15377.
17. Rashid S, et al. MLH1 deficiency leads to deregulated mitochondrial metabolism. Cell Death Dis. 2019;10:795.
18. Gu Y, et al. *BRCA*1-deficient breast cancer cell lines are resistant to MEK inhibitors and show distinct sensitivities to 6-thioguanine. Sci Rep. 2016;6:28217.
19. Ben-David U, et al. Genomic evolution of cancer models: perils and opportunities. Nat Rev Cancer. 2019;19(2):97–109.
20. Chang YH, et al. Ovarian cancer patient-derived organoids used as a model for replicating genetic characteristics and testing drug responsiveness: a preliminary study. Cell Transplant. 2024;33:9636897241281869.
21. Gantchev J, et al. Tools used to assay genomic instability in cancers and cancer meiomitosis. J Cell Commun Signal. 2022;16(2):159–77.
22. Xing H, Meng LH. CRISPR-Cas9: a powerful tool towards precision medicine in cancer treatment. Acta Pharmacol Sin. 2020;41(5):583–7.
23. Pacaud R, et al. Low dose DNA methyltransferase inhibitors potentiate PARP inhibitors in homologous recombination repair deficient tumors. Breast Cancer Res. 2025;27(1):8.
24. Xiao F, et al. Application of PARP inhibitors combined with immune checkpoint inhibitors in ovarian cancer. J Transl Med. 2024;22(1):778.
25. Klinakis A, et al. Targeting DNA repair in cancer: current state and novel approaches. Cell Mol Life Sci. 2020;77(4):677–703.

26. Plummer R, et al. First-in-human study of the PARP/tankyrase inhibitor E7449 in patients with advanced solid tumours and evaluation of a novel drug-response predictor. Br J Cancer. 2020;123(4):525–33.
27. Rundle S, et al. Targeting the ATR-CHK1 axis in cancer therapy. Cancers. 2017;9(5):41.
28. Bruyer A, et al. Combined inhibition of Wee1 and Chk1 as a therapeutic strategy in multiple myeloma. Front Oncol. 2023;13:1271847.
29. Strebhardt K, et al. The role of PLK1 in cancer exhibiting chromosomal instability. Mol Cell Oncol. 2018;5(6):e1485539.
30. Li C, et al. Combined Aurora Kinase A and CHK1 inhibition enhances radiosensitivity of triple-negative breast cancer. Int J Radiat Oncol Biol Phys. 2023;117(5):1241–54.
31. Ali JH, Walter M. Combining old and new concepts in targeting telomerase for cancer therapy: transient, immediate, complete and combinatory attack (TICCA). Cancer Cell Int. 2023;23(1):197.
32. Salmaninejad A, et al. Genomic instability in cancer: molecular mechanisms and therapeutic potentials. Curr Pharm Des. 2021;27(28):3161–9.
33. Bruin MAC, et al. Pharmacokinetics and pharmacodynamics of PARP inhibitors in oncology. Clin Pharmacokinet. 2022;61(12):1649–75.
34. Gorecki L, et al. Discovery of ATR kinase inhibitor berzosertib (VX-970, M6620): clinical candidate for cancer therapy. Pharmacol Ther. 2020;210:107518.
35. U.S. Food and Drug Administration. FDA website. 2024. https://www.fda.gov. Accessed on 5th Nov 2024.
36. Lord CJ, Ashworth A. PARP inhibitors: synthetic lethality in the clinic. Science. 2017;355(6330):1152–8.
37. Wang SSY, et al. PARP inhibitors in breast and ovarian cancer. Cancers. 2023;15(8):2357.
38. Shapiro GI, et al. Phase 1 study of the ATR inhibitor berzosertib in combination with cisplatin in patients with advanced solid tumours. Br J Cancer. 2021;125(4):520–7.
39. Jiang K, et al. Functions and inhibitors of CHK1 in cancer therapy. Med Drug Discov. 2024;22:100185.
40. Le DT, et al. Mismatch repair deficiency predicts response of solid tumors to PD-1 blockade. Science. 2017;357(6349):409–13.
41. Bhamidipati D, et al. PARP inhibitors: enhancing efficacy through rational combinations. Br J Cancer. 2023;129:904–16.
42. Nasioudis D, et al. Combination DNA damage response (DDR) inhibitors to overcome drug resistance in ovarian cancer. Cancer Treat Res. 2023;186:189–206.
43. Dias MP, et al. Understanding and overcoming resistance to PARP inhibitors in cancer therapy. Nat Rev Clin Oncol. 2021;18:773–91.
44. LaFargue CJ, et al. Exploring and comparing adverse events between PARP inhibitors. Lancet Oncol. 2019;20(1):e15–28.
45. Bielski CM, Taylor BS. Homing in on genomic instability as a therapeutic target in cancer. Nat Commun. 2021;12(1):3663.
46. Ranganathan K, et al. Exploring the clinical utility of liquid biopsy with cfDNA in cancer: a systematic review. J Liq Biopsy. 2024;5:100150.
47. Frasca M, et al. Artificial intelligence applications to genomic data in cancer research: a review of recent trends and emerging areas. Discov Anal. 2024;2:10.

Targeting Innate Immunity: A New Door to Cancer Therapy

Ujjayita Chowdhury and Rohan J. Khadilkar

1 Introduction

Immunotherapy in cancer is a booming discipline with various therapeutic strategies benefitting many patients over the last decade. It is a revolutionary therapeutic strategy to boost one's immune system to evade tumors. A few prominent immunotherapeutic approaches presently used as therapeutic strategies are checkpoint inhibitors, cell-based therapies, and targeted antibodies. These are techniques to rewire the immune system to combat cancer cells [1].

Immunotherapeutic strategies employing T-lymphocytes have garnered a lot of attention in recent times owing to their properties of antigen-directed cytotoxicity, wherein redirecting the T cells to tumors has opened up new horizons in cancer research from bench to bedside [2, 3]. However, with toxicities as side effects and few long-term benefits for patients in the case of primarily solid tumors, newer and more sophisticated approaches are the way forward. In vertebrates, the immune system is classically divided into two main branches: the innate arm and the adaptive arm. Immunological memory is a defining feature of the adaptive immune system in vertebrates wherein specialized immune cells like the B and T lymphocytes build a memory for the subsequent infection after recognizing a pathogen. The innate immune system is traditionally seen as nonspecific and lacking memory capabilities, primarily acting to respond quickly to a variety of pathogens and aid in early infection control [4]. However, researchers observed that specific ligands, which activate the immune system without antigen receptors (such as immunoglobulins or T-cell receptors), can induce a memory-like response. More recent studies

U. Chowdhury · R. J. Khadilkar (✉)
Stem cell and Tissue Homeostasis Lab, Cancer Research Institute, ACTREC-Tata Memorial Centre, Navi Mumbai, India

Homi Bhabha National Institute, Mumbai, India
e-mail: rkhadilkar@actrec.gov.in

A. S. Ramteke et al. (eds.), *Molecular Targets in Cancer Therapy*,
https://doi.org/10.1007/978-981-96-7336-0_11

show that innate immune cells, including myeloid and lymphoid cells, as well as non-immune cells like epithelial and endothelial cells, can retain the memory of previous exposure to microbes and inflammatory signals, becoming primed to launch a more robust, broad-spectrum response upon subsequent responses [2, 5]. This process, known as innate immune memory or trained immunity, is essential for the survival of organisms without adaptive immune systems (e.g., invertebrates and plants) and also functions within the more complex immune systems of vertebrates [6, 7].

2 History of Immunotherapy

Cancer immunotherapy dates back to the late nineteenth century when scientists harnessed the body's immune system to combat cancer. William B. Coley, widely called the "Father of Immunotherapy," noticed that tumors in sarcoma patients regress when treated with certain live and inactivated bacteria mixture. Some classical immunotherapeutic techniques involve oncolytic viruses, antibody-based cell therapy, and cancer vaccines [8, 9].

Oncolytic viruses are being used for the property of oncolytic viruses to lyse and kill specifically tumor cells, which are being modified now to be used as novel therapeutic strategies, used alone or in combination with other therapies [10]. Cancer vaccines, on the other hand, elicit antitumor responses against tumor-specific antigens [11]. Significant clinical outcomes have been observed in only a few cases in preventing cancers. Antibody-based cell therapy is another arm of classical immunotherapeutic strategy wherein antibodies block the proliferative signaling of tumor cells or engage with the immune cells to get the tumor killed by the immune cells [8]. The development of rituximab, a monoclonal antibody that binds to the CD20 protein present on the surface of immature B-cells, marks the beginning of antibody-based cell therapy against cancer. It was the first monoclonal antibody approved by the FDA for the treatment of non-Hodgkin's lymphoma [12].

The role of innate immune cells in cancer pathophysiology is as pertinent as it is to understanding adaptive immunity. The innate immune cells are highly heterogeneous and have multiple roles in tumorigenesis. These cells not only engage the adaptive immune arm to mount antitumor responses but also play paradoxical roles in tumor growth. Hence, these cells can be harnessed to remodel antitumor response and become meaningful candidates for therapeutic strategies [9]. Understanding the long-ignored innate arm will be beneficial in connecting the dots of cancer and immunity.

3 Innate Immune Response in Cancer

It is well known that the first cells that recognize a pathogen during infection are the macrophages and neutrophils. They secrete additional soluble chemical compounds to attract the cells of the adaptive arm, like T cells, to mount cytotoxic effects [4].

Tumor-promoting inflammation is one of the hallmarks of cancer pathophysiology [13]. Besides, tumors are heavily infiltrated by various immune cells. These responses represent the immune system's attempt to eliminate tumors, and in fact, there is mounting evidence for antitumoral responses to various tumor kinds [13]. However, it is also known that tumors evade immune responses successfully and continue to grow. An aberrant immune system supports cancer cell proliferation, inducing immune suppression [14].

The innate immune cells that infiltrate most solid tumors are highly heterogeneous and plastic, having a diverse array of phenotypes. Immune plasticity is shown by the innate immune cells, which can become either tumor-promoting or tumor-evading depending on the signals from the tumor and the microenvironment [15]. The heterogeneous population of myeloid cells, which are derived from a common myeloid progenitor in the bone marrow that infiltrates the tumor, include macrophages/monocytes, neutrophils, myeloid-derived suppressor cells (MDSCs), and dendritic cells (DCs) [15]. Various therapeutic strategies can be devised, remodeling the innate immune cells for antitumor effects (Fig. 1).

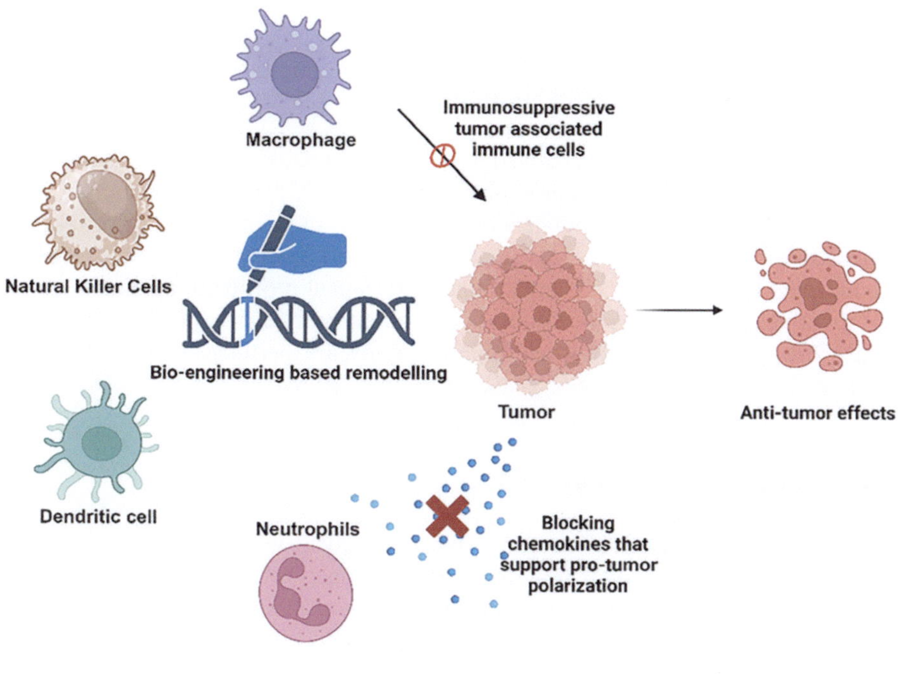

Fig. 1 An overview of bioengineering-based rewiring of various candidates of the innate immune arm, namely the macrophages, natural killer cells, dendritic cells, and neutrophils, can lead to antitumor responses. (Created in BioRender. Khadilkar, R. (2025) https://BioRender.com/l41w774)

4 Candidates of the Innate Immune Arm in Cancer Pathophysiology

4.1 The Majestic Macrophage/Monocyte

Macrophages make up a prominent population of leukocytes that infiltrate most solid tumors. These macrophages can directly kill the tumor cells in response to specific responses [16, 17]. Macrophages are classified into two main types: classically activated (M1) macrophages, which are antitumorigenic, and alternatively activated (M2) macrophages, which are pro-tumorigenic. M1 macrophages are associated with pro-inflammatory responses, whereas M2 macrophages exhibit immunosuppressive functions [18, 19]. The opposing functions that macrophages play in cancer render them important candidates for rewiring. However, it is of significance that a spectrum of phenotypes lies between the two polarized states.

As the tumor advances, signals from the tumor microenvironment prompt tumor-associated macrophages (TAMs) to adopt tumor-supportive roles. Initial studies revealed that this phenotypic transition could be driven by IL-4 and IL-13, often released by T lymphocytes. However, other factors, such as cytokines like IL-10 and TGF-β, metabolic products from tumor cells, hypoxic conditions, and immune complexes, can also promote the pro-tumorigenic polarization of TAMs [15–17, 20, 21].

4.2 The Not-So-Neutral Neutrophil

Neutrophils have long been perceived as neutral onlookers in the fight against cancer cells; however, recent literature proves otherwise [22]. The polarization of neutrophils leads to a highly heterogeneous population of immune cells that take up diverse roles depending on the specific tumor-derived ligands [22, 23]. Neutrophils are hijacked by the tumor cells in facilitating sustained proliferation, angiogenesis, and immunosuppression in the tumor microenvironment. Hence, polarization of the neutrophils and understanding neutrophil biology are pertinent to rewiring their plasticity and deriving antitumorigenic responses [22, 24].

4.3 The Underrated MDSCs

Physiologically, hematopoietic stem cells differentiate into various lineages and subsets of immune cells, which perform specific functions like phagocytosis, antigen presentation, and activating other immune cells. Myeloid-derived suppressor cells (MDSCs) are a subset of immature immune cells that showcase various immune suppressive and tumor-promoting effects, which is an outcome of an aberrant myelopoiesis [25]. A varied group of immature myeloid cells largely increase in number during pathological conditions like cancer, typically because of signals from tumors. They are categorized into two subtypes based on surface markers:

monocytic MDSCs (M-MDSCs) and granulocytic or polymorphonuclear MDSCs (PMN-MDSCs) [25–27].

4.4 The Deadly Dendritic Cells

Dendritic cells are the connecting link between innate and adaptive arms. The activation and sustainment of antitumor immune responses depend on the presentation of tumor antigens to naïve T cells, accompanied by functional signals from dendritic cells (DCs). As professional antigen-presenting cells (APCs), DCs play a crucial role in linking innate and adaptive immunity [28, 29]. Higher levels of DCs are strongly associated with better prognosis in cancer patients and improved outcomes from immune checkpoint inhibitors (ICIs) [29]. Technological advancements have provided valuable insights into the functional states of intra-tumoral dendritic cells (DCs) across different cancer types. These discoveries have deepened our understanding of the diversity of DC subtypes and their roles—both functional and dysfunctional—in cancer. This expanding knowledge underscores the potential of DCs as therapeutic targets, therapeutic agents, and predictors of patient responses to immunotherapies, such as immune checkpoint inhibitors (ICIs) [28].

4.5 The Defender NK Cell

Natural killer cells or NK cells are the innate immune cells that can kill target cells through cytotoxic mechanisms. NK cells are identified by CD56 receptors and the absence of surface TCR and associated CD3 molecules in human peripheral blood, bone marrow, and tissues [30]. With their non-MHC restrictive cytotoxic effects, NK cells release certain molecules like perforins and granzymes to kill damaged cells including tumor cells. Hence, NK cell-based therapy is proving to have more milestones to achieve than was thought so. NK cell-based therapy is gaining more and more attention with the latest approaches including the immune-checkpoint inhibition strategies, adoptive transfer of stem cell-derived NK cells, and CAR NK cell therapy wherein NK cell receptors are genetically engineered to have antitumor effects [31].

5 Macrophage as an Immunotherapeutic Target

CAR-T-cell therapy, which is gaining more and more popularity in clinics, are effective against most hematological malignancies. Yet, it has many limitations when it comes to solid tumors, as the infiltration by T cells is not that significant in all types of tumors [32]. The solid tumors are, however, densely infiltrated by the remarkable macrophages; hence, macrophages can be effective immune cells concerning cancer therapeutics. They are, in general, very plastic and are innately neither tumor-supportive nor tumor-evasive [21, 32].

Besides polarization of macrophages, there are several functions of activated macrophages, which are present in many different tissues of the human body with various morphology and gene signature. The tissue-resident nature of these macrophages gives them a unique advantage over other immune cells for tumor–host tissue cross talk [33]. This is another reason why macrophages can be deployed to rewire systemic inflammation, which is secondary effects of various tumors (Fig. 2). Targeting macrophages is intriguing to many as it can unravel insightful aspects in the field of immunotherapy [21, 32].

Macrophages easily localize to and remain within the tumor microenvironment. Macrophages are skilled at altering the extracellular matrix (ECM) [34, 35]. and have the ability to affect neighboring immune cells in both pro- and anti-inflammatory ways [21]. Macrophages are antigen-presenting cells, and they recruit T cells, presenting antigens, co-stimulating the immune system, and secreting cytokines that can start and intensify an adaptive immune response When combined, these effector

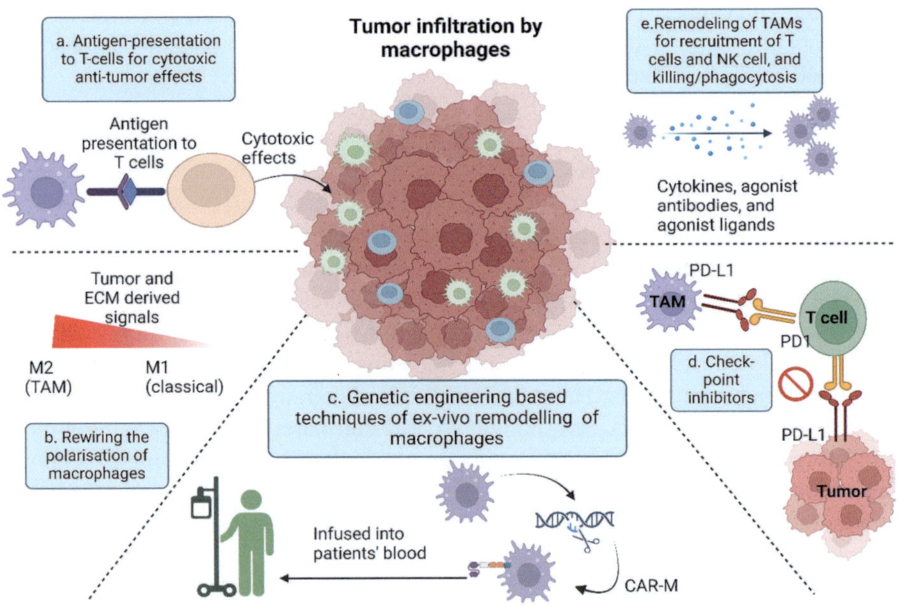

Fig. 2 Macrophages as a potent candidate for immunotherapeutic targets. (**a**) Classical antigen presentation properties of macrophages to T cells for tumor-killing effector functions. (**b**) Rewiring the polarization of macrophages from an immunosuppressive M2 phenotype to an antitumor M1 phenotype. (**c**) Ex vivo genetic engineering-based technique to remodel receptors for cytotoxic effector functions (e.g., CAR M therapy) and finally infuse the genetically engineered macrophages into the blood of the patients having cancer. (**d**) Various TAMs also showing PDL1 expression aiding in tumor escape can be blocked using the checkpoint inhibitors. (**e**) Robust generation of macrophages with antigen presentation properties and recruitment of T cells and NK cells for cytotoxicity using cytokine therapy and treatment with agonist antibodies and agonist ligands that promote tumor-killing properties. (Created in BioRender. Khadilkar, R. (2025) https://BioRender.com/m23s539)

roles facilitate the dissemination of epitopes and mitigate the difficulties caused by target antigen heterogeneity [21, 32, 36–38].

The immunotherapy techniques that have evolved over the years include: immune checkpoint inhibitors, adoptive cell therapy, and the use of monoclonal antibodies. Diverse therapeutic strategies have come up involving macrophages as targets. The most commonly used immune checkpoint inhibitors that exist in the clinic are anti-PD1 and anti-PDL1 therapies [21]. Studies suggest, that PD-1 and PD-L1 are expressed in TAMs, promoting tumor immune escape [39]. Hence, not just the adaptive system, but exploiting the PD-1 on TAMs will also be a therapeutic strategy to boost the responses of checkpoint inhibitors.

Adoptive cell therapy is an immunotherapeutic technique wherein immune cells are engineered in vitro having special effector functions before administration into patients with tumor. The immune cells can be derived either from the hosts themselves or from healthy donors. Macrophages, with their varied antitumor properties, are an established candidate for macrophage-based cell therapy [21, 40].

Several monoclonal antibodies have been effective in cancer immunotherapy. To highlight the role of macrophages as a target of therapy, one significant drug, Trastuzumab, targets the Human Epidermal growth factor Receptor 2 (HER2) protein on tumor cells and has shown efficient outcomes in breast cancer therapy [41]. It was also seen that by binding with Fcγ receptors on macrophages, Trastuzumab mediated macrophage phagocytic killing, and after increasing the expression of Fcγ receptors on macrophages, this effector response was broadened. Also, resistance to Trastuzumab was overcome by shifting the polarization of macrophages from the M2 to the M1 phenotype [42, 43]. These are a few of the therapeutic strategies from the many sophisticated immunotherapeutic strategies that exist using macrophages as a target [24].

6 NK Cell-Based Therapy

Various NK cell-based therapeutic approaches include cytokine-stimulated autologous NK cells and allogeneic NK cells, which are techniques of cytokine-stimulated robust NK cell infusion. Genetic modification of NK cells, including CAR-NK cells and TCR NK cells, also proved to be an important genetic engineering-based therapeutic strategy [30, 44]. These strategies include isolation and expansion of the immune cells using cytokines, and other strategies involve genetic engineering techniques to genetically modify NK cells to target tumor cells (Fig. 3).

Despite the remarkable success of CAR-T-cell therapy, several limitations with CAR-T-cell therapies include severe side effects such as cytokine-release syndrome (CRS) and neurological toxicities, which result in the severely triggered immune system and multiple organ failure, which might prove to be fatal and lead to devising alternative strategies such as the CAR NK-cell therapy primarily because they can potentially avoid many of these problems—possibly even the troublesome toxicities—while reiterating the dramatic anticancer benefits of CAR T cells. However,

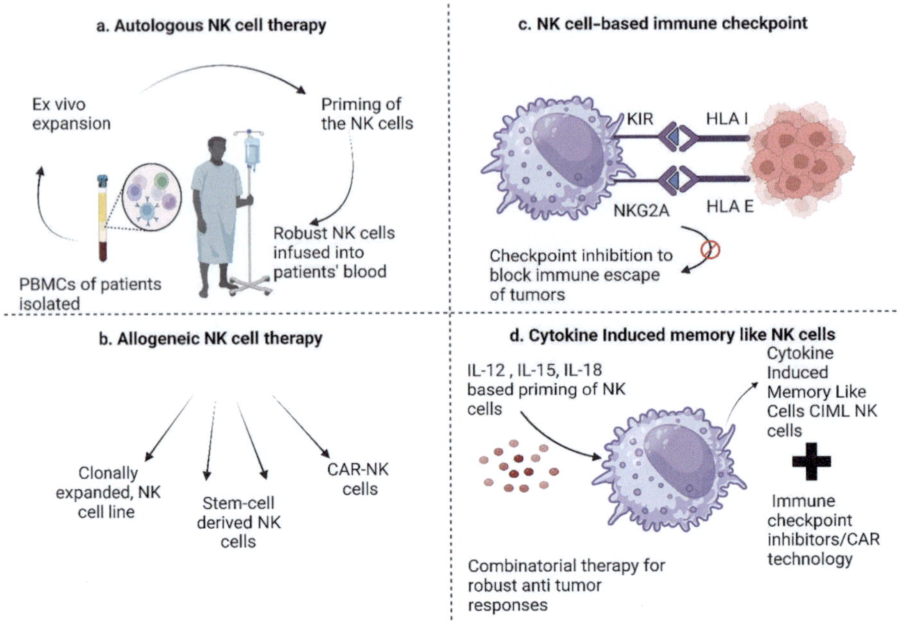

Fig. 3 Overview of various therapeutic strategies using NK cells as a target. (**a**) Autologous NK cell therapy wherein ex vivo expansion and priming of NK cells is done and reinfused into the patients' blood for antitumor responses. (**b**) Various allogeneic NK cell therapies that exist. (**c**) NK cell-based immune checkpoint inhibitors are used to avoid immune escape by tumors by presenting ligands for inhibiting receptors of NK cells. (**d**) Cytokine-induced memory-like (CIML-NK cells) NK cells as a new therapeutic strategy that, when used in combination with immune checkpoint inhibitors and CAR-NK cells/CAR-T cells, will produce robust responses. (Created in BioRender. Khadilkar, R. (2025) https://BioRender.com/z79v805)

more validation is required to bring them to clinics as a successful alternative of the CAR T-cell therapy.

7 Conclusion

Fine-tuning the immune system is a key to devising novel immunotherapeutic strategies. The innate immune cells being more complex might act as a double-edged sword in cancer progression. However, approaches like trained immunity and educating the immune cells are gaining more and more focus these days. As cancer is rightly termed as an "unhealed wound," chronic inflammation is a result of an aberrant inflammatory response that also triggers tumorigenesis [45]. Targeting inflammatory cells offers a therapeutic potential to impact cancer progression and enhance adaptive immune responses. However, the outcome of the inflammatory response is extremely context dependent. Further efforts must be made to identify the molecules that cause inflammatory cells to change into an anticancer phenotype. Various cytokines and chemokines are generally involved in the polarization of the immune cells and, hence, the outcome of either a pro-tumor or antitumor response.

Furthermore, as innate immunity cells affect adaptive immune cells and have bi-directional cross talk with each other, engaging the lymphocytes into antitumor responses by remodeling the immune cells will result in better tackling of tumor growth [46].

A holistic, systemic approach to managing and diving into relevant compartments of both the innate and adaptive immune system will not just help in controlling tumor growth but also tumor non-autonomous distant effects like muscle wasting and deregulated metabolism, which are as much important in contributing to the overall quality of life of the cancer patients [47–49].

A few questions as to how the tumors secrete factors to deploy the immune cells to regulate the host metabolism and still keep themselves safe from such factors is something to look into. What factors determine the outcome of the polarization of immune cells? Are they just tumor-derived signals or yet-undiscovered tumor microenvironment-derived factors as well? What drives the bias of myeloid cells? How do the tumor cells reprogram the host tissues, resulting in an emergent myelopoiesis to bring about tumor-immune escape? These questions hold pertinence and need to be addressed.

Innate immune signaling in cancer pathophysiology, with its intricacies in downstream signaling cascades, and extremely heterogeneous population is poorly understood, although it plays crucial roles in the antitumor immune responses by enhancing and priming the adaptive immune arm. Examining the activation of innate immune signaling cascades expanded the search for novel therapeutic adjuvants and highlighted its significance in cancer research. It holds immense promise as a therapeutic modality. Exploring the functioning and activation of innate immune signaling cascades will expand the search for novel therapeutic adjuvants and will highlight the significance and applicability of deploying innate immune cells as therapeutic interventions for dealing with cancer.

Acknowledgments We would like to thank CRI-ACTREC and Department of Atomic Energy for funding and support.

Declaration of Interest We hereby declare that there is no conflict of interest.

References

1. Waldman AD, et al. A guide to cancer immunotherapy: from T cell basic science to clinical practice. Nat Rev Immunol. 2020;20(11):651–68.
2. Rodrigues J, et al. Hemocyte differentiation mediates innate immune memory in Anopheles gambiae mosquitoes. Science. 2010;329(5997):1353–5.
3. Zhao C, et al. Macrophages in tissue repair and regeneration: insights from zebrafish. Cell Regen. 2024;13(1):12.
4. Delves PJ, et al. Roitt's essential immunology. Wiley; 2017.
5. Boutros M, et al. Sequential activation of signaling pathways during innate immune responses in Drosophila. Dev Cell. 2002;3(5):711–22.

6. Netea MG, et al. Trained immunity: a memory for innate host defense. Cell Host Microbe. 2011;9(5):355–61.
7. Hajishengallis G, et al. Innate immune memory, trained immunity and nomenclature clarification. Nat Immunol. 2023;24(9):1393–4.
8. Dobosz P, Dzieciątkowski T. The intriguing history of cancer immunotherapy. Front Immunol. 2019;10:2965.
9. Zhang Y, Zhang Z. The history and advances in cancer immunotherapy: understanding the characteristics of tumor-infiltrating immune cells and their therapeutic implications. Cell Mol Immunol. 2020;17(8):807–21.
10. Lin D, et al. Oncolytic virotherapy: basic principles, recent advances and future directions. Signal Transduct Target Ther. 2023;8(1):156.
11. Grimmett E, et al. Cancer vaccines: past, present and future; a review article. Discov Oncol. 2022;13(1):31.
12. Oiseth SJ, Aziz MS. Cancer immunotherapy: a brief review of the history, possibilities, and challenges ahead. J Cancer Metastasis Treat. 2017;3(10):250–61.
13. Hanahan D, Weinberg RA. Hallmarks of cancer: the next generation. Cell. 2011;144(5):646–74.
14. Gonzalez H, et al. Roles of the immune system in cancer: from tumor initiation to metastatic progression. Genes Dev. 2018;32(19–20):1267–84.
15. Maiorino L, et al. Innate immunity and cancer pathophysiology. Annu Rev Pathol. 2022;17(1):425–57.
16. Mantovani A, et al. Tumour-associated macrophages as treatment targets in oncology. Nat Rev Clin Oncol. 2017;14(7):399–416.
17. Mantovani A, et al. Macrophages as tools and targets in cancer therapy. Nat Rev Drug Discov. 2022;21(11):799–820.
18. Murray PJ, et al. Macrophage activation and polarization: nomenclature and experimental guidelines. Immunity. 2014;41(1):14–20.
19. Ginhoux F, et al. New insights into the multidimensional concept of macrophage ontogeny, activation and function. Nat Immunol. 2016;17(1):34–40.
20. Kim J, Bae JS. Tumor-associated macrophages and neutrophils in tumor microenvironment. Mediat Inflamm. 2016;2016(1):6058147.
21. Duan Z, Luo Y. Targeting macrophages in cancer immunotherapy. Signal Transduct Target Ther. 2021;6(1):127.
22. Mayadas TN, et al. The multifaceted functions of neutrophils. Annu Rev Pathol. 2014;9(1):181–218.
23. Huang X, et al. Neutrophils in cancer immunotherapy: friends or foes? Mol Cancer. 2024;23(1):107.
24. Jinushi M. Yin and yang of tumor inflammation: how innate immune suppressors shape the tumor microenvironments. Int J Cancer. 2014;135(6):1277–85.
25. Lasser SA, et al. Myeloid-derived suppressor cells in cancer and cancer therapy. Nat Rev Clin Oncol. 2024;21(2):147–64.
26. Hegde S, Leader AM, Merad M. MDSC: markers, development, states, and unaddressed complexity. Immunity. 2021;54(5):875–84.
27. Hicks KC, et al. Myeloid Cell–Derived Oxidized Lipids and Regulation of the Tumor Microenvironment. Cancer Res. 2022;82(2):187–94.
28. Heras-Murillo I, et al. Dendritic cells as orchestrators of anticancer immunity and immunotherapy. Nat Rev Clin Oncol. 2024;21(4):257–77.
29. Le Gall CM, et al. Dendritic cells in cancer immunotherapy. Nat Mater. 2018;17(6):474–5.
30. Vivier E, et al. Natural killer cell therapies. Nature. 2024;626(8000):727–36.
31. Shimasaki N, Jain A, Campana D. NK cells for cancer immunotherapy. Nat Rev Drug Discov. 2020;19(3):200–18.
32. Kashfi K, et al. Macrophage reprogramming and cancer therapeutics: role of iNOS-derived NO. Cells. 2021;10(11):3194.
33. Liu M, et al. The crosstalk between macrophages and cancer cells potentiates pancreatic cancer cachexia. Cancer Cell. 2024;42(5):885–903.

34. Meizlish ML, et al. Mechanosensing regulates tissue repair program in macrophages. Sci Adv. 2024;10(11):eadk6906.
35. Wynn TA, Vannella KM. Macrophages in tissue repair, regeneration, and fibrosis. Immunity. 2016;44(3):450–62.
36. Bied M, et al. Roles of macrophages in tumor development: a spatiotemporal perspective. Cell Mol Immunol. 2023;20(9):983–92.
37. Cendrowicz E, et al. The role of macrophages in cancer development and therapy. Cancers. 2021;13(8):1946.
38. Dey T. Tumor-macrophage crosstalk: how to listen. Integr Biol. 2020;12(12):291–302.
39. Gordon SR, et al. PD-1 expression by tumour-associated macrophages inhibits phagocytosis and tumour immunity. Nature. 2017;545(7655):495–9.
40. Anderson NR, et al. Macrophage-based approaches for cancer immunotherapy. Cancer Res. 2021;81(5):1201–8.
41. Maadi H, et al. Trastuzumab mechanism of action; 20 years of research to unravel a dilemma. Cancers. 2021;13(14):3540.
42. Shi Y, et al. Trastuzumab triggers phagocytic killing of high HER2 cancer cells *in vitro* and *in vivo* by interaction with Fcγ receptors on macrophages. J Immunol. 2015;194(9):4379–86.
43. Xu M, et al. Intratumoral delivery of IL-21 overcomes anti-Her2/Neu resistance through shifting tumor-associated macrophages from M2 to M1 phenotype. J Immunol. 2015;194(10):4997–5006.
44. Kang S, et al. The advances and challenges of NK cell-based cancer immunotherapy. Curr Oncol. 2021;28(2):1077–93.
45. Deyell M, et al. Cancer metastasis as a non-healing wound. Br J Cancer. 2021;124(9):1491–502.
46. Fenis A, et al. New immune cell engagers for cancer immunotherapy. Nat Rev Immunol. 2024;24:1–16.
47. Ogino S, et al. Cancer immunology-analysis of host and tumor factors for personalized medicine. Nat Rev Clin Oncol. 2011;8(12):711–9.
48. Ferrer M, et al. Cachexia: a systemic consequence of progressive, unresolved disease. Cell. 2023;186(9):1824–45.
49. Biswas AK, Acharyya S. Understanding cachexia in the context of metastatic progression. Nat Rev Cancer. 2020;20(5):274–84.

Metabolic Targeting in Cancer Therapy: Strategies, Challenges, and Personalized Approaches

Saurabh Pal, Sourin Adhikary, Saria Anjum, Santanu Chatterjee, and Kausar M. Ansari

1 Introduction

Cancer cell metabolism is a unique and complex process that makes cancer cells function differently from normal, healthy cells [1]. As a normal cell transforms into a cancer cell, it undergoes significant metabolic changes, a process known as metabolic reprogramming [2]. These metabolic adaptations allow cancer cells to grow and multiply rapidly. One of the most important metabolic changes in cancer is increased glucose consumption. Unlike normal cells, which depend on oxidative phosphorylation (OXPHOS) for energy, cancer cells depend mainly on glycolysis to produce ATP, even when oxygen is available. This shift allows them to generate important building blocks like nucleotides, amino acids, and lipids, which support their rapid growth and energy needs [3]. In the 1920s, Otto Warburg discovered that tumor cells prefer glycolysis over OXPHOS, even though glycolysis produces only 2 ATP molecules per glucose, compared to 36 ATP molecules produced by OXPHOS in normal cells [4]. This phenomenon, known as the Warburg effect, causes cancer cells to produce large amounts of lactate which creates an acidic tumor environment that helps cancer cells spread, invade tissues, and escape detection by the immune

S. Pal
Food Toxicology Laboratory, FEST Division, CSIR-Indian Institute of Toxicology Research, Lucknow, Uttar Pradesh, India

S. Adhikary · S. Anjum · K. M. Ansari (✉)
Food Toxicology Laboratory, FEST Division, CSIR-Indian Institute of Toxicology Research, Lucknow, Uttar Pradesh, India

Academy of Scientific and Innovative Research (AcSIR), Ghaziabad, Uttar Pradesh, India
e-mail: kmansari@iitr.res.in

S. Chatterjee
SRM Institute of Science and Technology, Kattankulathur Campus, Chengalpattu, Tamil Nadu, India

A. S. Ramteke et al. (eds.), *Molecular Targets in Cancer Therapy*,
https://doi.org/10.1007/978-981-96-7336-0_12

system. Because cancer cells have high metabolic demands, many modern therapies specifically target key metabolic pathways essential for survival, such as glucose metabolism, nucleotide synthesis, and oxidative stress regulation [5].

One of the most effective approaches has been the use of antimetabolites like methotrexate, 5-fluorouracil (5-FU), and gemcitabine. These drugs take advantage of cancer cells' high demand for nucleotides, which are essential for DNA replication. Although antimetabolites can be toxic, they are still widely used in chemotherapy and help to improve patient survival [6]. From a treatment perspective, cancer cells have a unique metabolism that offers many potential drug targets. This has increased interest in targeting cancer metabolism to improve treatment success. Emerging trends in cancer therapy now focus on exploiting metabolic reprogramming to disrupt cancer-specific vulnerabilities, such as targeting glycolysis, glutaminolysis, and fatty acid metabolism, while overcoming therapy resistance. These approaches are integrated with conventional therapies like chemotherapy, radiotherapy, and immunotherapy to enhance efficacy and improve patient outcomes. This chapter discusses the key metabolic changes in cancer cells, potential therapeutic targets, and strategies for more effective cancer treatments.

2 Aerobic Glycolysis and Mitochondrial Function in Cancer Metabolism

Cancer cells rely on altered metabolic pathways to support their rapid growth and survival, particularly under nutrient-limited conditions such as hypoxia. One hallmark of cancer metabolism is aerobic glycolysis where cancer cells preferentially use glycolysis for energy production instead of oxidative phosphorylation (OXPHOS), even in the presence of oxygen. This phenomenon is known as the Warburg effect. This metabolic shift allows cancer cells to generate ATP quickly and produce essential biosynthetic intermediates (such as nucleotides and lipids) necessary for cell proliferation. Cancer cells increase the expression of glucose transporters like GLUT1 and GLUT3 to facilitate glycolysis. Inhibiting key enzymes in glycolysis, such as GLUT1 inhibitors (e.g., BAY-876), HK2 inhibitors (e.g., lonidamine), and LDHA inhibitors (e.g., FX11), can block glucose uptake, impair ATP production, and induce cell death by increasing oxidative stress and disrupting energy homeostasis [7]. These inhibitors starve cancer cells of energy and biosynthetic precursors, making them promising candidates for cancer therapy. In addition to glycolysis, mitochondrial function and OXPHOS are crucial for cancer metabolism, supporting energy production, biosynthesis, and redox regulation. Cancer cells exhibit enhanced mitochondrial activity, adapting to fluctuating oxygen levels and nutrient availability. By switching between glycolysis and OXPHOS and utilizing alternative pathways like fatty acid oxidation (FAO), the tricarboxylic acid (TCA) cycle, and glutamine metabolism, cancer cells maintain metabolic flexibility, allowing them to survive and proliferate under stress. The TCA cycle helps generate energy carriers, while the mitochondria's electron transport chain (ETC) produces ATP. However, mutations in mitochondrial DNA (mtDNA) can lead to the accumulation of reactive oxygen species (ROS), which promote tumor growth but can also damage cells [8]. Some cancer cells manage high ROS levels by using antioxidant enzymes or shifting to glycolysis. Targeting mitochondrial metabolism, glycolysis, and ROS regulation with specific inhibitors offers a potential therapeutic approach to disrupt tumor adaptation, starve cancer cells, and reduce oxidative stress, thus inhibiting cancer progression.

2.1 Targeting Amino Acid Metabolism in Cancer

Amino acids are vital for cancer cell growth; they provide the building blocks for proteins and support critical metabolic processes like energy production, biosynthesis, and redox regulation. Rapidly proliferating cancer cells have an increased demand for amino acids such as glutamine and serine to support their growth and epigenetic regulation [9]. Some cancers also become dependent on external sources of nonessential amino acids, like arginine and asparagine, due to genetic mutations that impair their synthesis [10]. Therapeutic inhibitors targeting amino acid metabolism aim to deprive cancer cells of these crucial nutrients. Glutaminase inhibitors (e.g., CB-839) block glutamine metabolism, starving cells of energy and biosynthetic precursors [11]. Arginine deiminase (ADI) specifically targets tumors that cannot synthesize arginine, including melanoma and prostate cancer [12]. Serine biosynthesis inhibitors (e.g., NCT-503) disrupt serine production, essential for nucleotide synthesis and one-carbon metabolism. These inhibitors exploit cancer-specific vulnerabilities, effectively impairing tumor growth by limiting access to key amino acids and disrupting essential metabolic processes.

2.2 Targeting Fatty Acid Metabolism in Cancer

Lipid metabolism, particularly fatty acid synthesis (FAS) and fatty acid oxidation (FAO), plays a critical role in cancer cell survival, proliferation, and progression. Both FAS and FAO are frequently upregulated in cancer, enabling tumor cells to meet their high energy and biosynthetic demands. FAS converts nutrients into essential components for membrane biosynthesis, energy storage, and signaling molecules. A key enzyme in FAS is fatty acid synthase (FASN), which synthesizes palmitate from acetyl-CoA and malonyl-CoA. FASN is regulated by signaling pathways such as EGFR, HER2, and PDGFR, and its stability is influenced by post-translational modifications like ubiquitination, acetylation, and sumoylation [13–16]. Other lipogenic enzymes, including ATP-citrate lyase (ACLY) and acetyl-CoA carboxylase (ACC), are also crucial for de novo fatty acid synthesis. ACLY converts citrate to acetyl-CoA, a precursor for fatty acid and cholesterol synthesis, while ACC produces malonyl-CoA [17]. Inhibitors such as TVB-2640 (FASN inhibitor), SB-204990 (ACLY inhibitor), and ND-646 (ACC inhibitor) disrupt lipid production, increase oxidative stress, and impair cancer cell proliferation, making them promising therapeutic targets [18]. On the other hand, FAO breaks down fatty acids in the mitochondria to generate ATP, providing an alternative energy source under conditions of glucose scarcity or oxidative stress. Cancer cells often increase FAO to adapt to metabolic challenges in the tumor microenvironment, such as acidosis or nutrient deprivation. Epigenetic changes in enzymes like ACC2, which normally inhibits FAO, further enhance this metabolic shift. Inhibitors such as etomoxir and perhexiline target carnitine palmitoyltransferase 1 (CPT1), a key enzyme in FAO, blocking fatty acid transport into mitochondria and impairing energy production [19]. By switching between FAS and FAO, cancer cells maintain metabolic flexibility, supporting their survival, resistance to therapy, and metastasis. Targeting these pathways with inhibitors offers a promising strategy to disrupt cancer cell metabolism. For example, FASN inhibitors like TVB-2640 block lipid synthesis, while FAO inhibitors like etomoxir impair energy production, collectively starving cancer cells of essential resources [20]. Understanding the interplay between FAS and FAO provides valuable insights into potential therapeutic interventions to improve cancer treatment outcomes.

3 Hypoxia-Induced Metabolic Alteration in Cancer Cells

Hypoxia-induced metabolic reprogramming is a critical survival strategy that allows cancer cells to adapt and survive in the low-oxygen environments commonly found in solid tumors. The rapid proliferation of cancer cells exhausts oxygen more rapidly than the blood supply can replenish it [21]. In response, cancer cells undergo metabolic alterations, such as enhanced glycolysis, reduced oxidative phosphorylation, and changes in fatty acid and lipid metabolism, to sustain energy production and promote tumor growth. These adaptations not only support survival under hypoxia but also drive disease progression and therapy resistance. Understanding these mechanisms provides valuable insights into potential therapeutic targets for developing more effective cancer treatments.

3.1 Hypoxia-Driven Alterations in Glycolytic Pathway

HIF-1α plays a central role in the metabolic reprogramming of cancer cells under hypoxia, regulating key pathways such as glycolysis, lipid metabolism, and amino acid metabolism [22]. Under hypoxic oxygen, HIF-1α upregulates genes encoding glucose transporters (e.g., GLUT1, GLUT3) and glycolytic enzymes, including hexokinase 1/2, enolase 1, phosphoglycerate kinase 1, and pyruvate kinase M2, to enhance glucose uptake and glycolysis [23]. Moreover, it also promotes lactate dehydrogenase A (LDHA) expression, enabling the conversion of pyruvate to lactate and sustaining anaerobic glycolysis [24]. Additionally, HIF-1α facilitates glucose-to-glycogen conversion, providing an energy reserve for metabolic stress. This shift from oxidative phosphorylation to aerobic glycolysis supplies cancer cells with the energy and biosynthetic precursors needed for survival and proliferation in hypoxic tumor microenvironments [25].

3.2 Impact of Hypoxia on Mitochondrial Function and Oxidative Phosphorylation

Mitochondria play a central role in cancer cell metabolism, undergoing significant reconfiguration under hypoxia to adapt to low oxygen levels. The tricarboxylic acid (TCA) cycle, crucial for energy production and biosynthesis, is altered in hypoxic conditions. Pyruvate dehydrogenase kinase 1 (PDK-1) inhibits pyruvate dehydrogenase (PDH), reducing acetyl-CoA production and slowing the TCA cycle, leading to aspartate deficiency, which is vital for nucleotide synthesis and cancer cell proliferation [26]. Mutations in TCA cycle enzymes like succinate dehydrogenase (SDH), fumarate hydratase (FH), and isocitrate dehydrogenase (IDH1/IDH2) cause the accumulation of oncometabolites (e.g., succinate, fumarate, L-2-hydroxyglutarate), which inhibit prolyl hydroxylases (PHDs), stabilizing HIF-1α and promoting tumor progression [27]. Under hypoxia, α-ketoglutarate (αKG) is redirected to L-2HG synthesis via lactate dehydrogenase A (LDHA) and malate dehydrogenase (MDH1/MDH2), further stabilizing HIF-1α and enabling cancer cells to bypass oxidative metabolism [28].

The electron transport chain (ETC) is also reconfigured under hypoxia. Complex IV (cytochrome c oxidase, COX) adapts through HIF-1-mediated expression of COX4I2 and LON protease, replacing COX4I1 to enhance efficiency and reduce ROS production while maintaining ATP synthesis [29]. HIF-1 also upregulates NDUFA4L2, inhibiting Complex I, and regulates miR-210, which targets succinate dehydrogenase (SDH) and Complex IV subunits, fine-tuning ETC activity to sustain mitochondrial function under hypoxia [30].

3.3 Hypoxia-Altered Lipid Metabolism

Cancer cells exhibit elevated lipid levels, especially under hypoxic stress, wherein hypoxia-inducible factor 1 (HIF-1) directs a metabolic transition toward lipid accumulation and modified fatty acid metabolism. In hypoxic conditions, HIF-1 facilitates lipid droplet formation, modifies β-oxidation, and increases de novo fatty acid production. This is enabled via the overexpression of fatty acid-binding proteins, including FABP3, FABP7, and ADRP, which are essential for the stability and function of lipid droplets [31]. These changes enable cancer cells to alleviate oxidative stress, ensuring survival and continued proliferation after hypoxic exposure. Furthermore, HIF-1 inhibits mitochondrial fatty acid oxidation by downregulating essential enzymes, such as medium-chain acyl-CoA dehydrogenase (MCAD) and long-chain acyl-CoA dehydrogenase (LCAD). This suppression diminishes reactive oxygen species (ROS) formation and mitigates PTEN signaling, thus promoting tumor cell proliferation and survival [32]. Additionally, HIF-1α exerts a negative regulatory effect on c-Myc, a transcriptional activator of PGC-1β, crucial to producing MCAD and LCAD. Carnitine palmitoyl transferase 1A (CPT1A), the rate-limiting enzyme for mitochondrial fatty acid import, is concurrently inhibited by HIF-1 and HIF-2. This inhibition restricts fatty acid oxidation, increases lipid droplet accumulation, and facilitates tumor adaptability to metabolic stress. The coordinated modifications in lipid metabolism provide cancer cells with a survival advantage. These changes enhance their ability to survive in hypoxic environments.

The synthesis of fatty acids increases the production of neutral triacylglycerols (TAGs), stored as lipid droplets (LDs) for energy reserves. HIF-1 activates enzymes such as AGPAT2 and lipin-1, essential for lipid droplet formation and chemoresistance in hypoxic conditions [33]. AGPAT2 derivatives, lysophosphatidic acid (LPA), and phosphatidic acid (PA) are utilized to produce new membranes. Hypoxia stimulates the development of additional lipid droplet membrane components, including perilipin 2 (PLIN2), which is modulated by HIF-2α and facilitates the storage of triglycerides and cholesterol, hence preserving endoplasmic reticulum integrity and homeostasis during nutrition and oxygen deprivation [34]. HIF-1 enhances lipid accumulation in both cancerous and normal cells by promoting HIG2/HILPDA, which inhibits adipose triglyceride lipase (ATGL), hence obstructing intracellular lipolysis [35]. During hypoxia, the unsaturated fatty acid oleate is preferentially liberated from triglycerides in lipid droplets into the phospholipid pool to mitigate the accumulation of saturated lipids. Lipid signaling molecules like sphingosine kinase 1 also stimulate HIF-1 activity, further integrating lipid metabolism with cancer cell survival and proliferation [36].

3.4 Hypoxia-Driven Alterations in Amino Acid Metabolism

Hypoxic cancer cells reprogram glutamine and serine metabolism to support tumor growth, redox balance, and biosynthesis. Glutamine is a key carbon and nitrogen source necessary for ATP production and cellular equilibrium. It is upregulated under hypoxia via transporters like SLC1A5 and SNAT2/SLC38A2. It is converted to glutamate and then to α-ketoglutarate (αKG), which undergoes HIF-1-mediated reductive carboxylation to produce citrate for lipid synthesis [37]. This adaptation is critical in cancers like glioblastoma where citrate depletion or IDH2 inhibition suppresses growth. HIF-1 also promotes SIAH2-mediated degradation of αKG dehydrogenase and enhances reductive carboxylation. In VHL-mutated renal cell carcinoma, stabilization of HIF drives glutamine-derived aspartate production for pyrimidine synthesis [38]. Glutamine-derived glutamate further supports glutathione biosynthesis, transamination, and amino acid synthesis, ensuring tumor survival. Serine is crucial for cancer cell proliferation, supporting amino acid synthesis, phospholipid production, and one-carbon metabolism for nucleotide biosynthesis and redox balance. In hypoxia, PHGDH enhances serine biosynthesis by redirecting glycolytic intermediates. This pathway is often upregulated in cancers like non-small cell lung, cervical, colorectal, and breast cancer, especially under hypoxic conditions via HIF-1 [39]. HIF-1 also promotes cysteine metabolism and boosts glutathione production for antioxidant defense. Integrating serine metabolism and the one-carbon cycle maintains NADPH levels, which is vital for redox balance and mitigating oxidative damage.

4 Therapeutic Approaches Targeting Cancer Metabolism

4.1 Targeting Glycolysis

Targeting glycolysis in cancer cells disrupts energy production and biosynthetic pathways, reducing tumor growth and enhancing sensitivity to conventional therapies. Inhibitors of glycolytic enzymes or glucose transporters effectively starve cancer cells of energy and essential building blocks. This section summarizes current and emerging drugs targeting glucose metabolism, providing insights for developing innovative anticancer therapies.

4.1.1 Glucose Transporter (GLUT) Inhibitors

Cancer cells exhibit increased glucose uptake to support their high metabolic demands. Therefore, inhibiting GLUTs is a promising strategy to prevent glucose from entering cancer cells and disrupting the glycolytic pathway. WZB117 inhibits GLUT1, reducing glucose uptake and tumor growth in breast, lung, and pancreatic cancers [40]. However, further optimization is needed to improve its bioavailability and selectivity. BAY-876 is a selective inhibitor of GLUT1, disrupts glycolysis, and induces apoptosis in colorectal and breast preclinical models; it has shown potential in combination with chemotherapy to enhance its anticancer effects [7]. Phloretin is a natural compound that inhibits multiple GLUT isoforms, including GLUT1, GLUT2, and GLUT4 [41]. In preclinical studies, it reduces glucose uptake and glycolytic flux. However, phloretin's lack of specificity for GLUT1 limits its potential for clinical application.

4.1.2 Hexokinase (HK) Inhibitors

HK catalyzes the first step of glycolysis and converts glucose into glucose-6-phosphate. It is crucial for tumor glycolysis. Targeting HK is a promising cancer strategy. 3-Bromopyruvate (3-BP) inhibits HKII by alkylating its active site, depleting ATP, and inducing oxidative stress. Preclinical studies have demonstrated that 3-BP effectively induces cancer cell death by depleting cellular ATP levels and enhancing oxidative stress [42]. It has shown antitumor activity against liver, breast, and colon cancers. However, its toxicity and off-target effects have limited its clinical use.

Lonidamine is another HK2 inhibitor that disrupts HKII-mitochondria interaction, leading to reduced glycolytic flux and induction of apoptosis [43]. It has been tested in combination with other therapies, such as chemotherapy and radiotherapy, in clinical trials for gliomas and other solid tumors. However, its clinical efficacy has been modest, and further optimization is needed. 2-Deoxy-D-glucose (2-DG), a glucose analog, blocks glycolysis by competing for phosphorylation by HK [44]. While effective in preclinical studies in combination with radiotherapy, its clinical impact is limited by low potency and high-dose toxicity.

4.1.3 Pyruvate Kinase (PK) Inhibitors

Pyruvate kinase M2 (PKM2), the key enzyme in the final step of glycolysis, converts phosphoenolpyruvate (PEP) to pyruvate with the generation of ATP. PK exists in multiple isoforms, with pyruvate kinase M2 (PKM2) being most relevant in cancer metabolism, where it regulates the flow of metabolites through glycolysis. Unlike the liver and muscle forms of pyruvate kinase (PKM1), PKM2 exists in a less active dimeric form in cancer cells, which limits glycolytic flux and supports anabolic pathways for cell growth. Targeting PKM2 to restore its active tetrameric form and enhance glycolytic flux is a potential therapeutic strategy to inhibit cancer growth. Several small-molecule activators and inhibitors of PKM2 have been developed. Its active tetrameric form can enhance glycolysis and inhibit cancer growth.

TEPP-46 is a small-molecule activator that stabilizes the tetrameric form of PKM2, promoting glycolysis and reducing biosynthetic flux. It has shown antitumor activity in lung and colorectal cancer models [45]. Shikonin, a natural PKM2 inhibitor, disrupts glycolysis and reduces ATP production, inducing oxidative stress; it has shown anticancer effects in preclinical models of leukemia, breast cancer, and hepatocellular carcinoma [46].

4.1.4 Glyceraldehyde 3-Phosphate Dehydrogenase (GAPDH) Inhibitor

Targeting GAPDH, a key enzyme in the "pay-off phase" of glycolysis, is a promising approach to inhibit tumor glycolysis. One of the most promising inhibitors in this category, 3-Bromopyruvate (3-BrPA), a pyruvate analog, binds to GAPDH within cancer cells; it severely depletes ATP, leaving the cancer cells without energy [47].

4.1.5 Lactate Dehydrogenase (LDH) Inhibitors

LDH catalyzes the last step of the glycolytic pathway, converting pyruvate into lactate. FX11, an LDH inhibitor, reduces intracellular ATP levels, significantly enhances oxidative stress, and leads to the death of tumor cells [48]. Similarly, Oxamate, another LDH inhibitor, sensitizes resistant cancer cells to chemotherapy agents [49]. Therefore, inhibiting LDH has shown promising results in preclinical studies, and its future development relies on the results of clinical trials.

4.1.6 Monocarboxylate Transporter (MCT) Inhibitors

Lactate accumulation inside cells lowers intracellular pH. Therefore, the export of lactate into the extracellular space is essential. This process occurs actively through specific transporters called MCTs, including MCT1 and MCT4. α-cyano-4-hydroxy-cinnamic acid, an MCT inhibitor, disrupts tumor growth by causing intracellular lactate buildup, leading to cell acidification and cancer cell death [50]. However, additional research is needed to confirm the target specificity and therapeutic potential for clinical use.

4.2 Targeting Mitochondrial Metabolism

Given the essential role of OXPHOS in cancer cells, targeting mitochondrial metabolism, particularly OXPHOS, has emerged as a promising therapeutic strategy. Several inhibitors of mitochondrial function are in clinical or preclinical development, aimed at disrupting mitochondrial bioenergetics and inducing cancer cell death:

4.2.1 Phenformin and Metformin

Phenformin and Metformin, biguanide compounds initially developed for treating type 2 diabetes, inhibit Complex I of the ETC [51]. As shown in Fig. 1, Phenformin has been shown to reduce OXPHOS, increase ROS production, and induce apoptosis in cancer cells. Although it has limited use due to its toxicity, while Metformin is being tested in clinical trials for its anticancer properties.

4.2.2 IACS-010759

IACS-010759 is a potent inhibitor of Complex I that has shown promise in preclinical models of leukemia, glioblastoma, and other OXPHOS-dependent tumors [52]. This compound impairs ATP production and enhances ROS generation, leading to cancer cell death, especially in tumors reliant on OXPHOS (Fig. 1).

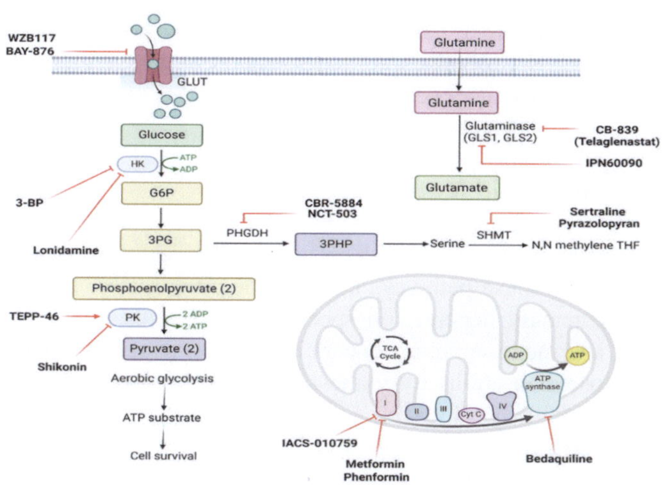

Fig. 1 Schematic depiction of metabolic inhibitors targeting glycolysis and amino acid metabolism pathways associated with cancer cell survival. Inhibitors are mentioned in red

4.2.3 Bedaquiline

Initially approved for tuberculosis treatment, Bedaquiline targets ATP synthase and inhibits mitochondrial ATP production and induces apoptosis in cancer cells (Fig. 1) [53].

4.2.4 Oligomycin

Oligomycin is an antibiotic that inhibits ATP synthase (Complex V) and reduces mitochondrial ATP production [54]. It has been used in experimental models to investigate mitochondrial dependency in cancer cells, although toxicity concerns limit its clinical utility.

4.2.5 Elamipretide (SS-31)

Elamipretide is a mitochondria-targeted peptide stabilizing mitochondrial membranes and preventing oxidative damage [55]. It is currently in clinical trials for various mitochondrial diseases and is being investigated for its potential to modulate mitochondrial function in cancer therapy, particularly in combination with other treatments.

4.3 Targeting Mitochondrial DNA (mtDNA) Mutations

Cancers with mutations in mitochondrial DNA, especially in genes encoding components of the ETC are vulnerable to therapies that exploit mitochondrial dysfunction. Targeting mutated mtDNA or enhancing ROS generation in such tumors can selectively induce cell death in cancer cells.

4.4 Targeting Amino Acid Metabolism

Given the critical role of amino acid metabolism in cancer, targeting key enzymes and transporters involved in these pathways has become an attractive therapeutic approach. Cancer cells heavily depends on specific amino acids, like arginine, asparagine, glutamine, and methionine; in cancer cells, due to genetic or metabolic alterations, cells often lose their ability to synthesize these nutrients.

4.4.1 Treatment of Amino Acid-Depleting Enzymes

Amino acid-depleting enzymes such as L-asparaginase, L-arginine deiminase, and L-methionine γ-lyase selectively inhibit cancer cell growth without harming normal cells [56]. The FDA-approved enzyme L-asparaginase treats acute lymphoblastic leukemia (ALL) by breaking down asparagine, which is essential for cancer cell survival.

4.4.2 Glutamine Biosynthesis Inhibitors

Inhibitors of the glutamine biosynthesis pathway are being developed to disrupt the metabolic dependencies of cancer cells. Glutaminase (GLS) is crucial for glutamine metabolism, converts glutamine to glutamate and has emerged as a key cancer target. CB-839 (Telaglenastat), an allosteric GLS inhibitor blocks glutamine breakdown and starves cancer cells of energy (Fig. 1) [57]. Currently, it is in clinical trials to treat various cancers as a chemotherapeutic agent. Another promising drug IPN60090 selectively inhibits GLS-1 without affecting GLS-2 and has demonstrated tumor suppression in preclinical studies, especially with TORC1/2 inhibitors [58]. Now in phase I trials, it holds promise for glutamine-dependent cancers.

4.4.3 Serine Biosynthesis Inhibitors

Targeting serine biosynthesis is a promising cancer therapy approach. Key enzymes like Serine Hydroxymethyltransferase (SHMT) and 3-Phosphoglycerate dehydrogenase (PHGDH) have been identified as potential drug targets. SHMT1/2 play a vital role in cancer growth. A pyrazolopyran compound selectively inhibits SHMT1, inducing apoptosis in lung cancer cells (Fig. 1) [59]. Additionally, the antidepressant Sertraline, repurposed as an SHMT inhibitor, shows selective toxicity in serine/glycine-dependent breast cancer and leukemia cells.

NCT-503 is a potent PHGDH inhibitor; by binding to the enzyme's active site, it shows a reduction in serine biosynthesis and inhibition of tumor cell proliferation in PHGDH-amplified cancers (Fig. 1) [60]. Similarly, CBR-5884 is another selective inhibitor for PHGDH. It is an allosteric inhibitor that disrupts PHGDH tetramerization and suppresses tumor growth in breast cancer and melanoma models (Fig. 1) [61].

4.5 Targeting Fatty Acid Synthesis

Given the crucial role of fatty acid metabolism in cancer, therapeutic strategies that target fatty acid synthesis and oxidation have gained increasing attention. Several inhibitors targeting key enzymes in these pathways are under development, and some have shown promising results in preclinical and clinical studies.

4.5.1 Fatty Acid Synthase (FASN) Inhibitors

FASN, a key enzyme in de novo fatty acid synthesis, is an attractive cancer target. TVB-2640, an oral FASN inhibitor, impairs lipid biosynthesis and tumor growth in breast, ovarian, and prostate cancers (Fig. 2) and is in clinical trials [62]. IPI-9119 also shows antitumor activity by inducing apoptosis and reducing proliferation. Orlistat, initially an anti-obesity drug, inhibits FASN and demonstrates anticancer potential in prostate, breast, and melanoma models. However, its poor bioavailability and off-target effects limit its therapeutic use.

4.5.2 Acetyl-CoA Carboxylase (ACC) Inhibitors

ACC is the rate-limiting enzyme in fatty acid synthesis, making it a key target for cancer therapy. PF-05175157 is a small-molecule inhibitor of ACC that has been tested in preclinical cancer models (Fig. 2) [63]. By inhibiting ACC, PF-05175157 reduces the production of malonyl-CoA, a precursor for fatty acid synthesis. This

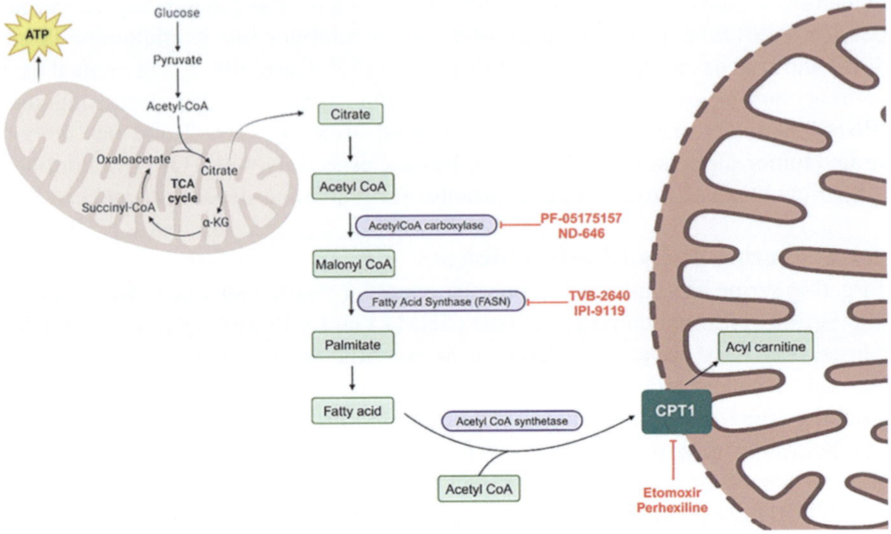

Fig. 2 Schematic illustration of inhibitors of key fatty acid metabolic pathway inhibitors under investigation in preclinical and clinical trials. Inhibitors are mentioned in red

disrupts lipid biosynthesis and impairs cancer cell growth. ND-646, a novel ACC inhibitor, has shown promise in preclinical studies of non-small cell lung cancer (NSCLC) and hepatocellular carcinoma [64]. It inhibits de novo fatty acid synthesis and enhances sensitivity to chemotherapy.

4.5.3 Carnitine Palmitoyltransferase 1 (CPT1) Inhibitors

CPT1 is the rate-limiting enzyme in FAO, transporting fatty acids into the mitochondria for oxidation. Inhibiting CPT1 reduces FAO and forces cancer cells to depend on glycolysis, thus increasing energy stress. Etomoxir, a CPT1 inhibitor, blocks FAO and reduces ATP production. Preclinical studies have shown that etomoxir inhibits tumor growth in breast cancer, glioblastoma, and leukemia models [65]. However, its clinical trials have been limited due to toxicity concerns and off-target effects.

Perhexiline, a CPT1 inhibitor that was initially developed for angina, has been repurposed for cancer therapy, showing efficacy in preclinical models of ovarian and pancreatic cancers [66]. By inhibiting FAO, it induces energy stress and apoptosis in cancer cells (Fig. 2).

4.6 Targeting AMPK/PPAR Pathways

AMPK and PPARs are key regulators of FAO. Inhibitors targeting these pathways can disrupt FAO and inhibit cancer cell survival under metabolic stress. AICAR is an AMPK activator that inhibits FAO by downregulating CPT1 expression. It has shown potential as an anticancer agent in preclinical studies, particularly in cancers that rely on FAO for survival, such as breast cancer and glioma. GW9662 is a PPAR-γ inhibitor that blocks FAO and has shown anticancer activity in preclinical models of colon and breast cancers. By inhibiting PPAR-γ, GW9662 reduces lipid metabolism and tumor growth.

5 Combination Therapies

In recent years, cancer treatment has advanced significantly, ranging from broad medications like chemotherapy and surgery to more targeted treatments like immunotherapy and metabolic therapies. There has been growing interest in combining metabolic inhibitors with conventional therapies such as chemotherapy, radiation, and immunotherapy to improve therapeutic efficacy. These combination treatments aim to minimize the resistance mechanisms of cancer cells and take advantage of cancer cells' weaknesses. Therefore, understanding cancer cell metabolism and combining it with conventional therapies will ultimately help to maximize patient outcomes.

5.1 Combination of Metabolic Inhibitors with Cancer Therapies

Metabolic inhibitors enhance chemotherapy, radiotherapy, and immunotherapy by targeting key metabolic pathways that support cancer cell survival and resistance.

5.1.1 Chemotherapy

Glycolysis inhibitors (e.g., 2-DG, 3-bromopyruvate) sensitize cancer cells, improving tumor regression and apoptosis when combined with doxorubicin or gemcitabine. Glutaminase inhibitors (e.g., CB-839) enhance chemotherapy effects, particularly in pancreatic cancer. AMPK activators (e.g., metformin) promote apoptosis and improve survival outcomes in ovarian and breast cancers. Autophagy inhibitors (e.g., chloroquine) enhance tumor sensitivity.

5.1.2 Radiotherapy

OXPHOS inhibitors (e.g., oligomycin) increase ROS and radiosensitivity in head and neck cancer. Glycolysis inhibitors impair DNA repair, improving tumor control. Hypoxia-targeting agents (e.g., PX-478) reduce radio resistance with ongoing clinical trials.

5.1.3 Immunotherapy and CAR T-Cell Therapy

Metabolic inhibitors combined with immune checkpoint inhibitors (e.g., anti-PD-1, anti-CTLA-4) improve tumor control and immune response. Targeting MDSCs enhances immunotherapy efficacy. In CAR T-cell therapy, metabolic inhibitors improve persistence and effectiveness against solid tumors. Preclinical and clinical trials demonstrate promising results for these combinations.

6 Challenges in Translating Metabolic Therapies to the Clinic

6.1 Complexity of Cancer Metabolism

The intricate and dynamic nature of cancer metabolism poses a significant challenge. Since metabolic pathways are interconnected, with compensatory mechanisms, for example, blocking glycolysis can trigger upregulation of glutaminolysis or fatty acid oxidation, enabling tumor survival. Understanding these adaptive responses is critical for designing combination therapies that prevent metabolic escape.

6.2 Patient Heterogeneity

Tumor metabolic heterogeneity, both within a single tumor and among patients, complicates the development of effective metabolic therapies. For instance, KRAS-mutant cancers often rely on glutamine metabolism, while PHGDH-amplified tumors depend on serine synthesis. Personalized approaches, such as metabolomic profiling and biomarker-driven therapy selection, are essential to address this variability.

6.3 Limited Clinical Trial Data

Although preclinical studies show promising preclinical results, robust clinical trial data demonstrating their safety and efficacy in diverse patient populations is still lacking. For example, while glutaminase inhibitors like CB-839 showed efficacy in preclinical models, their clinical success has been limited to specific subsets of renal cell carcinoma and hematologic malignancies. Expanding trials to diverse populations and integrating real-world evidence is crucial for validation.

6.4 Drug Delivery and Bioavailability

Effective delivery of metabolic inhibitors is a major difficulty, particularly for targeting brain tumors or hypoxic regions. For instance, the blood-brain barrier limits the efficacy of glycolysis inhibitors like 2-DG in glioblastoma. Advanced delivery systems, such as nanoparticle formulations (e.g., lipid-based carriers for FASN inhibitors) and prodrug strategies, are being explored to enhance bioavailability and target specificity.

6.5 Regulatory Challenges

The approval of metabolic therapies requires rigorous demonstration of safety and efficacy, which is complicated by the pleiotropic effects of metabolic inhibitors. For example, targeting AMPK can impact both tumor and normal cell metabolism, which raises concerns about off-target toxicity. Collaborative efforts between researchers, regulators, and industry are needed to streamline the approval process.

7 Personalized Approaches to Metabolic Therapy

Personalized medicine aims to tailor treatment strategies based on the individual characteristics of each patient's cancer. In the context of metabolic therapy, personalized approaches are gaining traction and may involve the following strategies:

7.1 Biomarker Identification

Identifying metabolic biomarkers such as specific metabolites (e.g., lactate, succinate), enzyme activity levels (e.g., LDHA, PHGDH), or genetic mutations (e.g., IDH1/2) can predict treatment response and guide therapy selection. For example, IDH1-mutant gliomas respond well to IDH inhibitors like Ivosidenib, while PHGDH-amplified breast cancers are vulnerable to PHGDH inhibitors like NCT-503. Integrating these biomarkers into clinical workflows enhances precision and improves outcomes.

7.2 Metabolomic Profiling

Integrating metabolomic profiling can provide a comprehensive insight into a patient's metabolic state, identifying vulnerabilities for targeted therapy. For instance, elevated serine levels in PHGDH-amplified tumors can guide the use of serine synthesis inhibitors. Advances in mass spectrometry and nuclear magnetic resonance (NMR) have made metabolomic profiling feasible in clinical settings, enabling real-time metabolic insights.

7.3 Adaptive Treatment Strategies

Given the dynamic nature of cancer metabolism, adaptive treatment strategies that allow for real-time monitoring of metabolic changes in response to therapy are essential. Liquid biopsies, which analyze circulating tumor DNA (ctDNA) and metabolites, offer a noninvasive way to track metabolic changes and treatment response. For example, rising lactate levels in blood samples may indicate glycolytic adaptation, prompting a switch to LDHA inhibitors. Real-time imaging techniques, such as hyperpolarized MRI, can visualize metabolic flux (e.g., pyruvate-to-lactate conversion) in tumors, enabling clinicians to adjust therapies based on metabolic activity.

7.4 Combination of Therapies

Personalized metabolic therapies could also be combined with other modalities (chemotherapy, radiotherapy, immunotherapy) to maximize efficacy in each patient. For instance, glutaminase inhibitors (e.g., CB-839) synergize with chemotherapy in KRAS-mutant pancreatic cancer, while glycolysis inhibitors enhance the efficacy of anti-PD-1 immunotherapy in melanoma. This individualized approach can potentially enhance therapeutic efficacy while minimizing side effects.

7.5 Engaging Multidisciplinary Teams

Developing personalized metabolic therapies will require collaboration among oncologists, metabolic researchers, molecular biologists, and pharmacologists. A multidisciplinary approach can facilitate the integration of metabolic insights into clinical practice and promote the development of innovative therapeutic strategies.

For example, combining AI-driven metabolomic analysis with clinician expertise can identify novel metabolic vulnerabilities and accelerate therapy development. By leveraging biomarkers, metabolomics, adaptive strategies, and multidisciplinary collaboration, personalized metabolic therapy offers a transformative approach to cancer treatment, addressing the unique metabolic needs of each patient.

8 Conclusion

The future of cancer therapy lies in the innovative targeting of cancer metabolism, offering a promising avenue to improve treatment efficacy and patient outcomes. Emerging trends, such as understanding metabolic heterogeneity, leveraging combination therapies, and integrating metabolomics and synthetic lethality, are driving the development of novel metabolic therapies. However, significant challenges still persist, including the complexity of cancer metabolism, patient heterogeneity, and regulatory hurdles.

To overcome these barriers, there is an urgent need for robust clinical trials to assess the safety and efficacy of metabolic inhibitors across diverse cancer types. For instance, trials evaluating glutaminase inhibitors (e.g., CB-839) and glycolysis-targeting agents (e.g., HK2 inhibitors) must expand to include larger, more diverse patient populations. Additionally, the integration of advanced technologies, such as CRISPR-Cas9 for identifying metabolic vulnerabilities, AI-driven drug discovery platforms (e.g., DeepMetabolism), and high-throughput screening for novel inhibitors, can accelerate the development of precision metabolic therapies.

Personalized approaches, guided by biomarker identification, metabolomic profiling, and adaptive treatment strategies, hold immense potential to enhance therapeutic precision. By tailoring therapies to individual metabolic profiles, clinicians can improve outcomes while minimizing side effects. The ongoing collaboration between researchers, clinicians, and regulatory bodies will be essential to translate these innovations from the laboratory to the clinic, offering new hope for patients facing this challenging disease.

Acknowledgments, Disclaimers, and Declaration of Interest This work was supported by grants from the Anusandhan National Research Foundation (CRG/2022/003866). Research findings and conclusions are those of the authors and do not necessarily represent the views of the CSIR-Indian Institute of Toxicology Research. The manuscript's CSIR-Indian Institute of Toxicology Research communication number is IITR/SECC-PME/MSS/2025/009. The authors report no declarations of interest. Figures reported in the book chapter have been prepared by BioRender (https://app.biorender.com).

References

1. Swift LH, Golsteyn RM. Genotoxic anti-cancer agents and their relationship to DNA damage, mitosis, and checkpoint adaptation in proliferating cancer cells. Int J Mol Sci. 2014;15(3):3403–31.
2. Ortmann BM. Hypoxia-inducible factor in cancer: from pathway regulation to therapeutic opportunity. BMJ Oncol. 2024;3:e000154.
3. Gray SP, Jandeleit-Dahm KAM. The role of NADPH oxidase in vascular disease-hypertension, atherosclerosis & stroke. Curr Pharm Des. 2015;21(41):5933–44.
4. Liao C, et al. Tumor hypoxia: from basic knowledge to therapeutic implications. Semin Cancer Biol. 2023;88:172–86.
5. Bergers G, Fendt SM. The metabolism of cancer cells during metastasis. Nat Rev Cancer. 2021;21(3):162–80.
6. Moldogazieva NT, Mokhosoev IM, Terentiev AA. Metabolic heterogeneity of cancer cells: an interplay between HIF-1, GLUTs, and AMPK. Cancers (Basel). 2020;12(4):862.
7. Liberti MV, Locasale JW. The Warburg effect: how does it benefit cancer cells? Trends Biochem Sci. 2016;41(3):211–8.
8. Miller ZA, et al. GLUT1 inhibitor BAY-876 induces apoptosis and enhances anti-cancer effects of bitter receptor agonists in head and neck squamous carcinoma cells. Cell Death Discov. 2024;10:339.
9. Smith ALM, et al. Mitochondrial DNA mutations in ageing and cancer. Mol Oncol. 2022;16(18):3276–94.
10. Lieu EL, et al. Amino acids in cancer. Exp Mol Med. 2020;52(1):15–30.
11. Bhutia YD, et al. Amino acid transporters in cancer and their relevance to "glutamine addiction": novel targets for the design of a new class of anticancer drugs. Cancer Res. 2015;75(9):1782–8.
12. Wang Z, et al. Targeting glutaminolysis: new perspectives to understand cancer development and novel strategies for potential target therapies. Front Oncol. 2020;10:589508.

13. Feun L, Savaraj N. Pegylated arginine deiminase: a novel anticancer enzyme agent. Expert Opin Investig Drugs. 2006;15(7):815–22.
14. Bai R, Cui J. Regulation of fatty acid synthase on tumor and progress in the development of related therapies. Chin Med J. 2024;137(16):1894–902.
15. Menendez JA, Lupu R. Fatty acid synthase regulates estrogen receptor-α signaling in breast cancer cells. Oncogenesis. 2017;6(2):e299.
16. Raab S, et al. Dual regulation of fatty acid synthase (FASN) expression by O-GlcNAc transferase (OGT) and mTOR pathway in proliferating liver cancer cells. Cell Mol Life Sci. 2021;78(13):5397–413.
17. Fhu CW, Ali A. Fatty acid synthase: an emerging target in cancer. Molecules. 2020;25(17):3935.
18. Zaidi N, et al. ATP-citrate lyase: a key player in cancer metabolism. Cancer Res. 2012;72(15):3709–14.
19. Mozolewska P, et al. Inhibitors of fatty acid synthesis and oxidation as potential anticancer agents in colorectal cancer treatment. Anticancer Res. 2020;40(9):4843–56.
20. Schlaepfer IR, Joshi M. CPT1A-mediated fat oxidation, mechanisms, and therapeutic potential. Endocrinology. 2020;161(2):bqz046.
21. Flavin R, et al. Fatty acid synthase as a potential therapeutic target in cancer. Future Oncol. 2010;6(4):551–62.
22. Chen Z, et al. Hypoxic microenvironment in cancer: molecular mechanisms and therapeutic interventions. Signal Transduct Target Ther. 2023;8(1):70.
23. Wang SF, et al. Role of mitochondrial alterations in human cancer progression and cancer immunity. J Biomed Sci. 2023;30(1):61.
24. Solaini G, et al. Hypoxia and mitochondrial oxidative metabolism. Biochim Biophys Acta. 2010;1797(6–7):1171–7.
25. Sinkler CA, et al. Tissue- and condition-specific isoforms of mammalian cytochrome c oxidase subunits: from function to human disease. Oxidative Med Cell Longev. 2017;2017:1534056.
26. Huang C, Freter C. Lipid metabolism, apoptosis and cancer therapy. Int J Mol Sci. 2015;16(1):924–49.
27. Deng X, et al. Pyruvate dehydrogenase kinase 1 interferes with glucose metabolism reprogramming and mitochondrial quality control to aggravate stress damage in cancer. J Cancer. 2020;11(4):962–73.
28. Wang Y, et al. The connection between tricarboxylic acid cycle enzyme mutations and pseudohypoxic signaling in pheochromocytoma and paraganglioma. Front Endocrinol (Lausanne). 2023;14:1274239.
29. Intlekofer AM, et al. Hypoxia induces production of L-2-hydroxyglutarate. Cell Metab. 2015;22(2):304–11.
30. Li Y, et al. Cytochrome c oxidase subunit IV is essential for assembly and respiratory function of the enzyme complex. J Bioenerg Biomembr. 2006;38(5–6):283–91.
31. Semenza GL. Hypoxia-inducible factor 1: regulator of mitochondrial metabolism and mediator of ischemic preconditioning. Biochim Biophys Acta. 2011;1813(7):1263–8.
32. Furuhashi M, Hotamisligil GS. Fatty acid-binding proteins: role in metabolic diseases and potential as drug targets. Nat Rev Drug Discov. 2008;7(6):489–503.
33. He J, et al. Short-chain fatty acids and their association with signalling pathways in inflammation, glucose and lipid metabolism. Int J Mol Sci. 2020;21(17):6356.
34. Triantafyllou EA, et al. Expression of AGPAT2, an enzyme involved in the glycerophospholipid/triacylglycerol biosynthesis pathway, is directly regulated by HIF-1 and promotes survival and etoposide resistance of cancer cells under hypoxia. Biochim Biophys Acta Mol Cell Biol Lipids. 2018;1863(9):1142–52.
35. Koizume S, Miyagi Y. Lipid droplets: a key cellular organelle associated with cancer cell survival under normoxia and hypoxia. Int J Mol Sci. 2016;17(9):1430.
36. Mylonis I, et al. Hypoxia-inducible factors and the regulation of lipid metabolism. Cells. 2019;8(3):214.
37. Chen J, Cui L, Lu S, Xu S. Amino acid metabolism in tumor biology and therapy. Cell Death Dis. 2024;15(1):42.

38. Jin J, et al. Targeting glutamine metabolism as a therapeutic strategy for cancer. Exp Mol Med. 2023;55(4):706–15.
39. Tedesco L, et al. von Hippel-Lindau mutants in renal cell carcinoma are regulated by increased expression of RSUME. Cell Death Dis. 2019;10(4):266.
40. Samanta D, Semenza GL. Serine synthesis helps hypoxic cancer stem cells regulate redox. Cancer Res. 2016;76(22):6458–62.
41. Zhang RS, et al. WZB117 enhanced the anti-tumor effect of apatinib against melanoma via blocking STAT3/PKM2 axis. Front Pharmacol. 2022;13:976117.
42. Habtemariam S. The molecular pharmacology of phloretin: anti-inflammatory mechanisms of action. Biomedicines. 2023;11(1):143.
43. Sun Y, et al. Mechanisms underlying 3-bromopyruvate-induced cell death in colon cancer. J Bioenerg Biomembr. 2015;47(4):319–29.
44. Nath K, Guo L, et al. Mechanism of antineoplastic activity of lonidamine. Biochim Biophys Acta. 2016;1866(2):151–62.
45. Dwarakanath B, Jain V. Targeting glucose metabolism with 2-deoxy-D-glucose for improving cancer therapy. Future Oncol. 2009;5(5):581–5.
46. Liu H, et al. The PKM2 activator TEPP-46 suppresses kidney fibrosis via inhibition of the EMT program and aberrant glycolysis associated with suppression of HIF-1α accumulation. J Diabetes Investig. 2021;12(5):697–709.
47. Huang B, et al. Shikonin ameliorated mice colitis by inhibiting dimerization and tetramerization of PKM2 in macrophages. Front Pharmacol. 2022;13:926945.
48. Ganapathy-Kanniappan S, et al. 3-Bromopyruvate: a new targeted antiglycolytic agent and a promise for cancer therapy. Curr Pharm Biotechnol. 2010;11(5):510–7.
49. Rellinger EJ, et al. FX11 inhibits aerobic glycolysis and growth of neuroblastoma cells. Surgery. 2017;161(3):747–52.
50. Yang Y, et al. Different effects of LDH-A inhibition by oxamate in non-small cell lung cancer cells. Oncotarget. 2014;5(23):11886–96.
51. Granja S, et al. The monocarboxylate transporter inhibitor α-cyano-4-hydroxycinnamic acid disrupts rat lung branching. Cell Physiol Biochem. 2013;32(6):1845–56.
52. Wheaton WW, et al. Metformin inhibits mitochondrial complex I of cancer cells to reduce tumorigenesis. elife. 2014;3:e02242.
53. Yap TA, et al. Complex I inhibitor of oxidative phosphorylation in advanced solid tumors and acute myeloid leukemia: phase I trials. Nat Med. 2023;29(1):115–26.
54. Goulooze SC, et al. Bedaquiline. Br J Clin Pharmacol. 2015;80(2):182–4.
55. Mackieh R, et al. Inhibitors of ATP synthase as new antibacterial candidates. Antibiotics (Basel). 2023;12(4):650.
56. Tung C, et al. Elamipretide: a review of its structure, mechanism of action, and therapeutic potential. Int J Mol Sci. 2025;26:944.
57. Pokrovsky VS, et al. Amino acid degrading enzymes and their application in cancer therapy. Curr Med Chem. 2019;26(3):446–64.
58. Harding JJ, et al. A phase I dose-escalation and expansion study of telaglenastat in patients with advanced or metastatic solid tumors. Clin Cancer Res. 2021;27(18):4994–5003.
59. Soth MJ, et al. Discovery of IPN60090, a clinical stage selective glutaminase-1 (GLS-1) inhibitor with excellent pharmacokinetic and physicochemical properties. J Med Chem. 2020;63(21):12957–77.
60. Marani M, et al. A pyrazolopyran derivative preferentially inhibits the activity of human cytosolic serine hydroxymethyltransferase and induces cell death in lung cancer cells. Oncotarget. 2016;7(4):4570–83.
61. Arlt B, et al. Inhibiting PHGDH with NCT-503 reroutes glucose-derived carbons into the TCA cycle, independently of its on-target effect. J Enzyme Inhib Med Chem. 2021;36(1):1282–9.
62. Falchook G, et al. First-in-human study of the safety, pharmacokinetics, and pharmacodynamics of first-in-class fatty acid synthase inhibitor TVB-2640 alone and with a taxane in advanced tumors. EClinicalMedicine. 2021;34:100797.

63. Marczyk M, et al. Comprehensive analysis of metabolic isozyme targets in cancer. Cancer Res. 2022;82(9):1698–711.
64. Svensson RU, et al. Inhibition of acetyl-CoA carboxylase suppresses fatty acid synthesis and tumor growth of non-small-cell lung cancer in preclinical models. Nat Med. 2016;22(10):1108–19.
65. Shim JK, et al. Etomoxir, a carnitine palmitoyltransferase 1 inhibitor, combined with temozolomide reduces stemness and invasiveness in patient-derived glioblastoma tumorspheres. Cancer Cell Int. 2022;22(1):309.
66. Dhakal B, et al. Perhexiline: old drug, new tricks? A summary of its anti-cancer effects. Molecules. 2023;28(8):3624.

MIX
Papier aus verantwortungsvollen Quellen
Paper from responsible sources
FSC® C105338

If you have any concerns about our products,
you can contact us on
ProductSafety@springernature.com

In case Publisher is established outside the EU,
the EU authorized representative is:
Springer Nature Customer Service Center GmbH
Europaplatz 3, 69115 Heidelberg, Germany

Printed by Libri Plureos GmbH
in Hamburg, Germany